Fortschritte der Chemie organischer Naturstoffe

Progress in the Chemistry of Organic Natural Products

37

Founded by L. Zechmeister
Edited by W. Herz, H. Grisebach, G. W. Kirby

Authors:
P. Albersheim, J. M. Brand, A. G. Darvill,
J. Häusler, M. McNeil, E. Öhler, H. Poisel,
U. Schmidt, R. M. Silverstein, J. Chr. Young

Springer-Verlag
Wien New York 1979

Dr. W. Herz, Professor of Chemistry, Department of Chemistry,
The Florida State University, Tallahassee, Florida, U.S.A.

Prof. Dr. H. Grisebach, Biologisches Institut II, Lehrstuhl für Biochemie der Pflanzen,
Albert-Ludwigs-Universität, Freiburg i. Br., Federal Republic of Germany

G. W. Kirby, Sc. D., Regius Professor of Chemistry, Chemistry Department,
The University, Glasgow, Scotland

With 8 Figures

© 1979 by Springer-Verlag/Wien
Softcover reprint of the hardcover 1st edition 1979
Library of Congress Catalog Card Number AC 39-1015

ISSN 0071-7886

ISBN-13: 978-3-7091-8547-6   e-ISBN-13: 978-3-7091-8545-2
DOI: 10.1007/978-3-7091-8545-2

Contents

The Structural Polymers of the Primary Cell Walls of Dicots. By M. McNeil, A. G. Darvill, and P. Albersheim 191

Dehydroamino Acids, α-Hydroxy-α-amino Acids and α-Mercapto-α-amino Acids. By
U. SCHMIDT, J. HÄUSLER, ELISABETH ÖHLER, and H. POISEL 251

List of Contributors

Albersheim, Prof. P., B. S., Ph. D., Department of Chemistry, University of Colorado, Boulder, CO 80309, U. S. A.

Brand, Dr. J. M., Biochemistry Department, University of Fort Hare, Alice, C. P. 5700, South Africa.

Darvill, A. G., B. S., Ph. D., Department of Chemistry, University of Colorado, Boulder, CO 80309, U.S.A.

Häusler, Dr. J., Institut für Organische Chemie, Universität Wien, Währinger Straße 38, A-1090 Wien, Austria.

McNeil, M., M. S., Department of Chemistry, University of Colorado, Boulder, CO 80309, U.S.A.

Öhler, Dr. Elisabeth, Institut für Organische Chemie, Universität Wien, Währinger Straße 38, A-1090 Wien, Austria.

Poisel, Dr. H., Gmundnerstraße 29, A-4800 Attnang-Puchheim, Austria.

Schmidt, Prof. Dr. U., Institut für Organische Chemie, Biochemie und Isotopenforschung der Universität Stuttgart, Pfaffenwaldring 55, D-7000 Stuttgart 80, Federal Republic of Germany.

Silverstein, Prof. Dr. R. M., Department of Chemistry, College of Environmental Science and Forestry, Syracuse Campus, State University of New York, Syracuse, NY 13210, U.S.A.

Young, Dr. J. Chr., Chemistry and Biology Research Institute, Canadian Department of Agriculture, Ottawa, Ontario, K1A 0C6 Canada.

Insect Pheromones: A Critical Review of Recent Advances in Their Chemistry, Biology, and Application

By J. M. Brand, Iowa City, Iowa, U.S.A.,
J. Chr. Young, Ottawa, Canada, and
R. M. Silverstein, Syracuse, New York, U.S.A.*

Contents

---

* CBRI Contribution No. 10,

I. Introduction

The chemical basis of insect behavior is firmly established and forms an integral part of regulatory biology. The many and varied studies on this topic constitute part of an overall attempt to understand behavior at the molecular level. A better understanding of this subject will only come about by interdisciplinary collaboration between chemists and biologists.

We have elected to review certain aspects of recent developments in pheromone chemistry. Scientific objectivity is difficult to achieve when summarizing the many, and sometimes conflicting, reports and opinions of investigators. We confess to personal biases based on our experiences, offer no apology but ask a rational tolerance.

A number of terms have been coined to designate the various kinds of chemical interactions between individuals and some are defined below. As a particular substance may be defined to act in more than one specific manner, depending on the context, these terms should not be considered as mutually exclusive.

The chemicals delivering the message are called *semiochemicals* (1). *Allelochemics* are those semiochemicals used for interspecific communication (2). These in turn are subdivided into *allomones*, which give adaptive advantage to the emitter (e.g. the defensive secretions of many arthropods or the spray of a skunk) or *kairomones,* which give adaptive advantage to the receiver (e. g. substances that enable predators to locate their prey) (*3*). Semiochemicals used for intraspecific communication are called *pheromones* (*4*).

Pheromones are classified according to the response they elicit. Chemical stimuli that trigger an immediate and reversible change in the behavior of the recipient are called *releasers,* whereas those inducing delayed, lasting responses are referred to as *primers* (*5*). These responses may be due to an individual chemical or, as is often the case, a mixture of chemicals. In the latter instance, the total mixture is the pheromone and the individual chemicals that make up the pheromone are termed *pheromone components.* In some semiochemical mixtures, all the components must be present to elicit a maximum response. If the total effect is greater than the sum of the effects of the individual components, the phenomenon is termed *synergism.* In some instances, synergism may occur from a combination of both insect and host produced chemicals (e. g. *6*). The major categories of semiochemicals are not mutually exclusive; the sex pheromones of some insects also serve as kairomones in that they attract predators (e. g. *7*).

The literature in the field of insect chemistry is vast; at the time of this writing, in excess of 10,000 scientific papers have been published. To make this review more manageable and meaningful and to avoid an encyclopedic approach, we have attempted to emphasize only the recent leading studies in one area, namely pheromones.

The last review of insect pheromone chemistry in this series was by EITER in 1970 (*8*). Since then, a number of other reviews have been published in the form of books (*9—16*) or review articles (*17—21*).

Sex pheromones are secreted by one sex to attract the other as an initial part of the mating process. A variety of chemicals have been identified by screening as attractive to one sex, but until these compounds have been isolated and identified from the opposite sex, they should be termed *sex attractants* or *parapheromones.* In some species, particularly among beetles (Coleoptera), the pheromones may attract both sexes and therefore serve more than one function. Such compounds are called *aggregation* or *recruitment* pheromones. Many social insects use *alarm pheromones* to alert other members of their species, or *trail pheromones* for foraging.

The importance of geometrical and optical isomers in the list of behavior-modifying chemicals is apparent, and we have taken this fact as a point of emphasis. We begin with a discussion of structure elucidation and chemical synthesis. This is followed by a section entitled "stereobiology" which includes recent reports on insect pheromones dealing with the occurrence of various isomers in insect species. Subsequently, studies on the biosynthesis of pheromones are covered in some detail; this section is followed by a brief mention of the application of pheromones to chemosystematics and speciation. Finally, we present a section on the present status of the practical applications of pheromones and offer some thoughts for the future.

A typical pheromone study involves intimate collaboration between a chemist and a biologist and follows a pattern (22):

Biologist

1. Selects an insect pest (often one of economic importance) and describes the behavior under pheromone control.

2. Develops a bioassay.

3. Collects source material containing the pheromone.

Chemist

1. Fractionates source material; each step must be monitored by the bioassay (biologist).

2. Isolates and identifies the compound(s) responsible for the behavior. In most cases, the insect uses a multicomponent pheromone system and a precise reproduction of the mixture may be necessary to attract or confuse; in many cases however, a single component of the blend (or a single parapheromone) may be sufficient to affect field behavior.

3. Synthesizes to confirm proposed structures and to furnish material for field studies.

4. Devises release systems for field studies.

Biologist

1. Confirms equivalent activity of isolated and synthesized compounds in the laboratory and in the field.

2. Develops protocol for survey and control.

At this point, large scale synthesis is turned over to the chemical industry and specialized industrial help is sought to develop efficient, slow-release formulation and dispersal systems. Only a handful of companies has been involved in these activities, and it is safe to say that returns have been marginal at best—in most cases developmental costs have exceeded returns.

II. Structure Elucidation

A. Isolation

1. Collection

The amount of semiochemical produced by an insect is highly variable and depends upon species and type of pheromone. Sex pheromones are usually produced in extremely small amounts [e. g. *ca.* 1 ng per female cabbage looper *(Trichoplusia ni)* *(23)*] whereas up to 20% of the body weight of the ant *Formica rufa* is formic acid *(24)* which functions mainly as a defensive allomone and partly as an alarm pheromone.

In choosing the method of isolation, one must consider how the insect stores and releases the pheromone. If it is stored in relatively large quantities ready for release, one can extract the whole insect or, preferably, the gland producing the chemical. If stored in small quantities, in precursor form, and/or released slowly over an extended period, it can be collected by passing air over the insects and through a conventional cold or cryogenic (liquid nitrogen) trap *(25)*, or through an absorbent such as Porapak Q *(26)*. Alternatively one can extract paper used to line the cages. Methods of collection have been reviewed *(15, 27, 28)*.

2. Bioassay

The collected material must now be fractionated and the pheromone components isolated in pure form. Each step in this process must be monitored by the bioassay which is a test for the presence or amount of a biologically active substance with the detector being an insect. The development, problems encountered, and methods of bioassay have been extensively reviewed by YOUNG and SILVERSTEIN *(27)*. One bioassay technique deserves special mention: the electroantennogram (EAG) developed by SCHNEIDER *(29)*. In this technique, electrodes are attached to the base and sensory hairs of an insect's antenna and one then measures the change in voltage potential in response to olfactory stimuli. The EAG procedure has been especially useful in studies on lepidopteran pheromones. In addition to identification of active fractions, EAGs have been useful for investigating physiological and environmental factors, molecular specificity, and other variables on antennal olfactory responsiveness. Responses have been obtained with as little as 0.1 pg of pheromone. For reviews on EAGs see *(28, 30, 31)*.

3. Fractionation

The choice of a fractionation method will depend upon the physical state, stability, and amount of pheromone available. To avoid decomposition or rearrangement, the mildest possible conditions should be used.

Traditional purification techniques such as distillation, sublimation, and recrystallization have been used only occasionally, due in part to the small amount of material available. Extraction and derivative formation has proved useful in many instances.

The technique of choice is chromatography. Column chromatography, including gel permeation, is often used in the first stages of purification. Since many pheromones are olefinic, silver nitrate impregnated adsorbants and ion exchange resins (*32*) have proved useful in the separation of *E*- and *Z*-isomers. The relatively new technique of high-performance liquid chromatography (HPLC) is finding increased use (e. g. *33—36*). Thin-layer chromatography (TLC) can be a versatile tool for preparative separations of relatively complex mixtures, for separation of *E*- and *Z*-olefins, and through the use of chromogenic reagents for demonstrating the presence or absence of certain functional groups.

Gas chromatography (GC) is almost universally used in pheromone studies because of its speed, resolution, sensitivity, precision of analysis, and simplicity, and because most pheromones are volatile and thus ideally suited for this technique. Among the major uses of GC in pheromone studies are: to fractionate mixtures and collect the various components, to determine the purity and homogeneity of a sample, to dertermine the amounts of various components in a mixture, to detect the presence of a specific compound, and to compare synthetic with natural material. Preparative GC involves diverting much of the effluent from the end of a column and trapping the various fractions (e. g. *37*). Purity is best determined by using several GC columns of high resolving power and differing polarity. Many of the lepidopteran sex pheromones contain mixtures of *E*- and *Z*-isomers with one predominating. In early studies, however, the minor component was not well resolved from the other, and its presence was often overlooked.

Fractionation of semiochemicals has been reviewed in detail (*15, 27*).

B. Identification

As natural products go, insect pheromones have relatively simple structures. This is due in part to the requirement for high volatility and rapid diffusion in air. Identification of these compounds is no easy matter, however, since the amounts produced by an individual are frequently at the nanogram level (or less). Only a decade ago, Tumlinson *et al.* (*38*) required 4.5 million boll weevils *(Anthonomus grandis)* to get enough material for identification. With the improvement in analytical instrumentation and microtechniques, rapid progress in the identification of pheromones has been made in the past few years. Since procedures for identification of pheromones have been reviewed at length (*15, 27, 34, 39*) and lists of identified pheromones have been adequately tabulated elsewhere (*15, 21, 40—43*), much of the emphasis here will be placed on examples to illustrate the useful techniques.

1. Spectrometric Methods

Structure determination rests heavily on information derived from mass (MS), nuclear magnetic resonance (NMR), infrared (IR), and ultraviolet (UV) spectra. Experienced chemists can frequently identify structures with no other information, using only 5—10 μg of material.

Modern mass spectrometers are capable of giving spectra from 10 ng or less of sample. In pheromone studies, the mass spectrometer is usually interfaced with a gas chromatograph (GC-MS), and, when the output is coupled to a computer, complex mixtures can be readily analyzed. The electron-impact MS sometimes gives the molecular weight, and, from the fragmentation pattern, one can often deduce the presence of functional groups and branching in molecules. High resolution mass spectrometers can give the elemental composition of the parent compound and its fragments. Chemical ionization mass spectrometry, from which the molecular weight can usually be determined, may use a reagent gas as the carrier gas with an interfaced GC. Computer searches of MS data banks may lead to identification.

With beam condensers and Fourier transform IR spectrometers or GC-IR combinations, spectra can be obtained from less than 1 ng (e. g. *44*). IR spectra are useful in identifying certain functional groups such as carbonyl or hydroxyl; the band at about 970 cm^{-1} is especially useful for confirming the *E*-configuration of the carbon-carbon double bonds.

UV spectrometry shows the absence or presence of conjugated systems and gives some indication of the type of conjugation. A number of pheromones have conjugated systems.

Applications of NMR spectrometry to pheromone identification have been restricted thus far to proton magnetic resonance. Until recently, the requirement for a relatively large sample size (100 μg) was a limiting factor in spectral identifications. TUMLINSON and HEATH (*34*) have recently reported good spectra from only 2 μg of grandisol (**33**, Chart 1), a pheromone component of the boll weevil. Complex molecules often produce spectra with overlapping peaks, thus making interpretation difficult. In some cases, this problem can be alleviated by use of lanthanide shift reagents such as Eu(fod)₃ (*45*). Chiral shift reagents have been used to determine the enantiomeric composition of several pheromone bicyclic ketals (**150, 195, 303 a**) (*46*) and alcohols (**1, 229, 232, 266a, 294a**) (*45*) (see Chart 1). The α-methoxy-α-trifluoromethylphenylacetyl derivatives of these alcohols were also used to determine enantiomeric composition (*45, 47, 48*). It is interesting to note that (**266 a**) from *Ips pini* and (**294 a**) from *Scolytus multistriatus* are 100% (−)-enantiomers, whereas, (**229**) from *Gnathotrichus sulcatus* is a 65 : 35 mixture of (+)/(−), (**232**) from *Dendroctonus frontalis* is 60 : 40 (+)/(−), and (**1**) from *Dendroctonus*

Chart 1

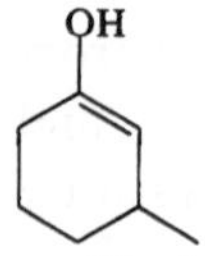

(1)

(2)

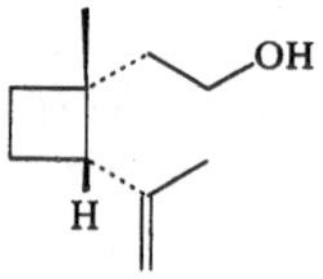

(33)
(+)-grandisol

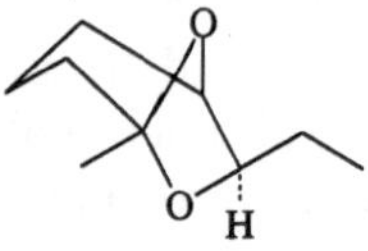

(150)
(+)-*exo*-brevicomin

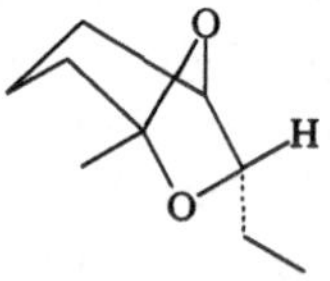

(151)
endo-brevicomin

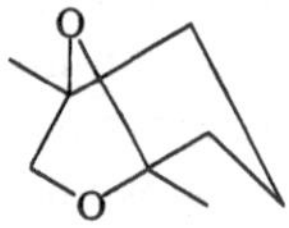

(195)
(−)-frontalin

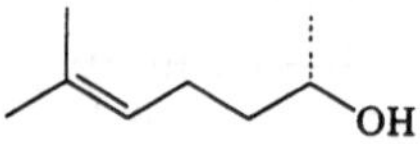

(229)
(+)-sulcatol

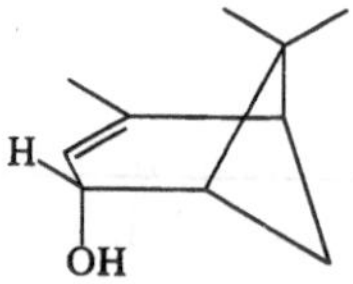

(232)
(−)-*trans*-verbenol

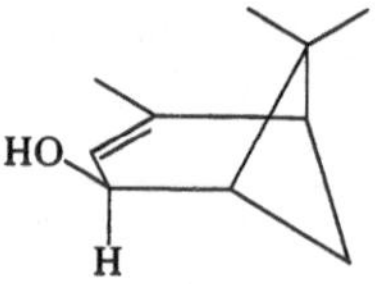

(234)
(+)-*cis*-verbenol

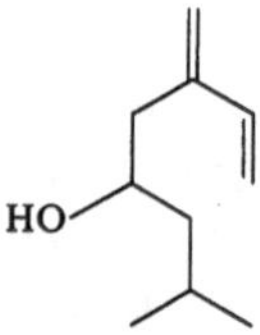

(240)
(±)-ipsenol

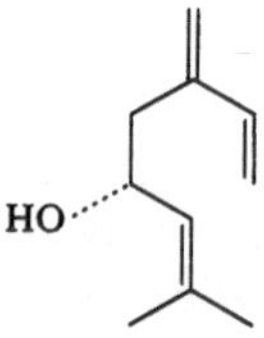

(266a)
(−)-ipsdienol

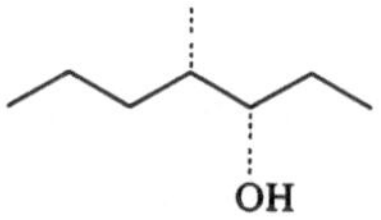

(294a)

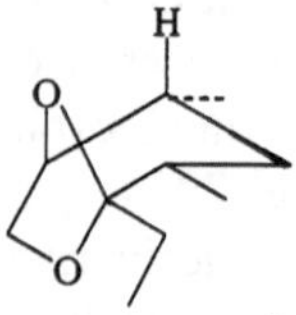

(303a)
α-(−)-
multistriatin

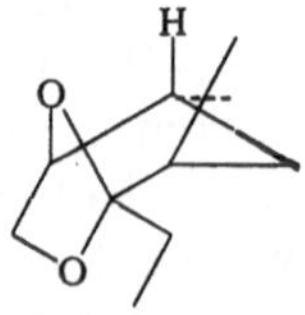

(303c)
γ-(−)-
multistriatin

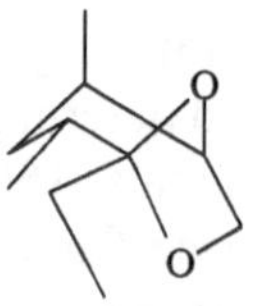

(303d)
γ-(+)-
multistriatin

pseudotsugae is 50 : 50 (+)/(−). These determinations were done on 5—200 μg of substrate. Carbon-13-NMR has become a powerful tool for structure identification. At present, large sample sizes (*ca.* 1 mg) are required, which limits its use in pheromone identification. However, it has proved useful in determination of optical purity of synthetic α-multistriatin (**303a,** Chart 1), an aggregation pheromone component of *Scolytus multistriatus* (*49*), and stereochemistry of the bicyclic ketals (**150**), (**151**), (**195**), (**303a**), (**303c**), and (**303d**) (*50*).

Optical rotatory dispersion (ORD) measurements have been made on some enantiomeric pheromones (*49*) and classical optical rotations are routinely determined. It should be noted that the sign of rotation may depend upon the solvent used: (1*S*,4*S*,5*S*)-*cis*-verbenol (**234**) is dextrorotatory in methanol and acetone (*48, 51*) and levorotatory in chloroform (*52*).

2. Chromatographic Methods

Use of the KOVATS system of GC retention indices (*53, 54*) can often provide considerable information on the functional groups and molecular size of an unknown compound. The retention index indicates where a compound will appear on a chromatogram with respect to straight-chain alkanes, using adjusted retention times. This method has been especially helpful with lepidopteran pheromones, which tend to be long, straight-chain unsaturated compounds with a single terminal functional group (e. g. *55*). From this and spectral evidence one can often choose model compounds for comparison. If two substances are resolved by GC, they can unambiguously be said to be different; however, identical retention times on a given column do not prove identity. YOUNG *et al.* (*56*) observed that the terpene alcohols (**2**) and (**240**) (Chart 1) had identical retention times on three of five columns investigated.

Similarly, comparison of known and unknown compounds by TLC and HPLC can be used to establish identity. Silver salt impregnated TLC is used to distinguish between *E*- and *Z*-isomers of closely related structures (e. g. *57*).

Reaction gas chromatography (*58—61*) has become one of the most convenient methods for performing chemical reactions at the microgram level. In this technique, the unknown compound is injected into the GC system and is retained or transformed, frequently at the injection port on a precolumn. The products that elute can be collected and analyzed.

One commonly employed technique is to use hydrogen as a carrier gas and to put an appropriate catalyst on a precolumn. On injection a compound may undergo a) hydrogenation to saturated analogs; b) dehydrogenation; or c) hydrogenolysis. Carbon-skeleton chromatography utilizes hydrogenolytic conditions to strip off all functional groups to give the parent hydrocarbon or a lower homolog. Determination of the structure of brevicomin (**150**) by SILVERSTEIN *et al.* (*62*) was facilitated when carbon-skeleton chromatography of (**150**) afforded nonane (**3**).

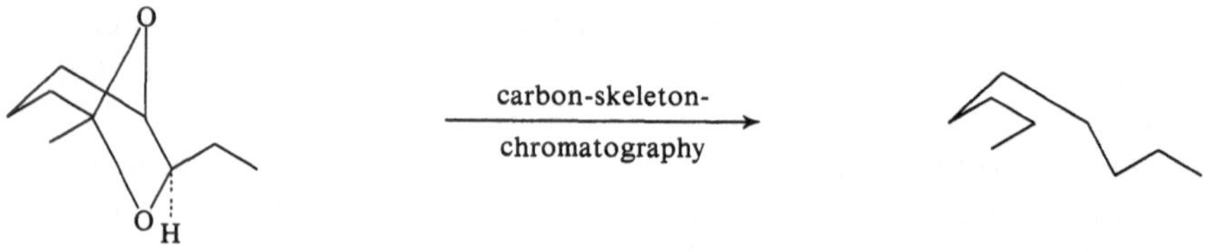

$$\xrightarrow{\text{carbon-skeleton-} \atop \text{chromatography}}$$

(**150**)   (**3**)

Compounds containing certain functional groups can be "subtracted" (retained) by a chemical placed in the GC pathway (*61, 63, 64*). The structure of the gypsy moth *(Lymantria dispar)* sex pheromone (**536**) was provided in part when it was subtracted by phosphoric acid, which is known to remove epoxides (*57*).

Functional group analyses can be performed by chemical reactions on TLC plates (*15*).

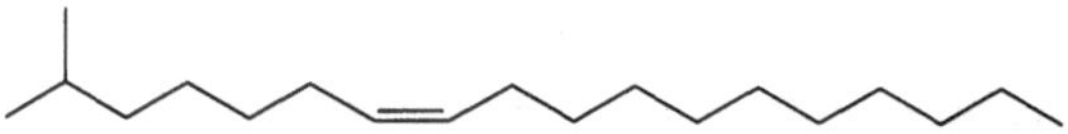

(**535**)

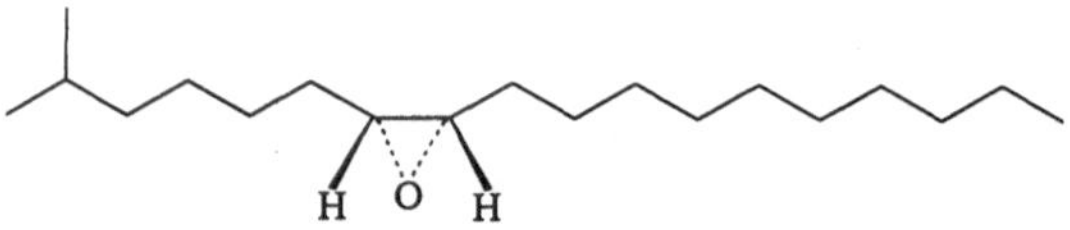

(**536**)

3. Microchemical Methods

When spectral and chromatographic data for an unknown compound are insufficient for elucidation of the total structure, it becomes necessary to perform chemical manipulations.

Hydrogenation by GC or in solution is frequently used to determine the number of olefinic bonds. The position of these bonds is usually determined by microozonolysis (followed by reduction) and examination of the aldehydic and/or ketonic fragments produced (*65—67*). The identification of fragments commonly encountered from ozonolysis of pheromones has been studied by MOORE and BROWN (68). Epoxidation of olefins has proved useful when there are two or more olefinic bonds and the products of ozonolysis are very volatile and difficult to analyze (34). When compounds containing two olefinic bonds are epoxidized, the reaction is monitored by GC and the reaction stopped at the mono-epoxide stage. Chemical ionization mass spectrometry helps to locate the position of the epoxide (69) and the configuration of the double bond is determined from the IR spectrum.

The nature of functional groups in a pheromone can be determined by chemical modification coupled with a sensitive bioassay of the converted material. For example, if a pheromone loses activity after saponification and regains it upon acetylation it likely contains an acetate group. Loss of activity after hydrogenation indicates unsaturation. INSCOE and BEROZA (15) tabulated many of the typical functional group tests that can be used in pheromone structure elucidation. These tests need not be restricted to purified material; they can be employed on crude extracts as well and the results may suggest purification methods to be used (or avoided).

Once the functionality has been identified, a search for precursors in the extracts may yield valuable information. When the sex pheromone of the gypsy moth was shown to be an epoxide, BIERL et al. (57) treated the monoolefin fraction of the extract and obtained biological activity. Identification of olefin (535) as the precursor led directly to (536) as the sex pheromone (see p. 10).

Use of solvents in the extraction of very volatile alarm pheromones can introduce impurities or mask short retention time components in GC analysis. These problems can be avoided by drawing glandular liquid into fine glass tubes and then introducing the pheromone into the GC by the solid sampling technique (70). Chemical pretreatment, reaction GC, or selective subtraction can also be conducted to obtain useful information.

4. Electroantennogram Methods

Difficulties with traditional bioassay techniques include development of olfactometers, maintenance of sufficient numbers of insects, synergism, replication, and time taken to get significant results. The EAG technique has none of these. This method has been widely used to define chain length, functional group, and the position and configuration of double bonds in lepidopteran pheromones. After determining the first two parameters, one screens a series of standards differing only in the olefin portion. Subject to field testing, that compound showing the greatest response is assumed to be the pheromone or a pheromone component. On the assumption that the response from a doubly unsaturated compound will be the summation of the responses from two monounsaturated molecules, Roelofs et al. (71) correctly predicted that (E,Z)-7,9-dodecadien-1-yl acetate (490) was the sex pheromone of the European vine moth (Lobesia botrana) after obtaining significant responses to the E-7 and Z-9 isomers of dodecen-1-yl acetate. Some caution in interpretation of results should be exercised since chemicals other than the pheromone may elicit an EAG response.

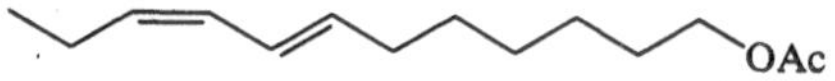

OAc

(490)

5. Screening Methods

As the number of identified pheromones increases, we are getting a clearer picture of the types of chemicals that are used by each order and family of insect. Structural variations of known lepidopteran phoromones have been tested in field traps and have attracted a wide variety of insects. It should be noted that attraction to a particular synthetic chemical or mixture does not constitute rigorous proof that the compound(s) is (are) a pheromone for that species; independent isolation and identification is required.

6. Examples

Three examples of how some of the above methods have been successfully combined to elucidate pheromone structures are given below.

a) Wild silkmoth (Antheraea polyphemus)

Kochansky et al. (72) purified extracts of the abdominal tips from about 1800 females by column chromatography on Florisil and 25% AgNO$_3$-silicagel. All bioassays were conducted by the EAG technique. GC analysis on polar and nonpolar columns suggested a major and minor component.

Comparison of retention times of the major component with that of hexadecyl acetate suggested that it might be an unsaturated C$_{16}$ acetate. A sample purified by preparative GC was examined by TLC; comparison of R$_f$ values of the sample and its mercuric acetate adduct with those of suitable standards confirmed the presence of two double bonds. Saponification and LiAlH$_4$ reduction destroyed EAG activity, which was regained on reacetylation. Acetylation had no effect and bromination destroyed activity. Hydrogenation also destroyed activity and gave a product that cochromatographed with hexadecyl acetate. MS analysis confirmed a diunsaturated C_{16} acetate. EAG analyses of a series mono-unsaturated C_{12}, C_{14}, C_{16}, and C_{18} acetates, alcohols, and aldehydes revealed the greatest responses to be from E-6 and Z-11 C_{16} acetates, which suggested (E,Z)-6,11-hexadecadienyl acetate as the major phero-mone component. Microozonolysis on 250 ng of material confirmed the positions of the double bonds.

GC data suggested that the minor component was a C_{16} aldehyde, and (E,Z)-6,11-hexadecadienal seemed a reasonable first guess. This guess was confirmed when the purified minor component was reduced with LiAlH$_4$, the resultant alcohol preparatively collected at the C_{16} alcohol GC retention time, and the material then acetylated to give a substance having chromatographic, chemical, and EAG properties identical to those of the major component.

Field bioassay with authentic synthetic material confirmed that a 9 : 1 mixture of acetate and aldehyde was highly attractive and thus constituted the sex pheromone.

b) Japanese beetle *(Popillia japonica)*

TUMLINSON *et al.* (73) purified the rinses of glass vessels used to hold virgin females by gel permeation liquid chromatography and sequential preparative GC on five columns of differing polarity. Bioassays were conducted by pouring fractions into petri dishes, which were placed on a golf-course fairway. The number of males responding in five minutes was counted and compared with the number responding to three virgin females in a small cage placed near by during the same time.

MS, IR, and NMR data suggested a γ-lactone of a 14-carbon hydroxy acid with one double bond of possibly Z configuration. Microozonolysis yielded a major compound having MS and GC properties identical to those of nonanal. The racemic Z- and E-isomers plus the saturated analog were synthesized and their spectrometric, chromatographic, and chemical properties were compared with those of material obtained from the insects. All three compounds were present in the extracts, but only the Z-isomer (144) corresponded to the active compound.

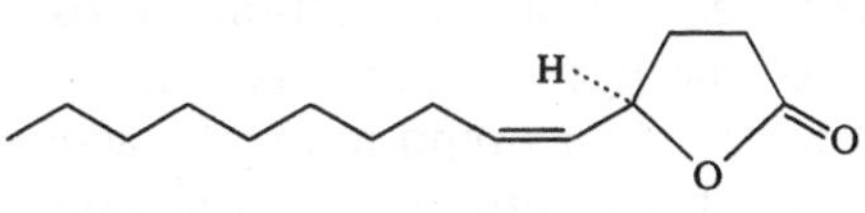

(144)

Since the racemic synthetic material was inactive in the field bioassay, the optically active enantiomers were synthesized. The R,Z enantiomer was attractive and as little as 1% of the S,Z enantiomer significantly reduced the number of males captured by traps baited with pure R,Z enantiomer.

c. California red scale *(Aonidiella aurantii)*

ROELOFS *et al.* (*74, 75*) obtained the crude sex pheromone by passing air over scale-infested potatoes and through a Porapak Q trap, and then extracting the absorbent with pentane. Two active compounds were isolated by Florisil column chromatography followed by HPLC on two different columns and preparative GC.

Both compounds were shown to be acetates when activity was lost on reduction with LiAlH₄ and regained after treatment of the products with acetyl chloride. Hydrogenation and further reduction to a hydrocarbon skeleton of both compounds afforded identical products. MS spectra were indicative of doubly and triply unsaturated C_{14} acetates and only the latter component gave a product, 3-ketobutan-1-yl acetate, on ozonolysis. Finally, NMR spectra were obtained, and the information was used to derive structures (**332a**) and (**333**) for the sex pheromones. These structures were confirmed by synthesis.

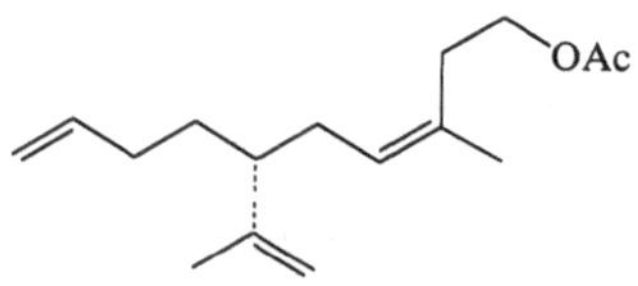

(**332a**)

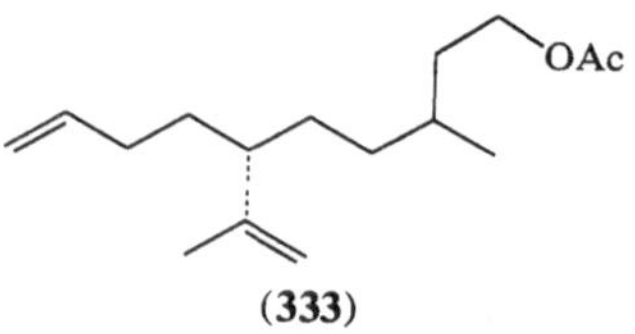

(**333**)

7. Problems

The ultimate proof of structure is unambiguous synthesis followed by demonstration of equivalent biological activity of synthetic material in the field. However, despite all precautions, errors still may occur. The sex pheromone of the gypsy moth was first postulated as (**4**) (*76*) and some synthetic material was reported to be active in the field (*77*). Later studies (*57*) showed that although (**4**) was present in extracts, the active material was (**536**) (see p. 10). Similarly, the sex pheromone of the pink boll worm moth *(Pectinophora gossypiella)* was first reported as (**5**) with synthetic material active in the laboratory (*78*) and later (*79*) shown to be a mixture of Z,Z- and Z,E-isomers of 7,11-hexadecadienyl acetate (**529a** and **529b**, respectively). A number of other sex pheromones have been misidentified, including those for the American cockroach

(Periplaneta americana) *(80)*, the codling moth *(Laspeyresia pomonella)* *(81)*, and the oak leaf roller moth *(Archips semiferanus)* *(82)*. EITER *(83)* offers critical comments on some pheromone identifications.

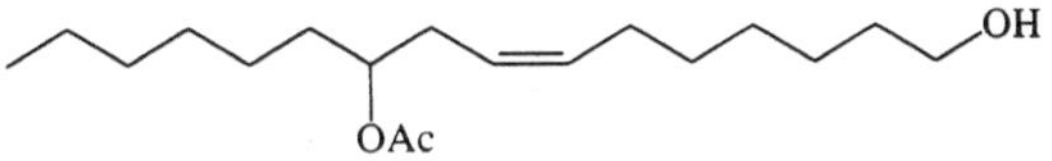

(4)

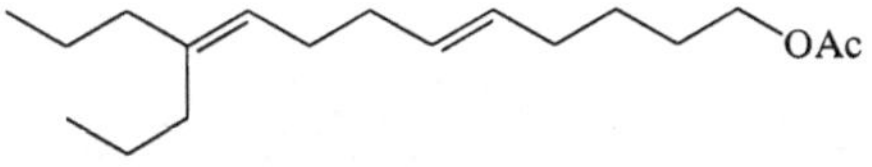

(5)

III. Synthesis

In addition to confirmation of an assigned structure, synthesis of enantiomers or *E/Z*-isomers can provide material to establish which of the compounds is actually produced by an insect or to determine the absolute stereochemistry of a chiral molecule.

As noted above, insect pheromones have relatively simple structures. However, many of the compounds contain double bonds or possess chiral centers. Furthermore, some insects can distinguish between isomers and small amounts of the "wrong" isomer in a synthetic preparation may inhibit responses to the "correct" one. Thus the requirement for stereospecificity renders many syntheses non-trivial. Synthetic chemists have responded to the challenge and many new reactions of widespread utility have been developed; notable among these are reagents for stereospecific generation of *E* and *Z* double bonds.

The remainder of this section is devoted to a survey of novel syntheses and is organized alphabetically by Order and Family. Certain aspects of insect pheromone synthesis have been recently reviewed *(83a, 89, 90)*.

A. Coleoptera

1. Bruchidae

The sex pheromone **(11)** of the dried bean beetle *(Acanthoscelides obtectus)* contains the conjugated carbomethoxyeneallene system. LANDOR et al. *(84)* generated the allene by lithium aluminium hydride reduction of the α-hydroxy-α′-tetrahydropyranyloxy acetylene **(8)**. Subsequent oxidation and coupling with the modified Wittig reagent, trimethyl phosphenoacetate, afforded the racemic ester (Scheme 1). DESCOINS et al. *(85)* prepared the masked functional groups **(14)** and by addition of lithium dioctyl cuprate, in one step produced the conjugated allene and completed the carbon skeleton (Scheme 2). Formation of the conjugated

DHP= Dihydropyran　　　THP = Tetrahydropyranyl

Scheme 1

Scheme 2

allene was accomplished by BAUDOUY and GORE (*86*) in a single step by addition of 2-hydroxypropyne to the α-mesyloxy acetylene (**17**) (Scheme 3). Reduction of the conjugated acetylene with methoxy lithium aluminium hydride was followed by standard methods to give (**11**).

A novel synthesis by MICHELOT and LINSTRUMELLE (*87*) is outlined in Scheme 4. Treatment of the lithium dialkyl cuprate (**22**) of the substituted allene (**21**) with methylpropynoate afforded (**11**) in 94% overall yield.

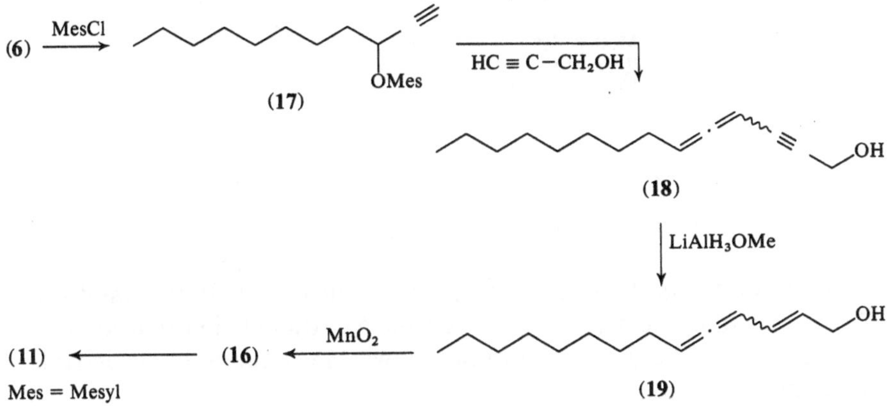

Scheme 3

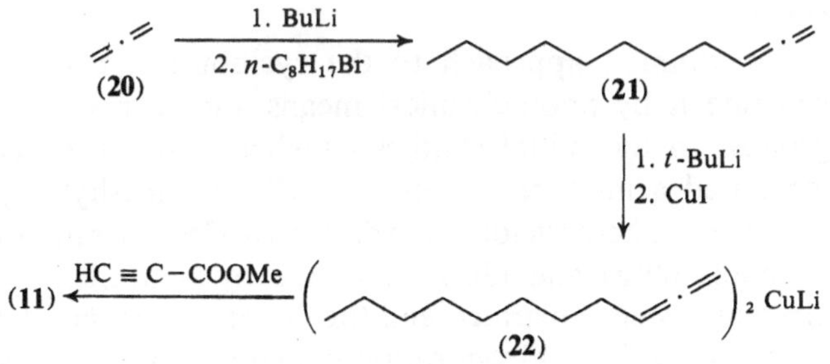

Scheme 4

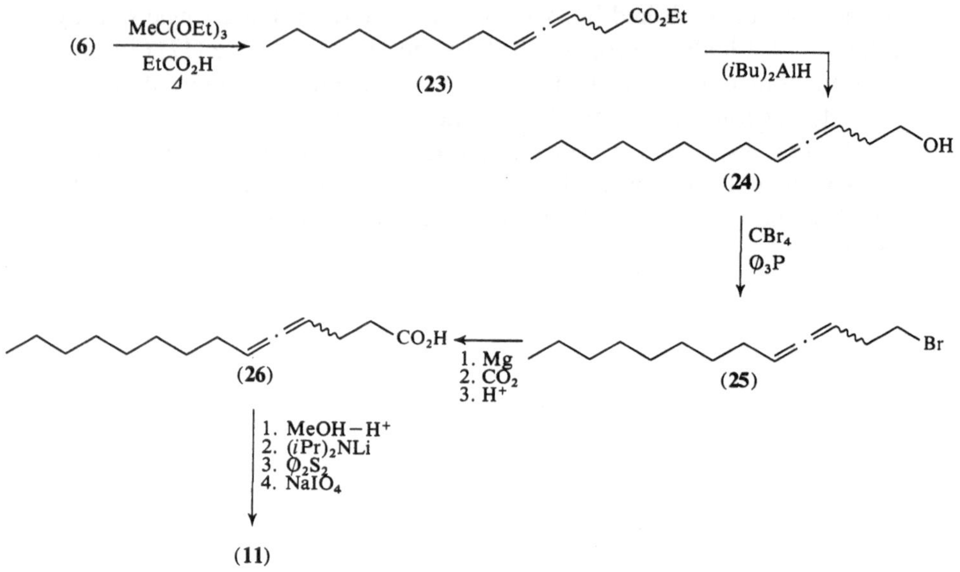

Scheme 5

Kocienski *et al.* (*88*) generated the β-allenic ester (**23**) by a modified Claisen rearrangement of ynol (**6**) (Scheme 5). After chain extension to a γ-allenic ester, the *E* α,β-unsaturation was introduced by selenoxide elimination.

2. Curculionidae

Male boll weevils (*Anthonomus grandis*) produce a mixture of alicyclic terpenes (**33, 110b, 111a,** and **111b**) which synergistically serves as an aggregation and sex pheromone. The unsaturated cyclobutane alcohol (**33**) (grandisol) has generated much interest and activity among synthetic chemists. Since many of the syntheses have been reviewed in detail by Katzenellenbogen (*89*) and Hendrick (*90*) the treatment here will be cursory.

A common approach to the cyclobutane ring system has been to generate it by photochemical means and then elaborate the functional groups. In the initial synthesis (Scheme 6), Tumlinson *et al.* (*38, 91*) irradiated a mixture of isoprene (**27**) and methylvinyl ketone (**28**) and obtained the keto cyclobutane derivative (**29**) as a mixture of diastereomers. Alcohols (**30a**) and (**30b**), obtained by treatment of (**29**) with methyl Grignard, were separated and the desired *cis*-isomer (**30a**) was converted to grandisol by hydroboration-oxidation followed by dehydration and hydrolysis.

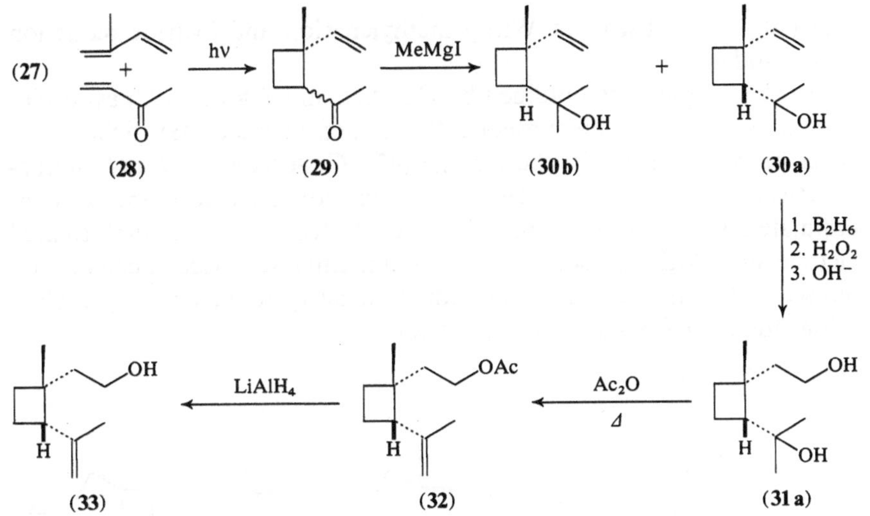

Scheme 6

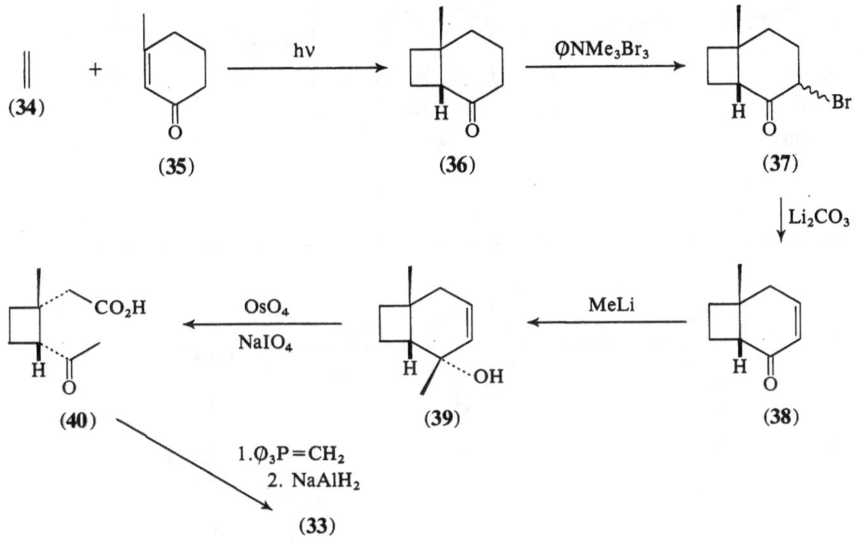

Scheme 7

By generating a fused bicyclic ring system, ZURFLÜH *et al.* (*92*) were able to preform the required *cis*-configuration at the cyclobutane ring (Scheme 7). Cyclohexanone (**36**) was transformed to a cyclohexenone (**38**). Allylic alcohol (**39**) was cleaved with osmium tetroxide-periodate to keto acid (40), and subsequent Wittig methylenation and hydride reduction gave grandisol.

Keto acid (40) was obtained by Cargill and Wright (93) via cyclopentanone (42) and the subsequently formed ozonide (45) (Scheme 8). After preparation of bicyclic lactone (47), Gueldner et al. (94) stereoselectively generated diol (31a), which was converted to grandisol contaminated with isomeric (44) (Scheme 9). Kosugi et al. (95) treated lactone (47) with the sodium salt of dimethyl sulfoxide (Scheme 10). Subsequent aluminium amalgam reduction, acetylation, Wittig methylenation, and hydrolysis afforded grandisol.

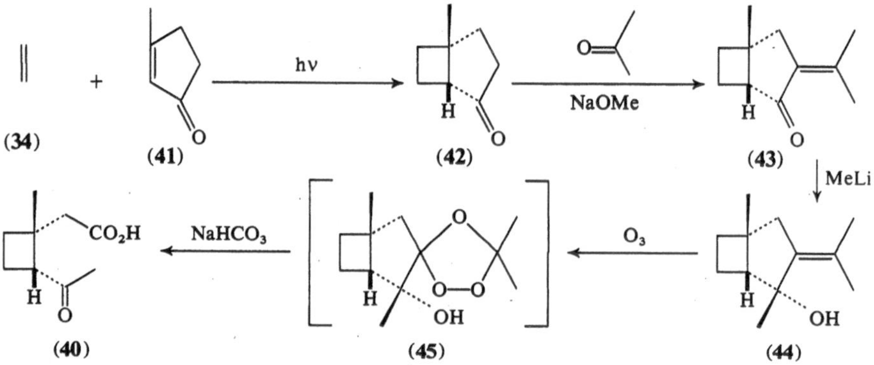

Scheme 8

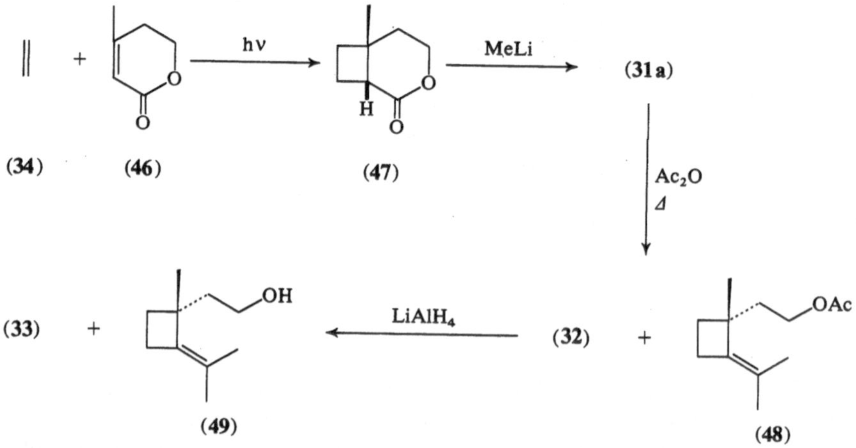

Scheme 9

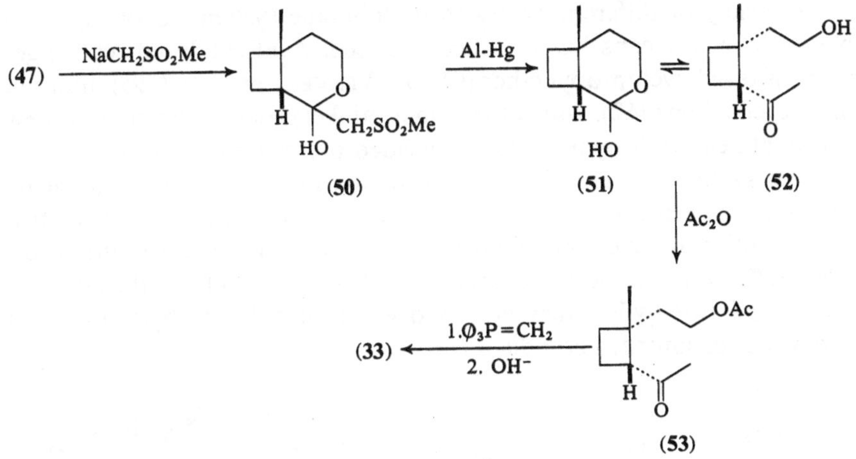

Scheme 10

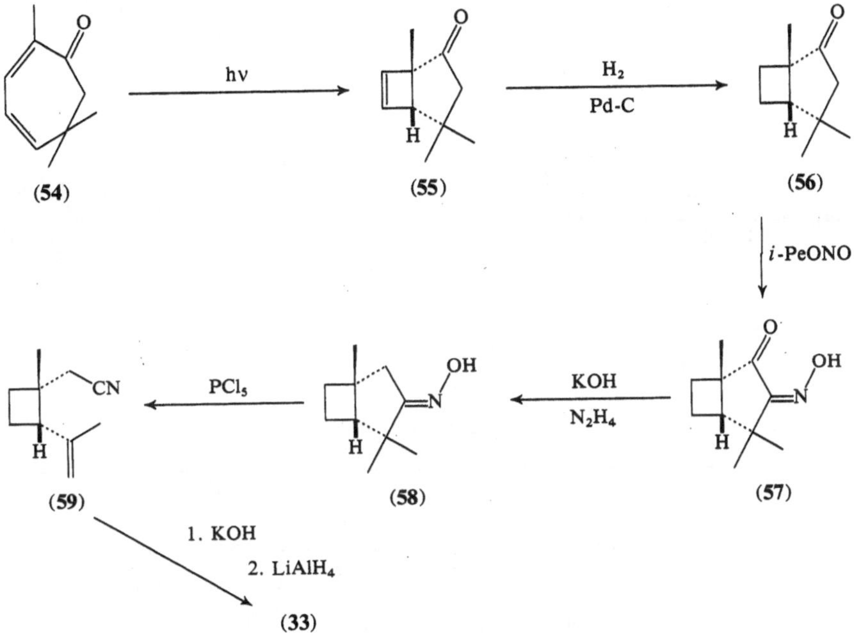

Scheme 11

An intramolecular photocyclization of eucarvone (**54**), which contains all of the required carbon atoms, was used by AYER and BROWNE (96) to give cyclopentanone (**55**) (Scheme 11). Transposition of the keto group, Beckman fragmentation of the resultant oxime (**58**), hydrolysis and reduction completed the synthesis.

A variety of different routes to cyclobutane systems involving non-photochemical means has also been devised. Acid catalyzed rearrangement of fused cyclopropyl ether (61) by Wenkert et al. (97, 98) afforded the bicyclic dione (62), which through a thioketal-desulfurization process and treatment with hydroxylamine yielded oxime (58) (Scheme 12). The shortest synthesis of grandisol is that reported by Billups et al. (Scheme 13) (99). Dimerization of isoprene in the presence of a zero-valent bis-cyclooctadienyl-nickel-phosphite complex gave the cis-cyclobutane di-olefin (65), which could be separated in 12—15% yield from the complex product mixture by low-temperature distillation. Selective hydroboration and oxidation afforded grandisol.

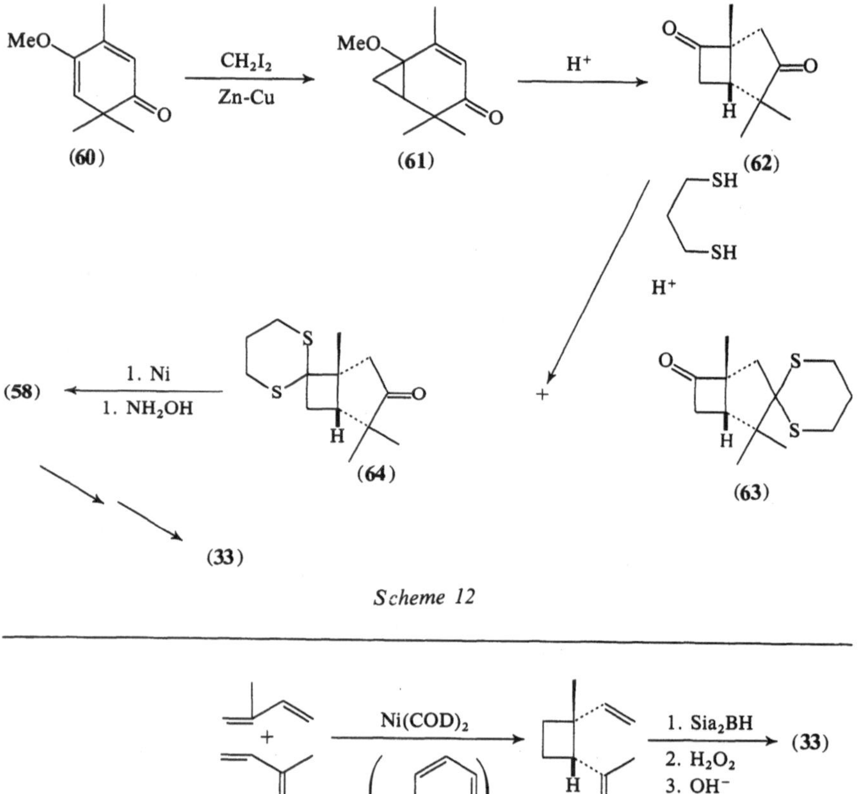

Scheme 12

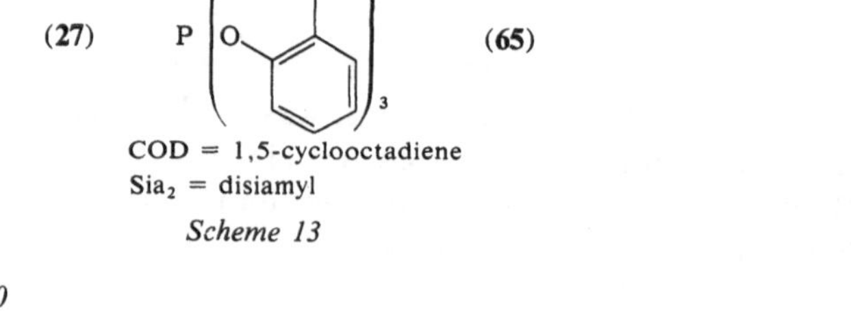

COD = 1,5-cyclooctadiene
Sia₂ = disiamyl

Scheme 13

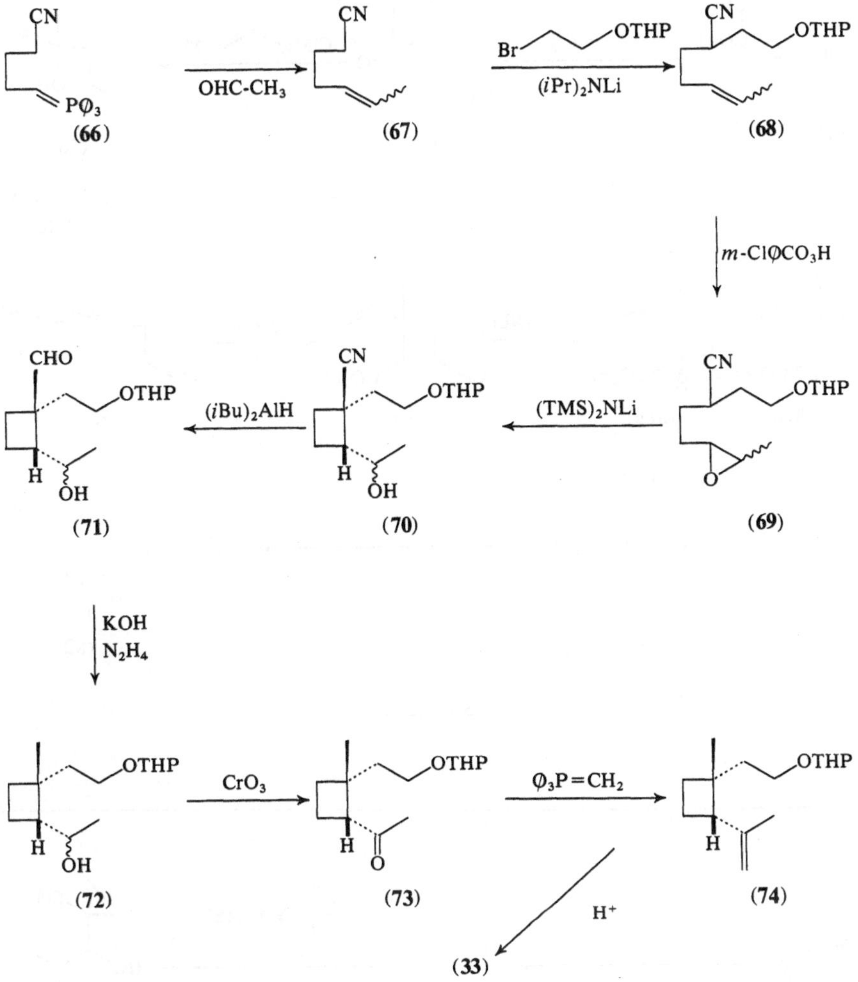

(66) → OHC-CH₃ → (67) → Br⟋⟋OTHP / (iPr)₂NLi → (68)

m-ClØCO₃H

(71) ← (iBu)₂AlH ← (70) ← (TMS)₂NLi ← (69)

KOH
N₂H₄

(72) → CrO₃ → (73) → Ø₃P=CH₂ → (74)

(33)
H⁺

(TMS)₂NLi = Lithium hexamethyldisilazane

Scheme 14

STORK and COHEN (*100*) gained entry into the *cis*-cyclobutane system by stereoselective cyclization of epoxynitrile (**69**) (Scheme 14). Reduction of the nitrile to a methyl group and elaboration of the isopropenyl group completed this synthesis. Cyclization of δ-chloroester (**80**) by BABLER (*101*) gave a 65 : 35 mixture of *cis*- and *trans*-cyclobutane olefin esters (**81 a** and **81 b,** respectively), which were carried through to a mixture of grandisol (**33**) and fragrantol (**82**), respectively (Scheme 15).

Scheme 15

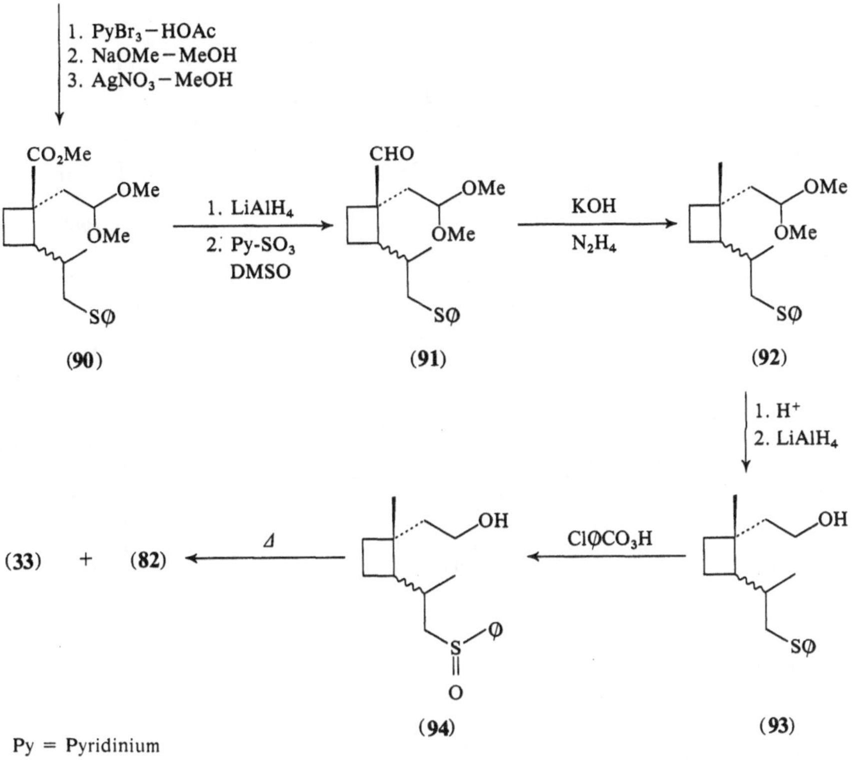

1. PyBr₃–HOAc
2. NaOMe–MeOH
3. AgNO₃–MeOH

CO₂Me
OMe
OMe
SØ
(90)

1. LiAlH₄
2. Py-SO₃
DMSO

CHO
OMe
OMe
SØ
(91)

KOH
N₂H₄

OMe
OMe
SØ
(92)

1. H⁺
2. LiAlH₄

OH
SØ
(93)

ClØCO₃H

OH
S
Ø
O
(94)

Δ

(33) + (82)

Py = Pyridinium

Scheme 16

In a novel approach, Trost and Keeley (*102*) used the annelation reagent lithiocyclopropylphenyl sulfide (**85**) to prepare first cyclobutanone (**88**) and then the fused biscyclobutane spiro system (**89**) as a mixture of diastereomers with the desired conformer in four fold excess (Scheme 16). Haloform-type cleavage afforded cyclobutane derivative (**90**), which upon transformation of the various functionalities yielded grandisol.

All of the above syntheses gave racemic grandisol. By starting with optically active (−)-β-pinene (**95**), which has a suitably substituted cyclobutane ring, Hobbs and Magnus (*103, 104*) were able to obtain (+)-(1*R*,2*S*)-grandisol (**33**) (Scheme 17). The ethanol side chain (in **100**) was elaborated by oxidation of the *endo*-methyl group, Wittig elongation and selective hydroboration-oxidation. Allylic oxidation of (**102**) and subsequent reduction yielded ketone (**104**). Upon irradiation this ketone underwent a Norrish type I cleavage to afford aldehyde (**105**), which contained the desired isopropenyl group. Rhodium catalyzed decarbonylation and hydrolysis afforded enantiomerically pure (+)-grandisol.

Scheme 17

Syntheses of the olefinic dimethylcyclohexane derivatives (**110b, 111a** and **111b**) have been relatively straightforward. 3,3-Dimethyl cyclohexanone (**106**) (Schemes 18 (*38, 91*), 19 (*105*), 20 (*106*), and 21 (*107*)), isophorone (**116**) (Schemes 22 (*108*) and 23 (*109*)) and geranic acid (**120**) (Scheme 24 (*110*)) have served as starting materials.

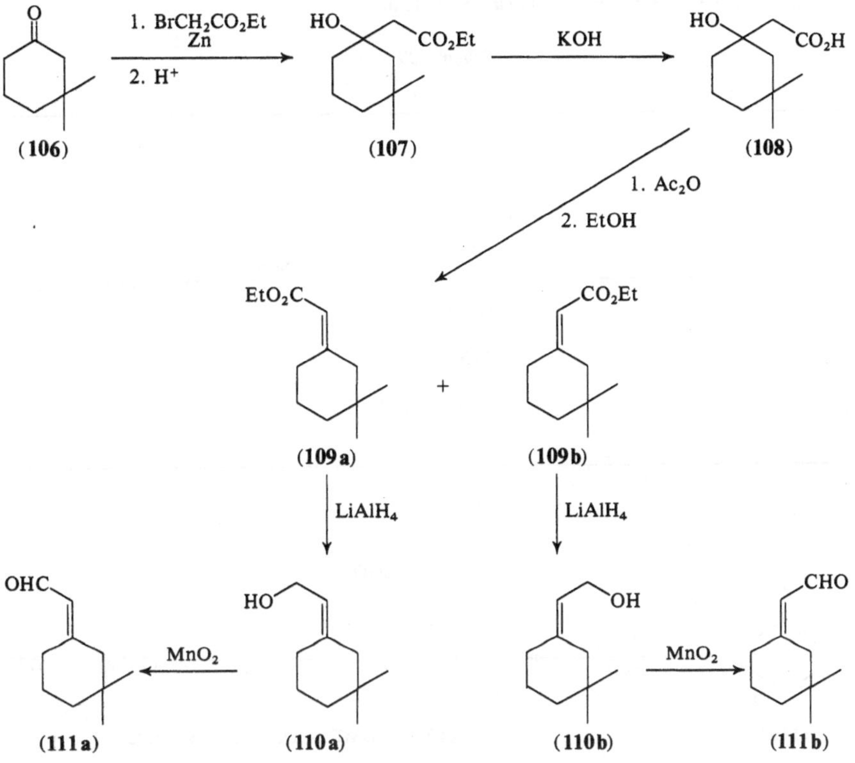

Scheme 18

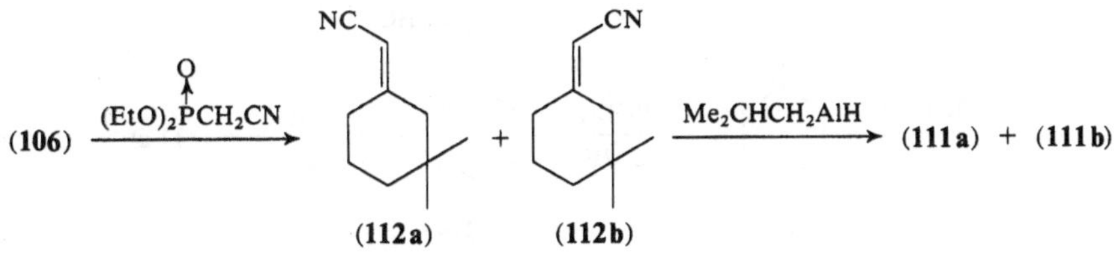

Scheme 19

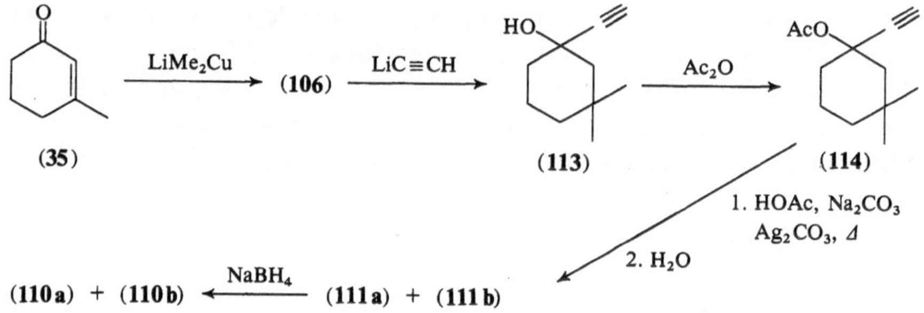

(35)   (106)   (113)   (114)

1. HOAc, Na$_2$CO$_3$
   Ag$_2$CO$_3$, Δ

(110a) + (110b) $\xleftarrow{\text{NaBH}_4}$ (111a) + (111b)   2. H$_2$O

Scheme 20

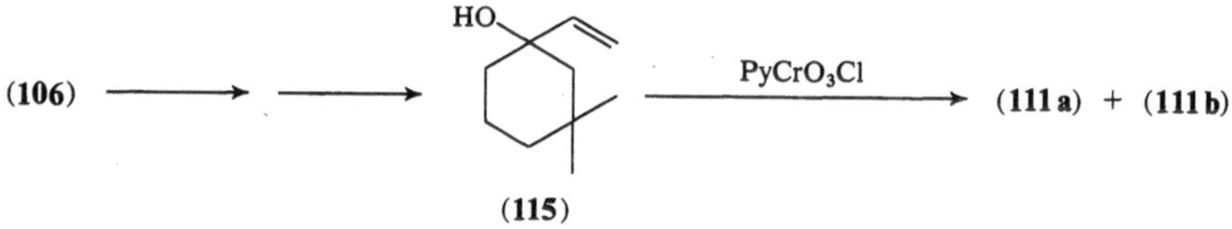

(106) $\longrightarrow$ $\longrightarrow$

(115)

$\xrightarrow{\text{PyCrO}_3\text{Cl}}$ (111a) + (111b)

Scheme 21

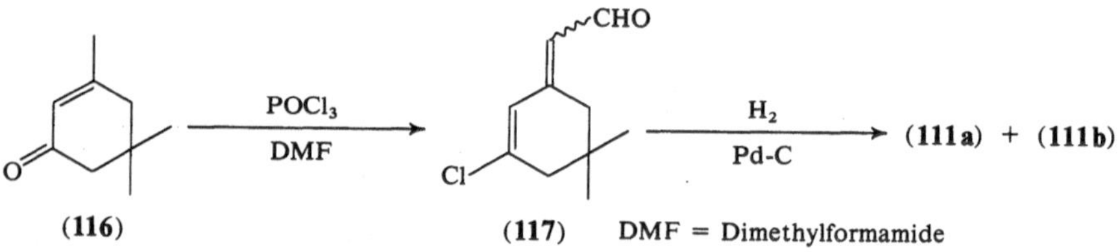

(116)   (117)   DMF = Dimethylformamide

POCl$_3$ / DMF

H$_2$ / Pd-C $\longrightarrow$ (111a) + (111b)

Scheme 22

(116) $\xrightarrow{\text{Na(OR)}_2\text{AlH}_2}$ (118) $\xrightarrow[\text{DMF}]{\text{POCl}_3}$ (119) $\xrightarrow[\text{Pd-C}]{\text{H}_2}$ (111a) + (111b)

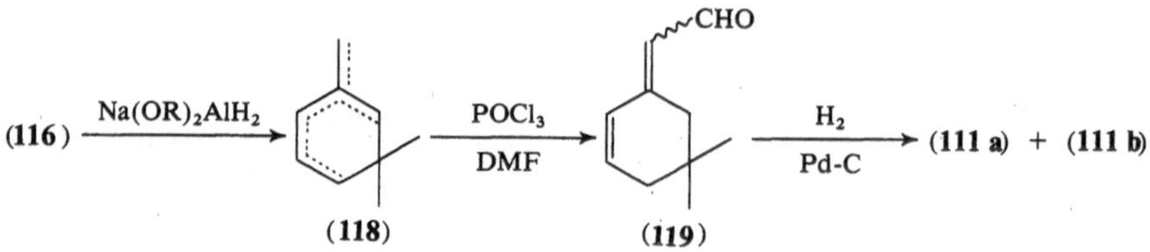

Scheme 23

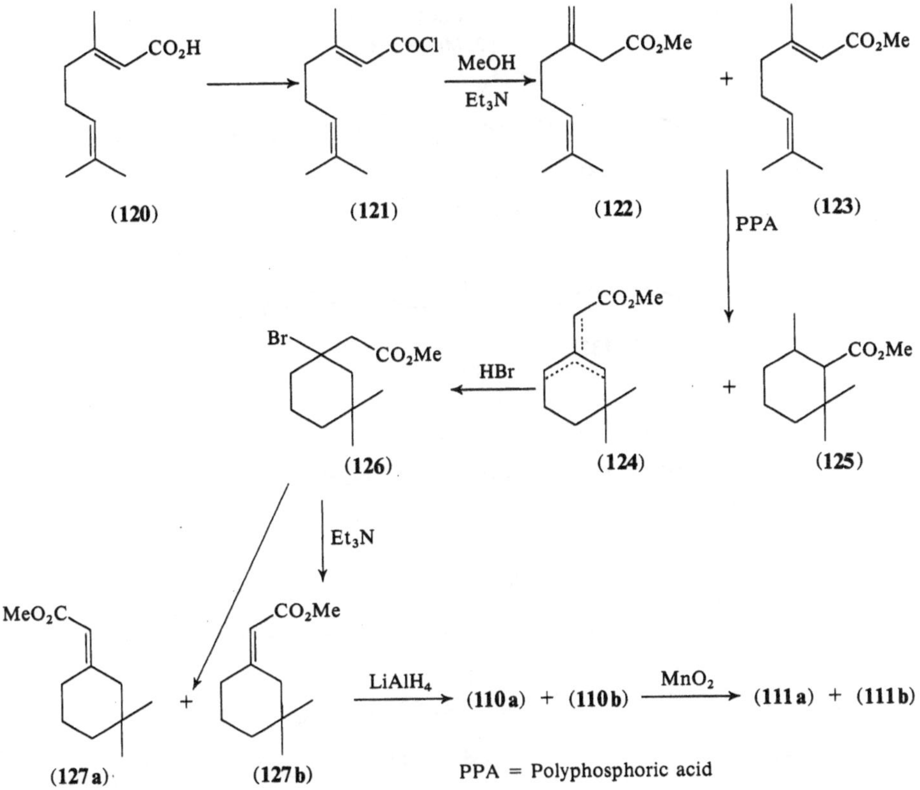
CO₂H
COCl
MeOH
Et₃N
CO₂Me
+
CO₂Me
(120)
(121)
(122)
(123)
PPA
CO₂Me
Br
CO₂Me
HBr
+
CO₂Me
(126)
(124)
(125)
Et₃N
MeO₂C
CO₂Me
+
LiAlH₄
(110a) + (110b)
MnO₂
(111a) + (111b)
(127a)
(127b)
PPA = Polyphosphoric acid

Scheme 24

3. Dermestidae

As shown in Scheme 25, SILVERSTEIN *et al.* (*111, 112*) began elaboration of the conjugated *E,Z* configuration of the black carpet beetle *(Attagenus megatoma)* sex pheromone (**133**) by simultaneous allylic

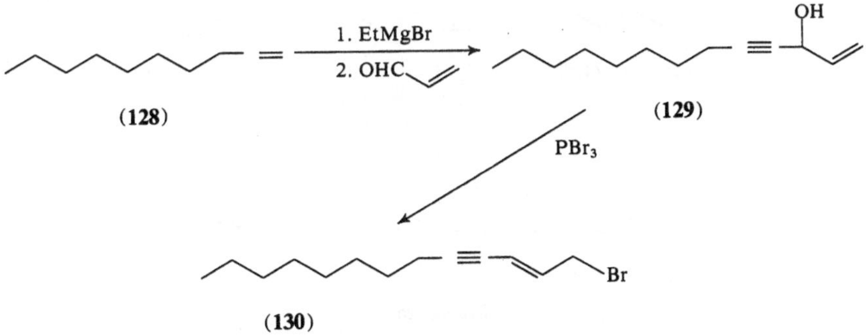
OH
1. EtMgBr
2. OHC
(128)
(129)
PBr₃
Br
(130)

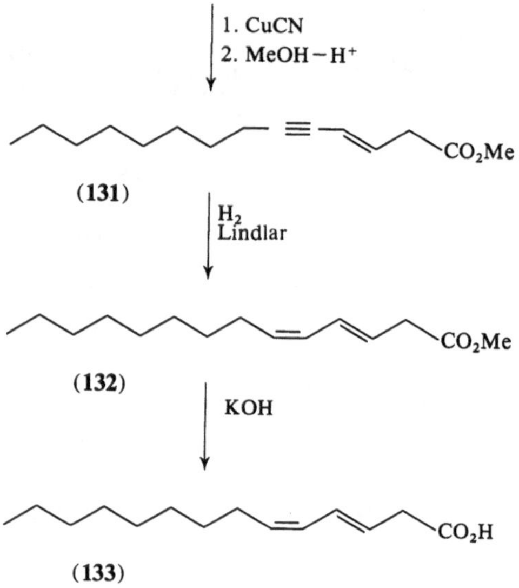
1. CuCN
2. MeOH−H+
CO2Me
(131)
H2
Lindlar
CO2Me
(132)
KOH
CO2H
(133)

Scheme 25

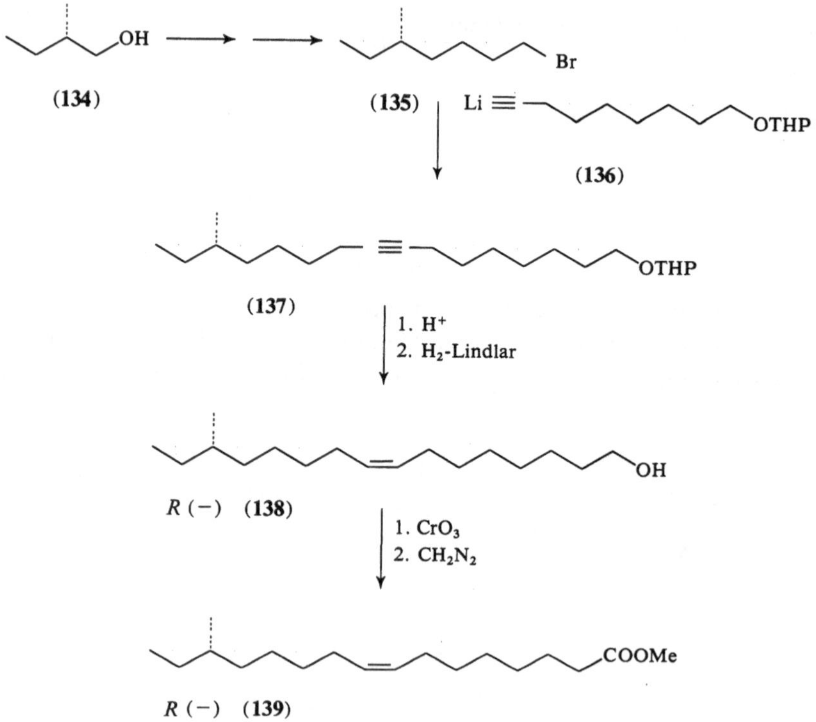
OH
(134)
Br
(135)
Li
OTHP
(136)
OTHP
(137)
1. H+
2. H2-Lindlar
OH
R (−) (138)
1. CrO3
2. CH2N2
COOMe
R (−) (139)

Scheme 26

References, pp. 157—190

rearrangement and bromination of the diunsaturated alcohol (**129**) to the bromide (**130**), which contained the desired *E*-isomer in two fold excess. Subsequent catalytic reduction of the triple bond afforded the diene.

The sex pheromone of *Trogoderma inclusum* is a mixture of the branched unsaturated ester (**139**) and its corresponding alcohol (**138**) and aldehyde (Scheme 26). Two syntheses giving racemic material (*113, 114*) used the Wittig reaction to form the *Z* double bond. Optically active products were obtained by MORI (*115*) *via* the alkyl bromide-lithium alkyne coupling reaction followed by catalytic reduction. In a similar manner, ROSSI and CARPITA (*116*) prepared optically active *S* (**138**), its *E*-isomer and the corresponding *Z*- and *E*-aldehydes.

4. Scarabidae

The Japanese beetle *(Popillia japonica)* sex pheromone (**144**) is a lactone with an unsaturated side chain (Scheme 27). Optically active (*R*)-(−)-glutamic acid (**140a**) served as the starting material in the synthesis by TUMLINSON *et al.* (*73*) shown in Scheme 27. After ring closure with retention of configuration and formation of the aldehyde, inverse addition of the Wittig reagent afforded the desired *R,Z* configuration.

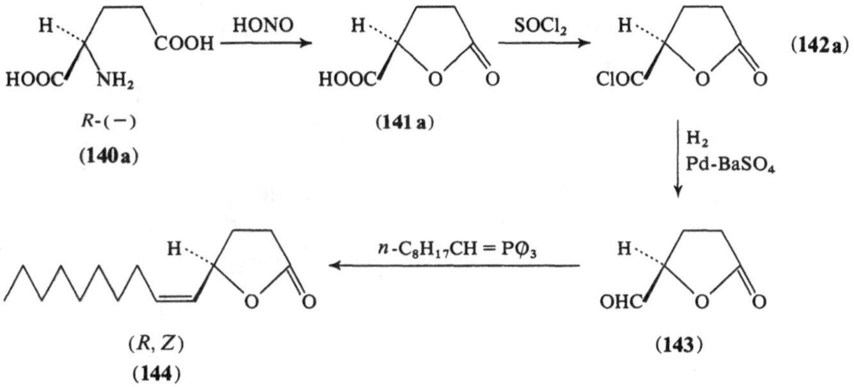

Scheme 27

5. Scolytidae

Beetles of the genus *Dendroctonus* use the bicyclic ketals *exo*-brevicomin (**150**, formulas in Chart 1) and frontalin (**195**) and the terpene *trans*-verbenol (**232**) as part of their pheromonal communication systems.

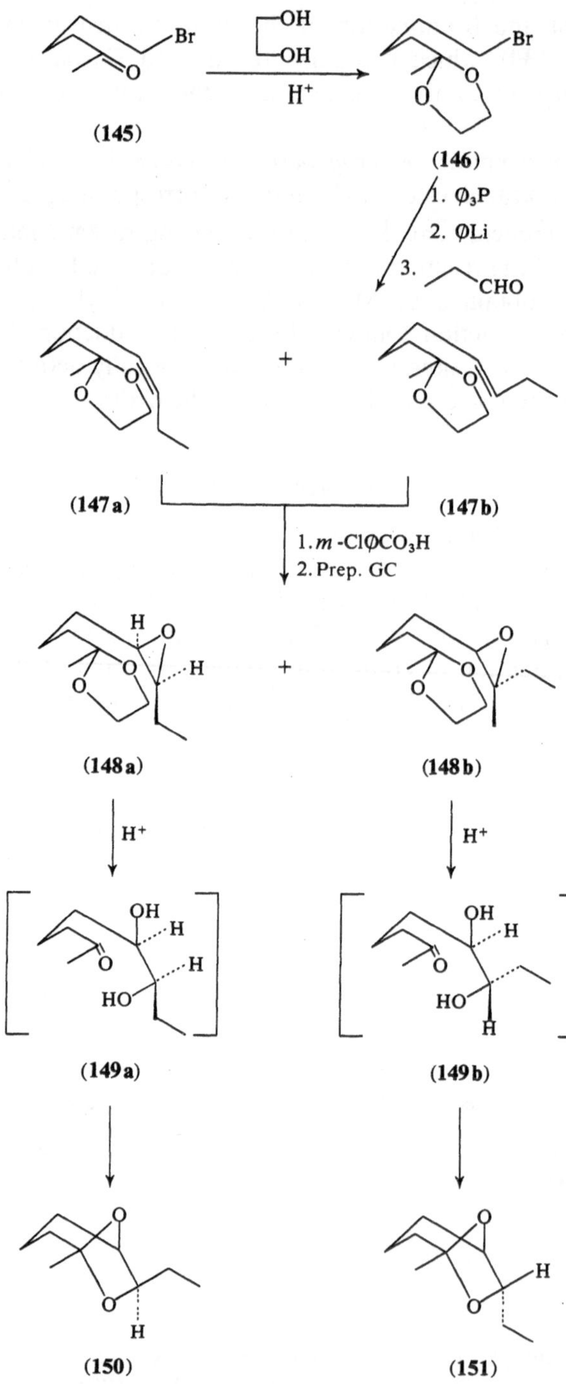

J. M. Brand, J. Chr. Young, and R. M. Silverstein:

(145)

(146)

1. $\varnothing_3$P
2. $\varnothing$Li
3. ⌒CHO

(147a) + **(147b)**

1. m-Cl$\varnothing$CO$_3$H
2. Prep. GC

(148a) + **(148b)**

H$^+$ H$^+$

(149a)   **(149b)**

(150)   **(151)**

Scheme 28

Syntheses of *exo*-brevicomin and its *endo*-epimer (**151**) have followed two general approaches. The first involves generation of the ketodiol intermediates (**149a** and **149b**) of Scheme 28, which on treatment with acid give *exo*- and *endo*-brevicomin respectively, or the ketoepoxy intermediates (**184a** and **184b**) of Scheme 35, which are thermally cyclized to give (**150**) and (**151**), respectively. The second approach begins with Diels-Alder cyclizations. Syntheses of frontalin have followed the same approaches. For a discussion of *trans*-verbenol, see *Ips* spp. below.

A mixture of epimeric epoxyketals (**148a** and **148b**), separable by preparative GC, was obtained by Silverstein *et al.* (*62, 117*) upon treatment of the isomeric mixture of (**147a** and **147b**) with *m*-chloroperbenzoic acid (Scheme 28). Acid hydrolysis of (**148a**) gave the intermediate ketodiol (**149a**), which spontaneously cyclized to racemic *exo*-brevicomin (**150**). Similarly, (**148b**) afforded racemic *endo*-brevicomin (**151**). Stereoselective syntheses of (**147a**) and/or (**147b**) have been achieved by Bellas *et al.* (*117*), Kocienski and Ostrow (*118*), and Mori (*119*) (Schemes 29—31, respectively). Knolle and Schaefer (*120*) prepared the carbon skeleton (**162a**) in one step *via* Kolbe electrolysis (Scheme 32).

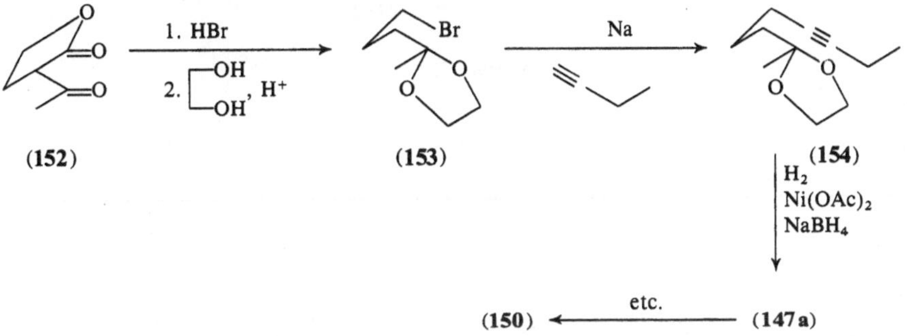

Scheme 29

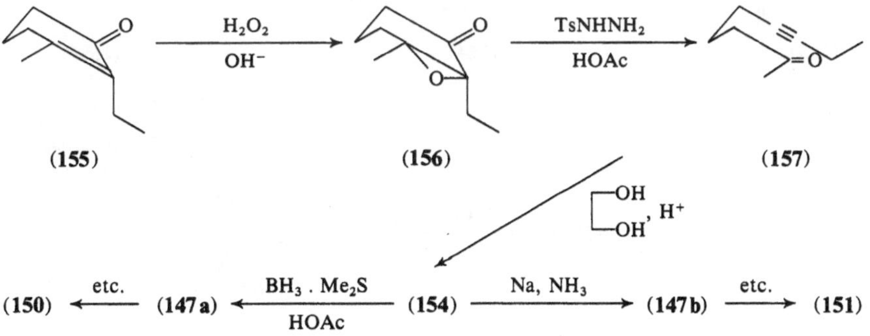

Ts = *p*-Toluenesulfonyl

Scheme 30

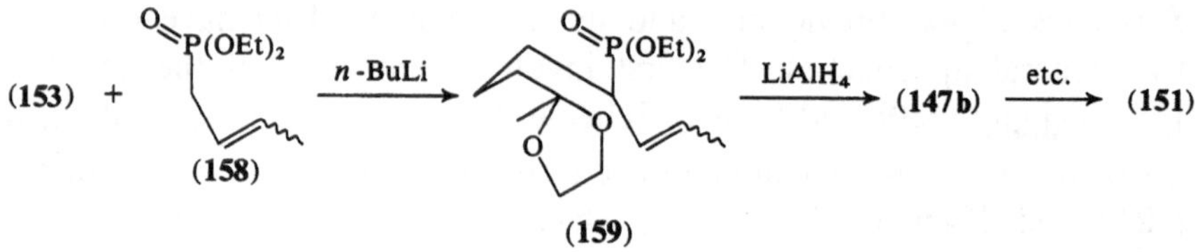

Scheme 31

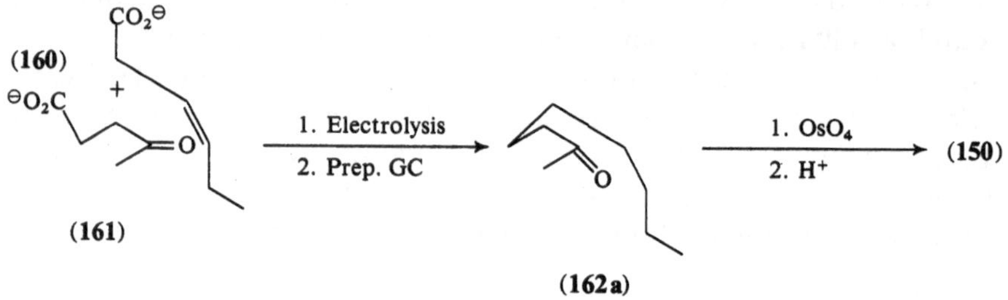

Scheme 32

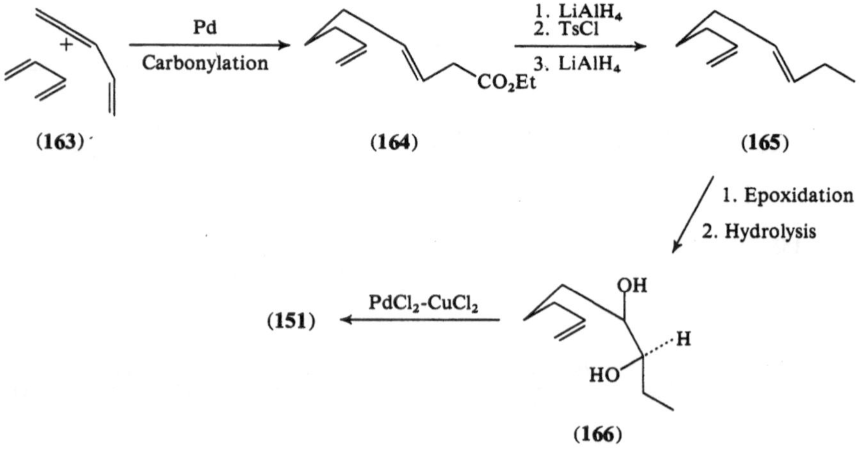

Scheme 33

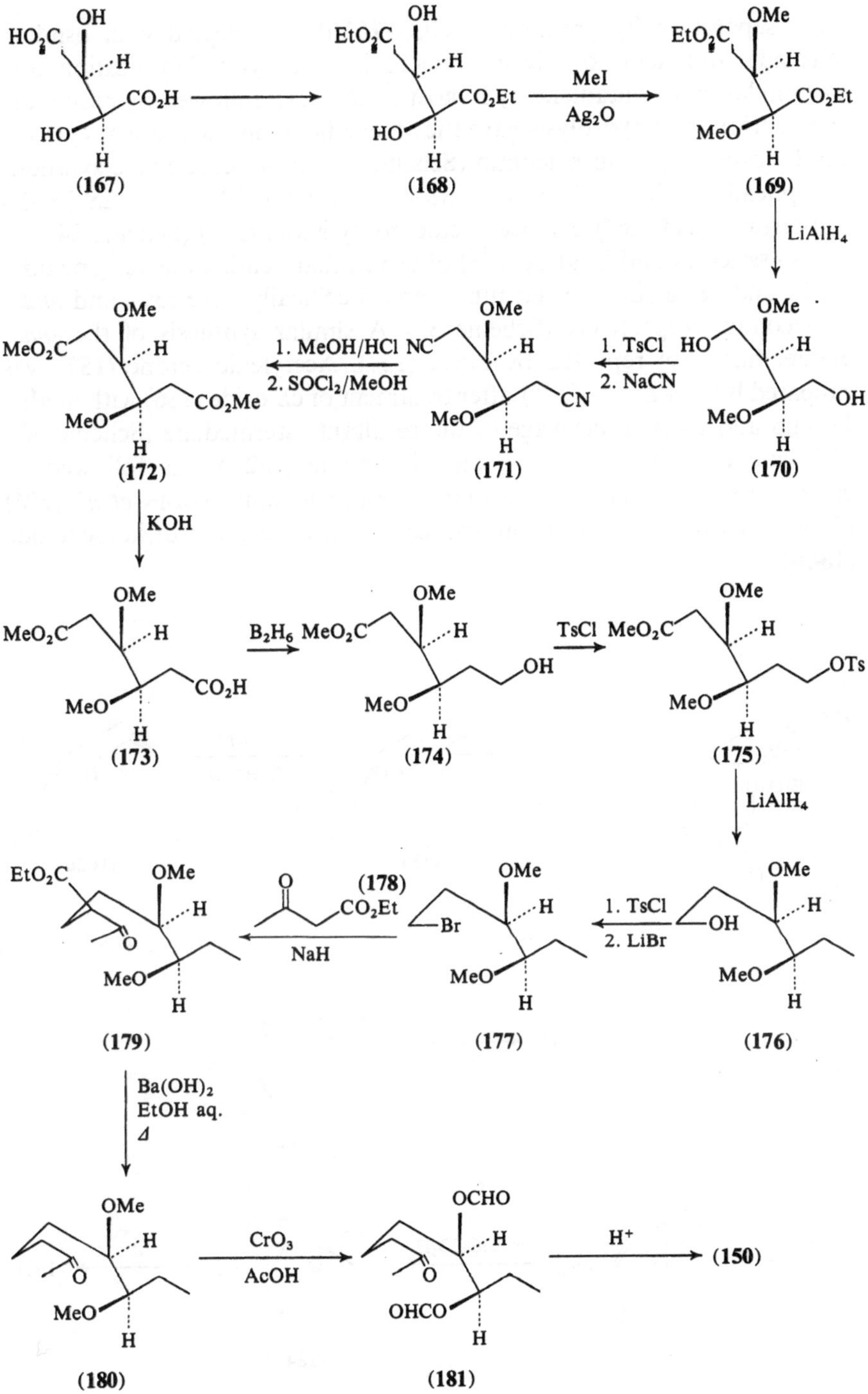

(167)
(168)
(169)
MeI
Ag₂O
LiAlH₄
1. TsCl
2. NaCN
(170)
(171)
1. MeOH/HCl
2. SOCl₂/MeOH
(172)
KOH
(173)
B₂H₆
(174)
TsCl
(175)
LiAlH₄
(176)
1. TsCl
2. LiBr
(177)
(178)
NaH
(179)
Ba(OH)₂
EtOH aq.
Δ
(180)
CrO₃
AcOH
(181)
H⁺
(150)

Scheme 34

After separation by preparative GC, (162a) was treated with osmium tetroxide and acid to afford brevicomin. Catalytic dimerization and carbonylation of butadiene by Byrom et al. (121), followed by reduction, epoxidation, and hydrolysis gave the alkenediol (166), which was cyclized catalytically to endo-brevicomin (Scheme 33). A stereoselective synthesis of optically active (1R,7R)-(+)-exo-brevicomin (150) from (2S,3S)-D-(−)-tartaric acid (167) has been achieved by Mori (122) (Scheme 34).

Wasserman and Barber (123) observed that heating the ketoepoxides (184a and 184b) in a sealed tube stereospecifically gave exo- and endo-brevicomin, respectively (Scheme 35). A similar synthesis of the endo-epimer has been reported by Look (124). Acetylenic ketone (157) was prepared by Coke et al. (125) after treatment of chloride (186) with methyl lithium and thermal cleavage of the resultant intermediate (Scheme 36). Catalytic reduction to the E olefinic ketone (162a) was followed by epoxidation and thermolysis to give exo-brevicomin. Rodin et al. (126) obtained exo-brevicomin from acid catalyzed cyclization of ketoepoxide (184a).

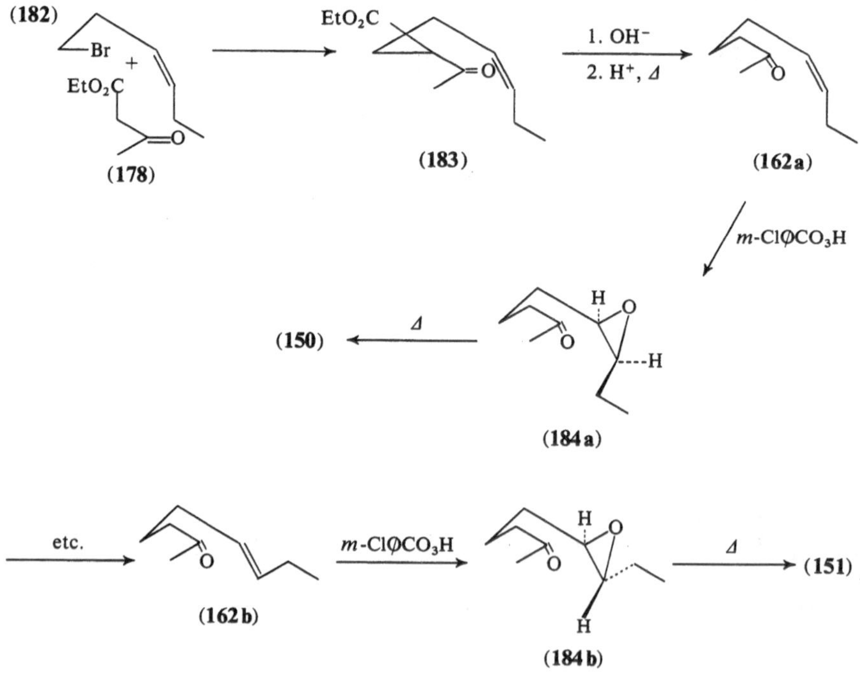

Scheme 35

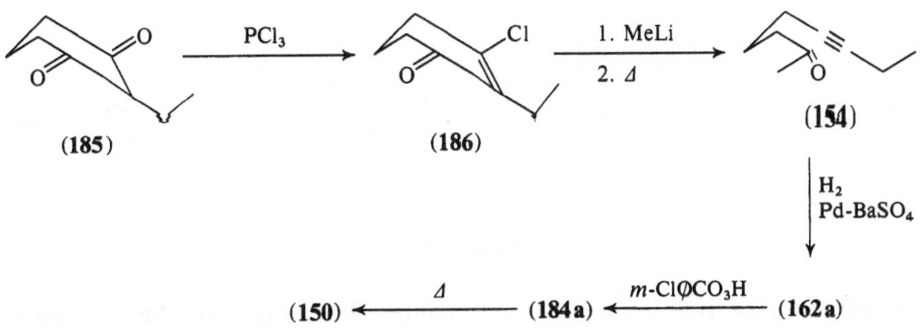

Scheme 36

As shown in Scheme 37, MUNDY *et al.* (*127*) treated the Diels-Alder adduct (**188**) of acrolein (**187**) and methylvinyl ketone (**28**) with ethyl-magnesium bromide to obtain an epimeric hydroxydihydropyran (**189**), which cyclized in the presence of mercuric acetate to a mixture of *exo-* and *endo*-brevicomin. Difficulties in purifying (**188**) precludes large-scale

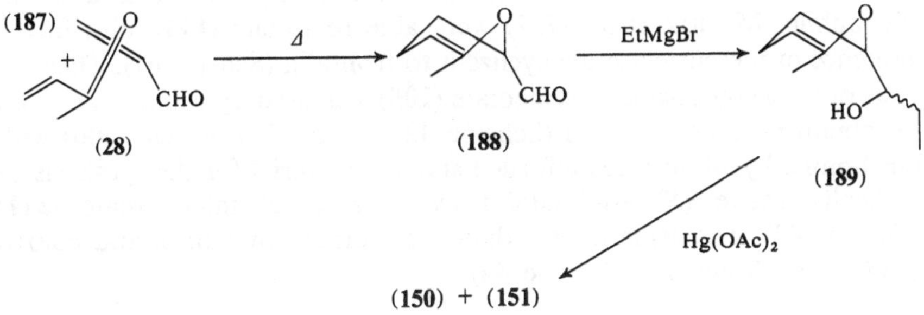

Scheme 37

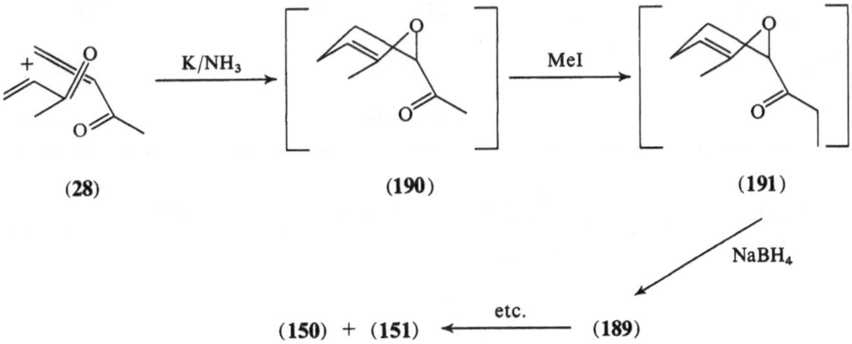

Scheme 38

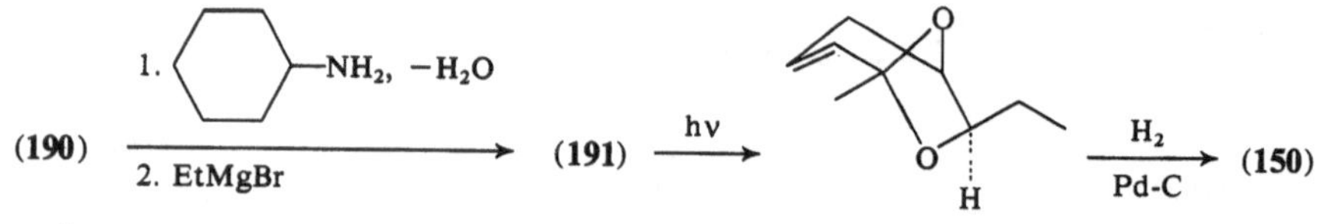

Scheme 39

preparation of brevicomin by this route.. In a later synthesis, Lipko-witz et al. (*128*) prepared (**189**) from the methylvinyl ketone dimer (**190**) (Scheme 38). Chaquin et al. (*129*) recently reported the stereospecific photolytic cyclization of keto dihydropyran (**191**) to an *exo*-ethylenic ketal (**192**), which readily yielded *exo*-brevicomin on catalytic reduction (Scheme 39).

The first synthesis of racemic frontalin (**195**) was accomplished in one step by Kinzer et al. (*130*) with a Diels-Alder reaction postulated to proceed *via* the hydroxy-dihydropyran intermediate (**194**) (Scheme 40). D'Silva and Peck (*131*) improved the synthesis by substituting formal-dehyde and acetone for methylvinyl ketone. By use of a different dienophile, Mundy et al. (*127*) were able to isolate (**194**), and in the presence of mercuric acetate cyclize it to frontalin (Scheme 41). Thermo-lysis of the ketoepoxide intermediate (**205**) was used by Mori et al. (*132*) to obtain racemic frontalin (Scheme 42). Resolution of acid (**206**) with cinchonine by Mori (*133*) afforded starting material for the synthesis of optically active (*S*)-(−)-frontalin *via* a ketodiol intermediate (**211**) (Scheme 43). The same intermediate was formed by Ohrui and Emoto (*134*) from *D*-glucose (Scheme 44).

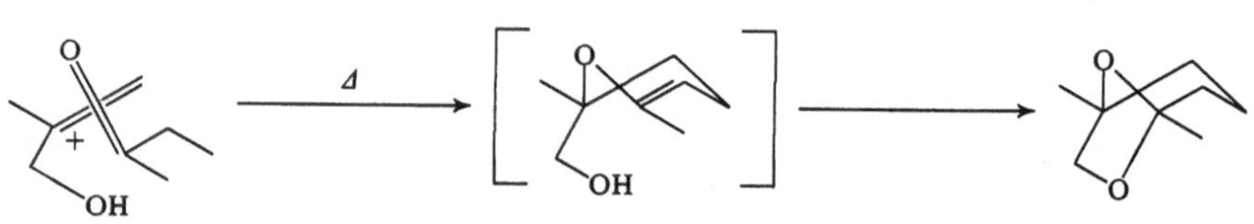

Scheme 40

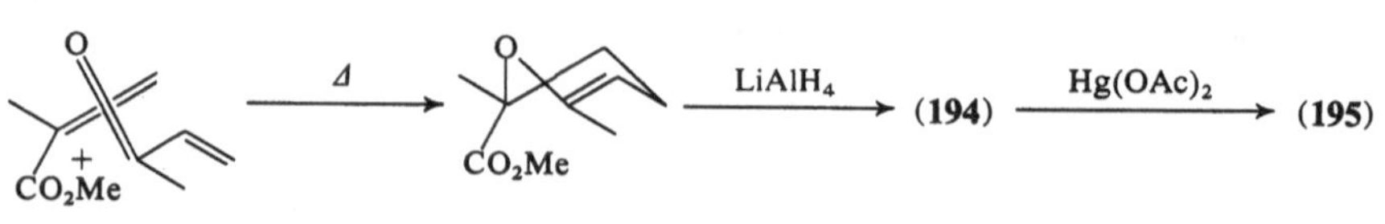

Scheme 41

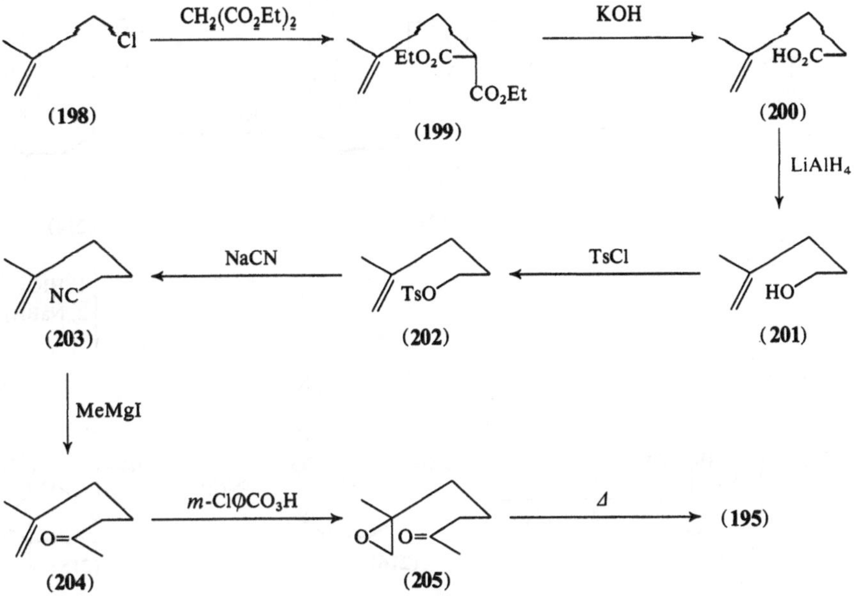

Scheme 42

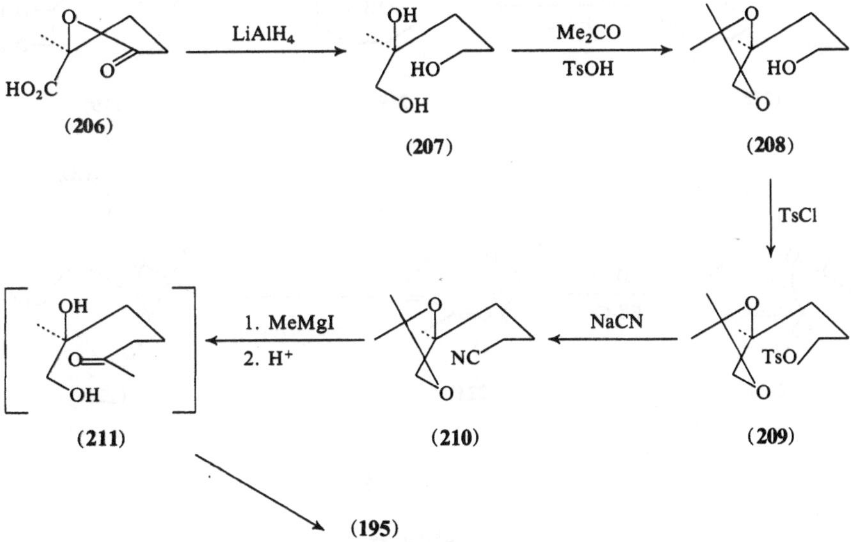

Scheme 43

Scheme 44

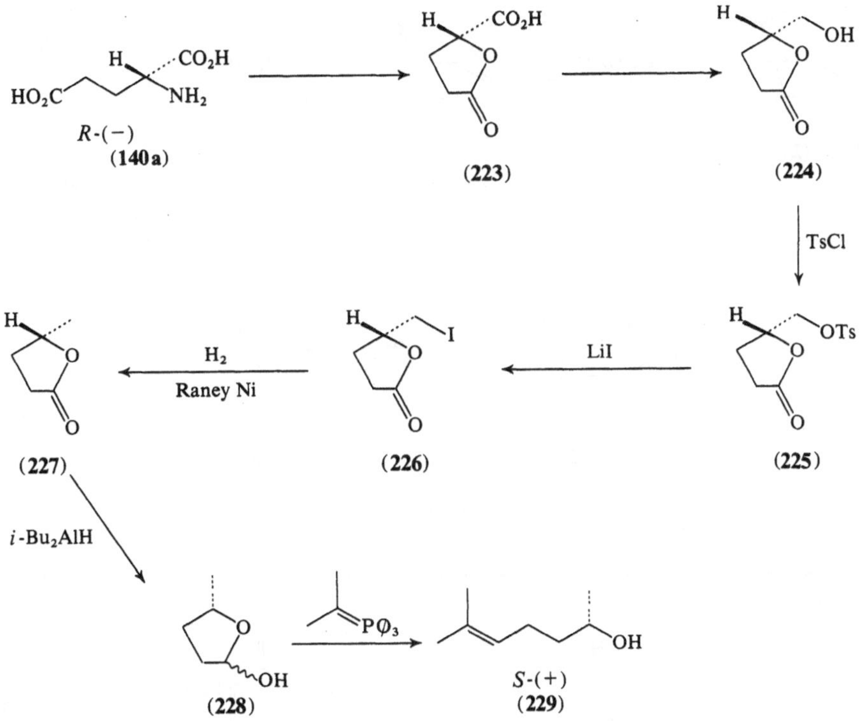

Scheme 45

Sulcatol (**229**), the sex pheromone of *Gnathotrichus sulcatus*, has been synthesized in its optically active (*S*)-(+)- and (*R*)-(−)-forms from (*R*)-(−)- and (*S*)-(+)-glutamic acid (**140**), respectively, by MORI (*135*) (Scheme 45).

Ips spp. bark beetles use a variety of terpene alcohols for communication. Among these are *cis*-verbenol (**234**, Scheme 46), ipsenol (**240a**, Scheme 50), and ipsdienol (**266a**, Scheme 52). *cis*-Verbenol was first synthesized by SILVERSTEIN *et al.* (*51, 136*) by stereospecific reduction of verbenone (**233**) with sodium borohydride. MORI (*47*) synthesized *trans*-verbenol (**232**), a *Dendroctonus* spp. attractant, by lead tetraacetate oxidation of α-pinene (**230**) followed by hydrolysis of the resultant acetate (Scheme 46). Repeated recrystallization of the 3β-acetoxyetienic acid ester afforded optically pure *trans*-verbenol. Oxidation (*48*) of the optically pure *trans*-epimer to verbenone and reduction with lithium aluminium hydride yielded optically pure *cis*-verbenol.

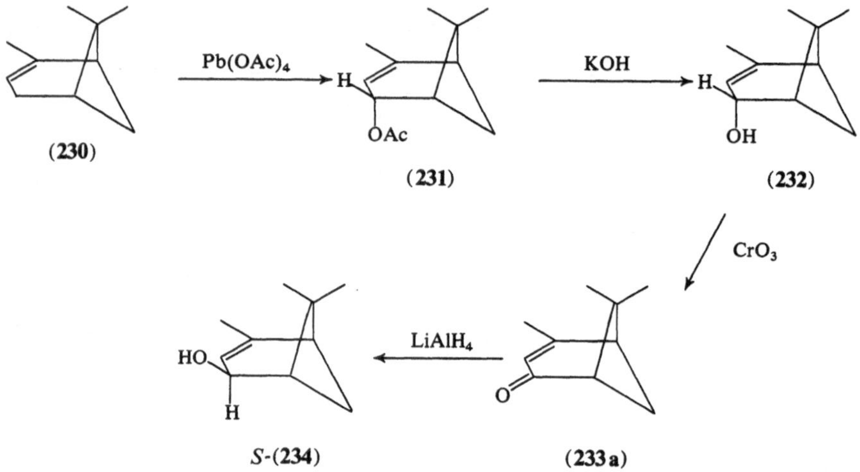

Scheme 46

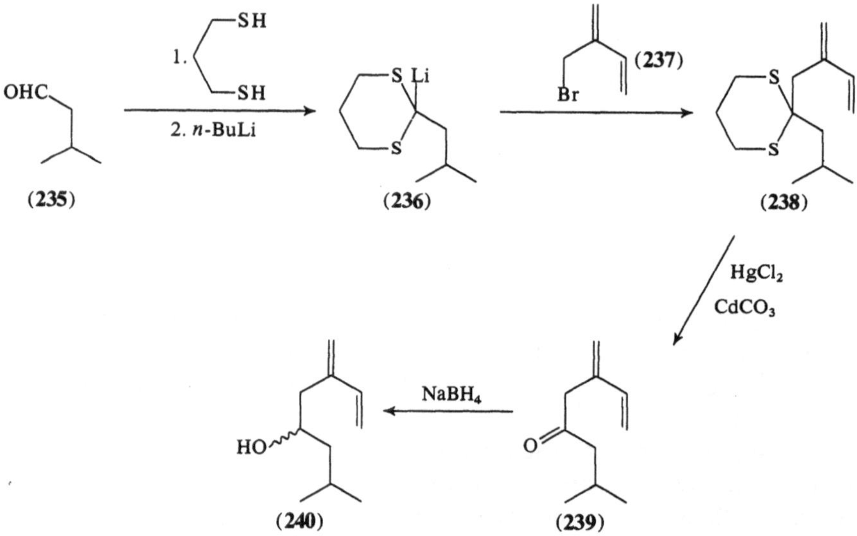

Scheme 47

SILVERSTEIN *et al.* (*51, 136*) began their synthesis of racemic ipsenol by coupling 2-bromomethyl butadiene (**237**) with the anion of the dithiane derivative (**236**) of 3-methylbutanal (**235**) (Scheme 47). Desulfurization of (**238**) and reduction of ketodiene (**239**) gave the desired alcohol (**240**).

KATZENELLENBOGEN and LENOX (*137*) were able to obtain ipsenol in one step by coupling (235) with either (237) in the presence of zinc or 2-mesyloxymethylbutadiene in the presence of lithium.

A thermal sigmatropic rearrangement of the allenic vinyl ether (243) by KARLSEN et al. (*138*) afforded 3-methylene-4-pentenal (244), which was treated with *iso*-butyl magnesium bromide (245) to give ipsenol (Scheme 48). CLINET and LINSTRUMELLE (*139*) prepared ketone (239) by treating the conjugated allenic ketone (247) with vinyl cuprate (Scheme 49). (S)-(+)-Leucine (248) served as the starting material for MORI's stereoselective synthesis (*140, 141*) of (S)-(−)-ipsenol (240 a) (Scheme 50). Epoxide (255) was condensed with diethylmalonate, cyclized, and dehydrated to give the unsaturated lactone (257). After protection of the methylene group as a phenylseleno derivative and reduction to a hemiacetal (*259*), the second methylene was added by treatment with a Wittig reagent.

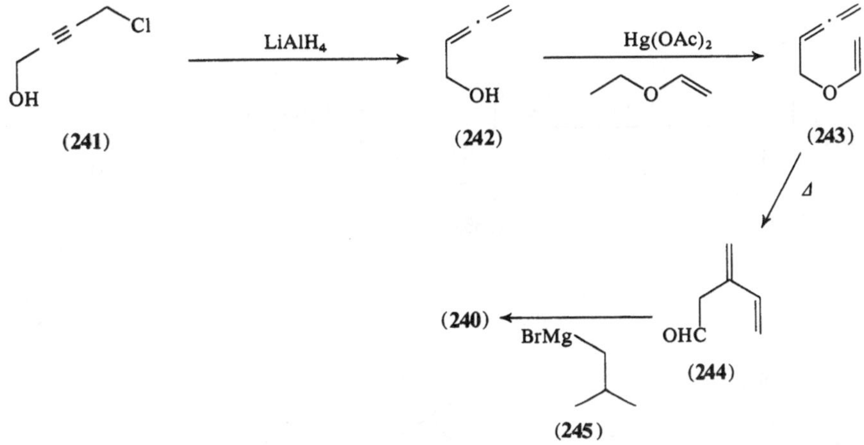

Scheme 48

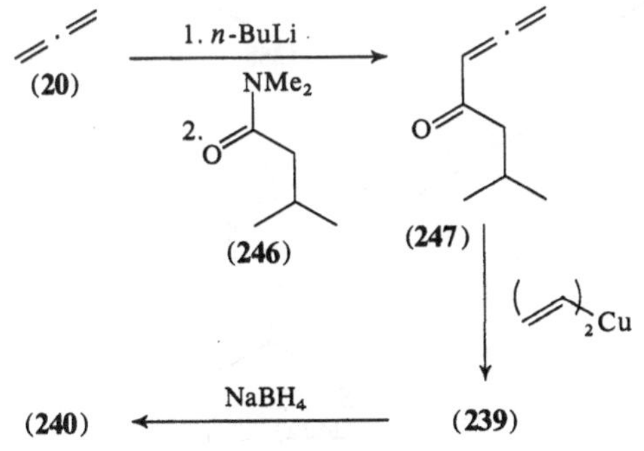

Scheme 49

Scheme 50

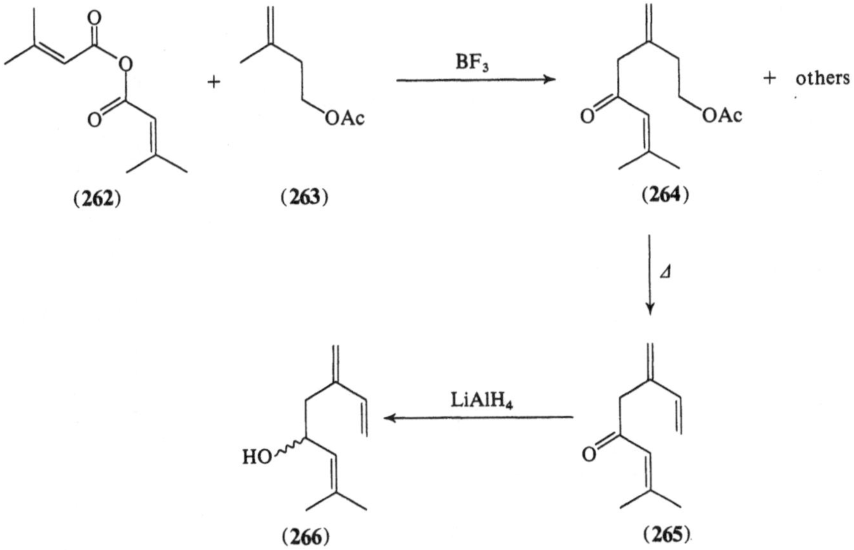

(262) (263) (264) (266) (265)

Scheme 51

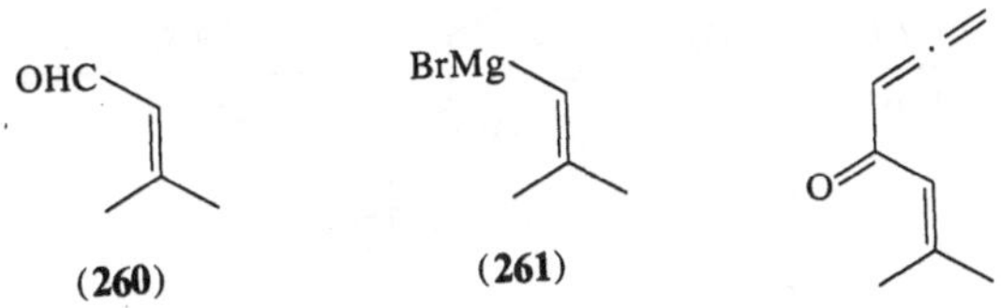

(260) (261) (267)

Syntheses of racemic ipsdienol (**266,** Scheme 51) have tended to follow the approaches used for ipsenol. SILVERSTEIN *et al.* (*51, 136*) began with the dithiane derivative of 3-methyl-2-butenal (**260**) and proceeded as in Scheme 47. RILEY *et al.* (*142*) coupled (**260**) with (**237**) in the presence of zinc, and KARLSEN *et al.* (*138*) coupled (**244**) with the Grignard reagent (**261**). Condensation of senecioic anhydride (**262**) with acetate (**263**) by GARBERS and SCOTT (*143*) gave a mixture of ketoacetates, with (**264**) predominating (Scheme 51). Pyrolytic elimination of acetic acid followed by hydride reduction of ketone (**265**) afforded racemic ipsdienol. By using the cross conjugated ketone (**267**) and vinyl cuprate, CLINET and LINSTRUMELLE (*139*) prepared ketone (**265**). MORI's synthesis (*144*) of (*R*)-(−)-ipsdienol (**266a**) began with *D*-mannitol (**268**) (Scheme 52) and followed essentially the same path as that of (*S*)-(−)-ipsenol (Scheme 50).

Scheme 52

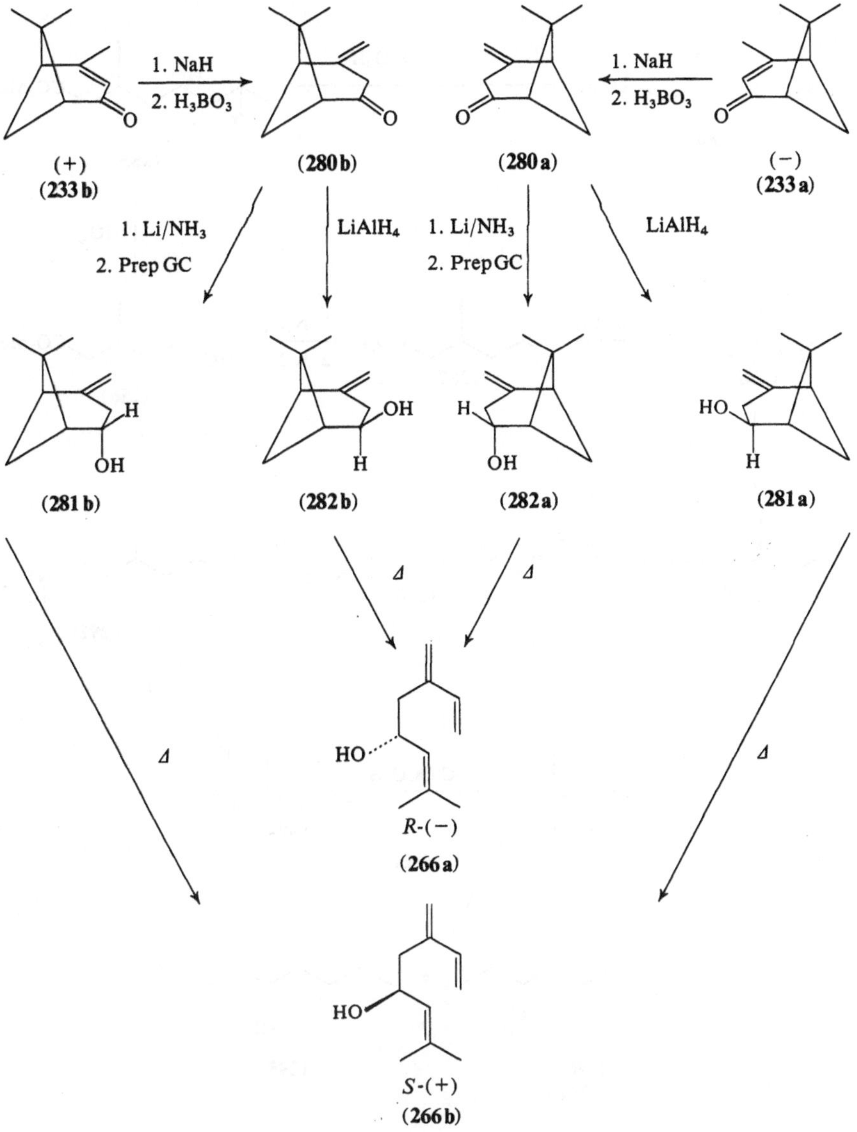

Scheme 53

The best route to (*R*)- or (*S*)-ipsdienol is that devised by OHLOFF and
GIERSCH (*145*), which commences with the optically active verbenones
(**233 a** and **233 b**) (Scheme 53). The a, β-unsaturated ketones were decon-
jugated by sequential treatment with sodium hydride and boric acid to
give (**280 a**) and (**280 b**). Reduction with lithium aluminium hydride

Scheme 54

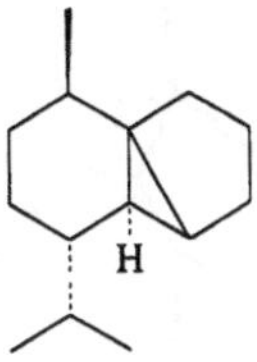

(283)

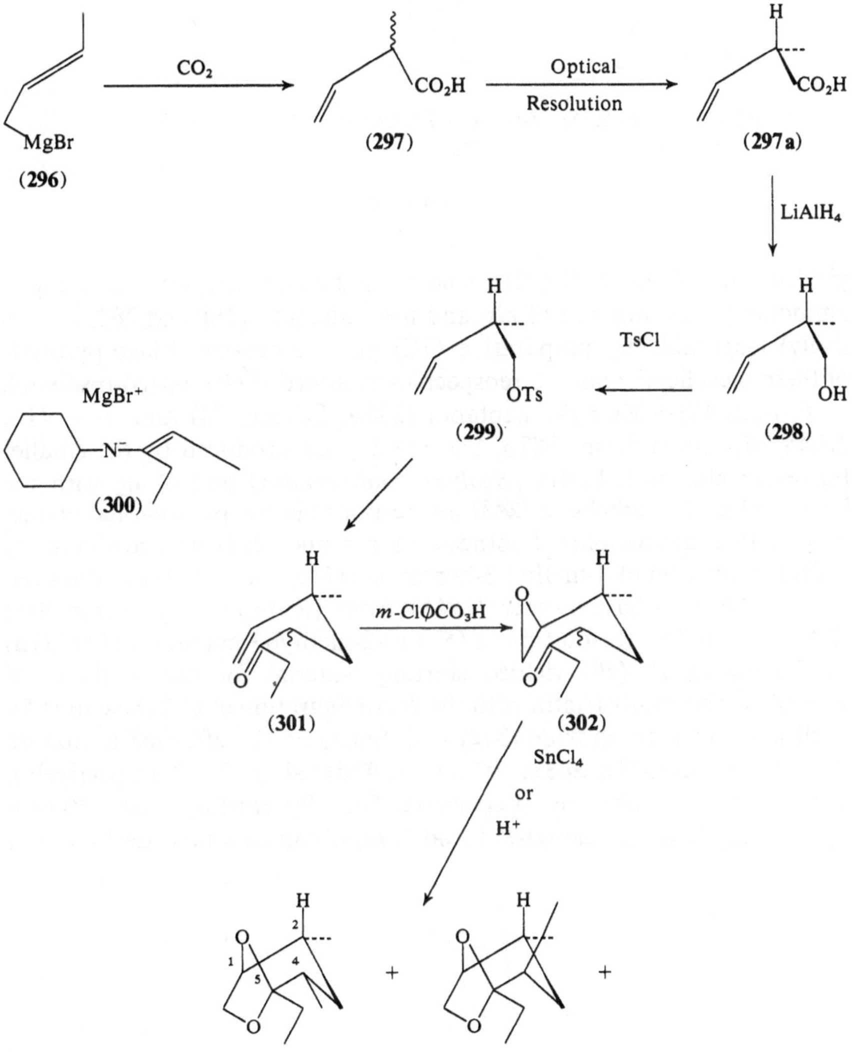

(296)

CO₂

(297)

Optical
Resolution

(297a)

LiAlH₄

(300)

(299)

TsCl

(298)

(301)

m-ClØCO₃H

(302)

SnCl₄

or

H⁺

α-(−)-(1S,2R,4S,5R)

(303a)

γ-(−)-(1S,2R,4R,5R)

(303c)

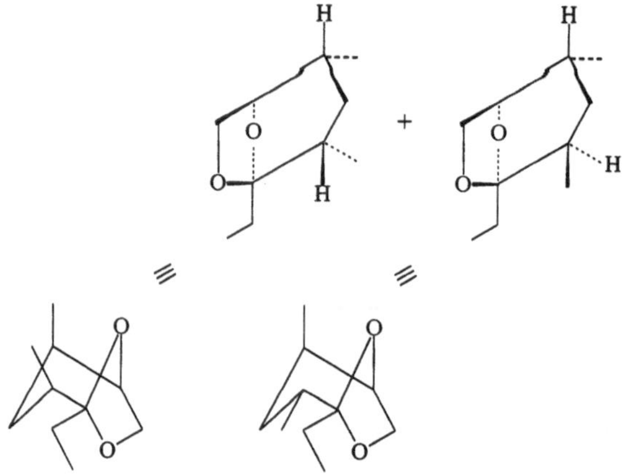

β-(+)-(1R,2R,4S,5S)   δ-(+)-(1R,2R,4R,5S)
(303 b)   (303 d)

Scheme 55

yielded pure *cis*-alcohols (**281**), whereas reduction with lithium in liquid ammonia gave a mixture of *cis*- and *trans*-alcohols (**281** and **282**, respectively) (separable by preparative GC) plus verbenone. Flash-pyrolysis of these bicyclic alcohols stereospecifically afforded (*R*)- or (*S*)-ipsdienol.

(−)-(3*S*,4*S*)-4-Methyl-3-heptanol (**294 a**, Scheme 54) and α-(−)-(1*S*, 2*R*,4*S*,5*R*)-multistriatin (**303 a**, Scheme 55) are produced by the smaller European elm bark beetle *(Scolytus multistriatus)* and along with the host-produced α-cubebene (**283**) are responsible for pheromonal attraction in this species (*46*). Synthesis of racemic (**294**) was achieved by hydride reduction of 4-methyl-3-heptanone (*146*). Mori (*147*) synthesized the (3*R*,4*R*)-(+)-diastereomer (**294 b**) from (*R*)-(+)-methyl citronellate (**284**) (Scheme 54). Resolution of (*S*)-(+)-2-methyl-3-butenoic acid (**297 a**) by Pearce et al. (*49*) yielded starting material for the synthesis of optically active multistriatin with the *R* configuration at C-2 (Scheme 55). Cyclization of keto epoxide (**302**) with SnCl₄ or H⁺ afforded a mixture of multistriatins (**303 a, 303 b, 303 c,** and **303 d**) (34 : 1 : 7 : 58, respectively), which were separable by preparative GC. By starting with (*R*)-(+)-glyceraldehyde acetonide (**269**) Mori (*148*) obtained a mixture (in about

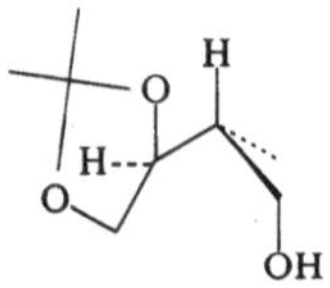

(**307 a**)

the same ratio as PEARCE *et al.* (*49*)) of optically active multistriatins having the *S* configuration at C-1 (Scheme 56). In a similar synthesis, ELLIOTT and FRIED (*149*) were able to obtain a racemic mixture of (**303a**) and (**303c**) (85 : 15) in 73% overall yield from (*Z*)-2-butene-1,4-diol *via* (**307a**).

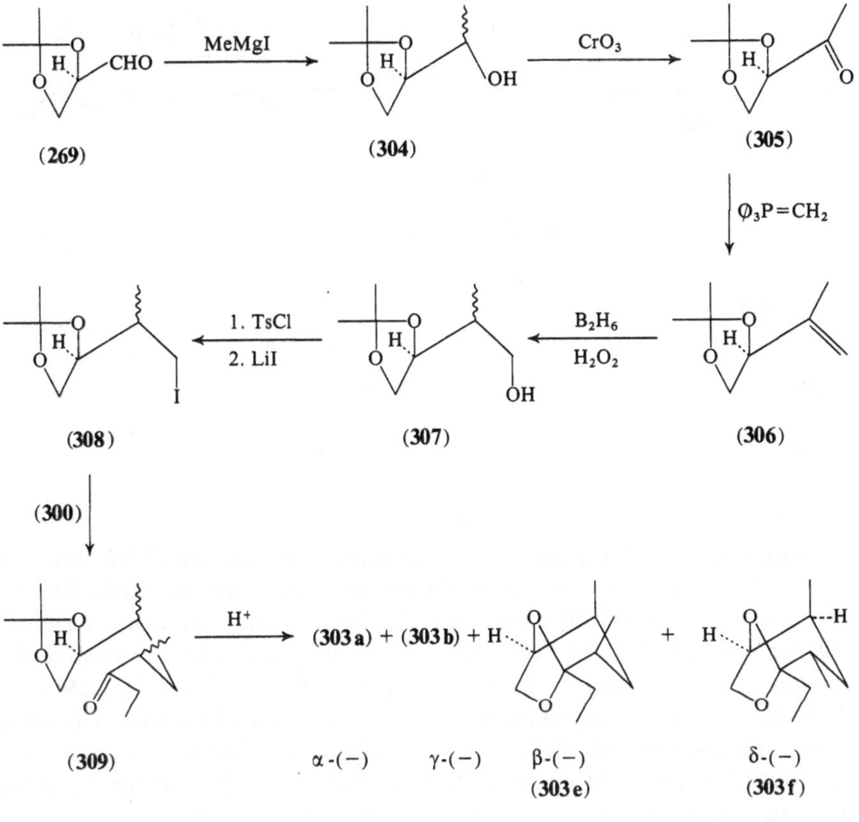

Scheme 56

B. Diptera

Synthesis of (*E*)-non-6-en-1-ol (**315**), the sex attractant of the Mediterranean fruit fly (*Ceratitus capitata*) was achieved by JONES *et al.* (*150*) through photochemical addition of propanal and 1,3-cyclohexadiene (**311**) and subsequent thermal cycloreversion of the bicyclic oxetane intermediate (**313**) (Scheme 57). The critical cycloreversion step was >95% stereoselective. This synthesis is significantly shorter than that reported by JACOBSON *et al.* (*151*). ROSSI (*152*) has also reported a convenient synthesis.

4*

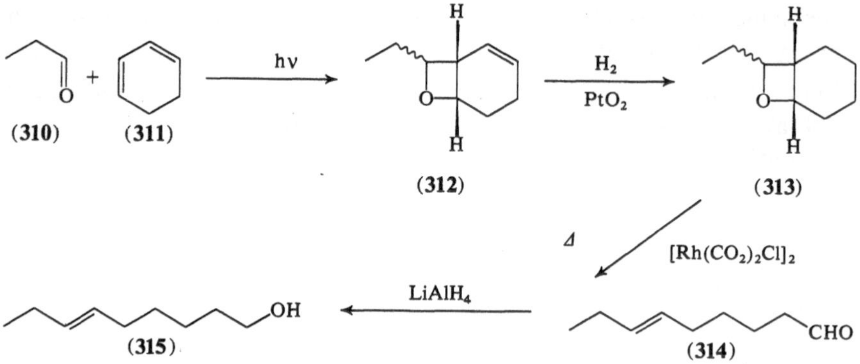

Scheme 57

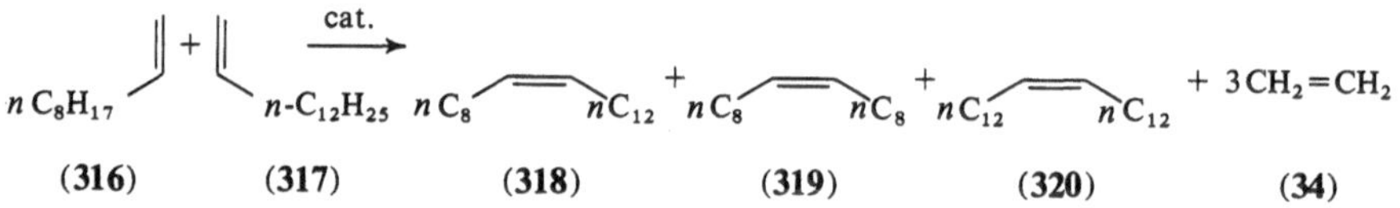

Scheme 58

The housefly *(Musca domestica)* uses the hydrocarbon (Z)-9-tricosene (318 in Scheme 58) as the major component of its sex pheromone. Simple syntheses involve Wittig coupling (153, 154) or alkylation of a terminal alkyne and subsequent reduction (155, 156) to afford the desired Z-isomer in *ca.* 95% purity. Erucic acid (157, 158) or oleic acid (159, 160) have served as starting material of known Z stereochemistry. The transition metal-catalyzed olefin cross-metathesis reaction has been applied by Rossi (161) to synthesize (318) as a mixture of E/Z-isomers together with the other possible C_{18} and C_{28} olefins (Scheme 58).

C. Homoptera

Females of the California red scale *(Aonidiella aurantii)* release a mixture of the branched unsaturated acetates (332a) and (333) to attract males. Synthesis of the R,Z component (332a) was achieved by Roelofs *et al.* (74, 75) starting from (S)-(+)-carvone (321) (Scheme 59). The final product consisted of a mixture of the Z and E isomers, which were easily separable by GC. The absolute stereochemistry of the 3-methyl group in (333) is unknown.

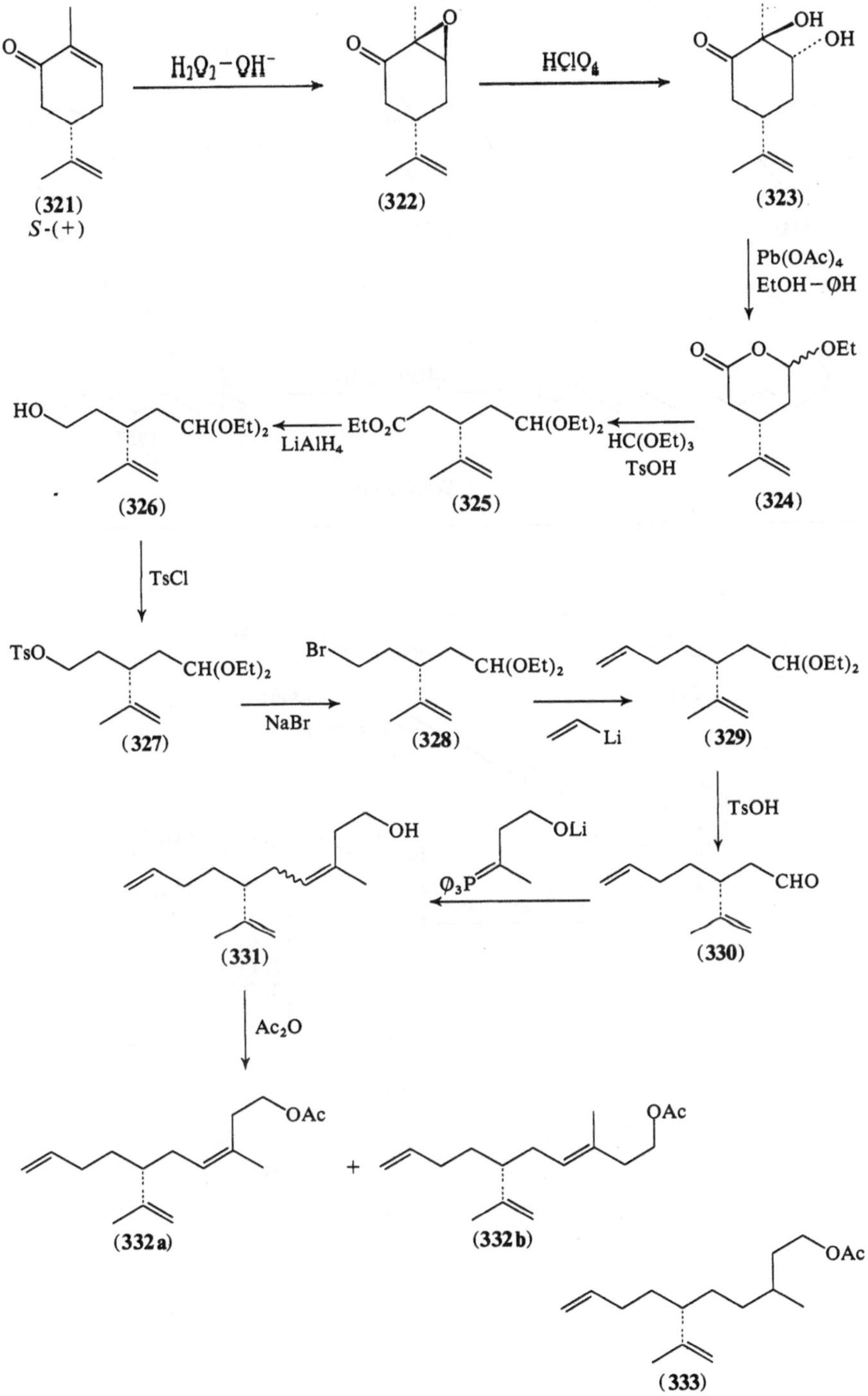

(321)
S-(+)

$H_2O_2-OH^-$

(322)

$HClO_4$

(323)

$Pb(OAc)_4$
$EtOH-\phi H$

(324)

$HC(OEt)_3$
TsOH

(325)

EtO_2C $CH(OEt)_2$

LiAlH$_4$

(326)

HO $CH(OEt)_2$

TsCl

(327)

TsO $CH(OEt)_2$

NaBr

(328)

Br $CH(OEt)_2$

$\triangle$ Li

(329)

$CH(OEt)_2$

TsOH

(330)

CHO

ϕ_3P OLi

(331)

OH

Ac$_2$O

(332a)

OAc

+

(332b)

OAc

(333)

OAc

Scheme 60

Scheme 61

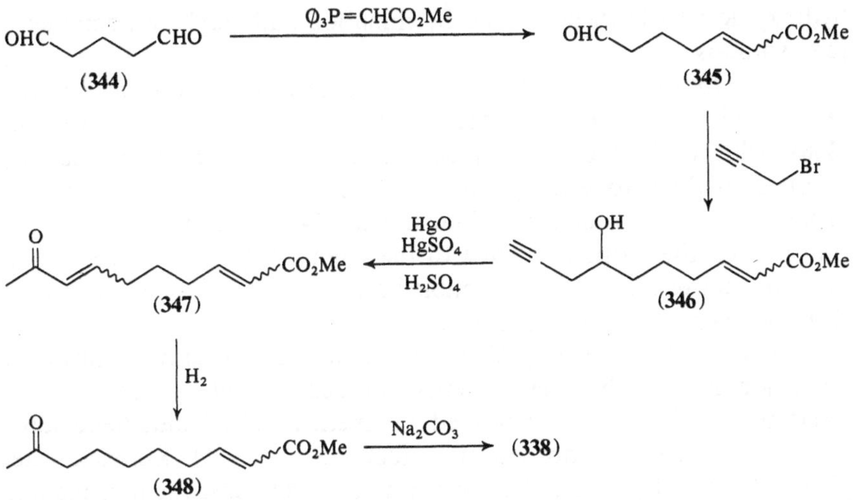

OHC⌇⌇⌇CHO $\xrightarrow{\varphi_3P=CHCO_2Me}$ OHC⌇⌇⌇⌇CO₂Me

(**344**)   (**345**)

$\downarrow \equiv\!\!\diagup\!\!\diagdown Br$

O   OH

⌇⌇⌇⌇CO₂Me $\xleftarrow[\text{H}_2\text{SO}_4]{\begin{array}{c}\text{HgO}\\\text{HgSO}_4\end{array}}$ $\equiv$⌇⌇⌇⌇CO₂Me

(**347**)   (**346**)

$\downarrow$ H₂

O

⌇⌇⌇⌇CO₂Me $\xrightarrow{\text{Na}_2\text{CO}_3}$ (**338**)

(**348**)

Scheme 62

$\equiv$⌇⌇⌇⌇I $\xrightarrow{\text{Me}_3\text{N}\longrightarrow\text{O}}$ $\equiv$⌇⌇⌇CHO

(**349**)   (**350**)

$\downarrow$ CH₂(CO₂H)₂

(**338**) $\xleftarrow[\text{HCO}_2\text{H}]{\text{HgSO}_4}$ $\equiv$⌇⌇⌇⌇CO₂H

(**351**)

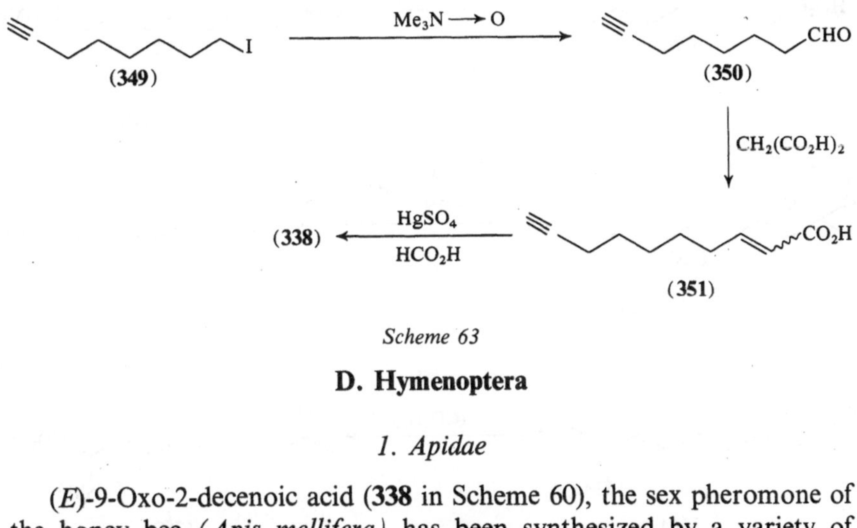

Scheme 63

D. Hymenoptera

1. Apidae

(*E*)-9-Oxo-2-decenoic acid (**338** in Scheme 60), the sex pheromone of the honey bee *(Apis mellifera)* has been synthesized by a variety of methods. In a lengthy sequence starting from azaleic acid, BUTLER et al. (*162*) dehydrohalogenated an α-bromomethyl ester and converted the other acid to a methyl ketone. Ozonolysis of 1-methyl-1-cycloheptene (**336**) by BARBIER et al. (*163*) afforded ketoaldehyde (**337**), which was condensed with malonic acid to give (**338**) (Scheme 60). JAEGER and ROBINSON (*164*) obtained (**337**) from 7-oxooctanoic acid *via* the corresponding acid chloride. Acidification of dihydropyran by KENNEDY et al. (*165*) and chain extension with malonic acid yielded the unsaturated hydroxy acid (**341**) (Scheme 61). Bromination and alkylation completed the synthesis.

Eiter (*166*) introduced the unsaturated carboxy moiety *via* a Wittig reaction with pentanedial and then chain extended to a β-hydroxyalkyne (**346**) (Scheme 62). Mercuric ion-catalyzed rearrangement and reduction afforded methylketo ester (**348**). In a relatively high yield synthesis, Sisido *et al.* (*167*) oxidized iodoalkyne (**349**), chain extended and transformed the terminal alkyne to the methyl ketone (Scheme 63). In all of the above examples, the double bond is assumed to be in the more thermodynamically stable *E* form.

Trost and Salzmann (*168*) were able to stereoselectively introduce an *E* double bond by their sulfenylation-dehydrosulfenylation method (Scheme 64). Palladium catalyzed telomerization of butadiene (**358**) by Tsuji *et al.* (*169*) afforded diene (**360**) (Scheme 65). Ketoacid ester (**363**), obtained by terminal olefin oxidation and subsequent reduction, was treated with diphenyldiselenide and the resultant phenylselenyl group oxidatively removed to give the *E* α,β-unsaturated keto ester (**365**).

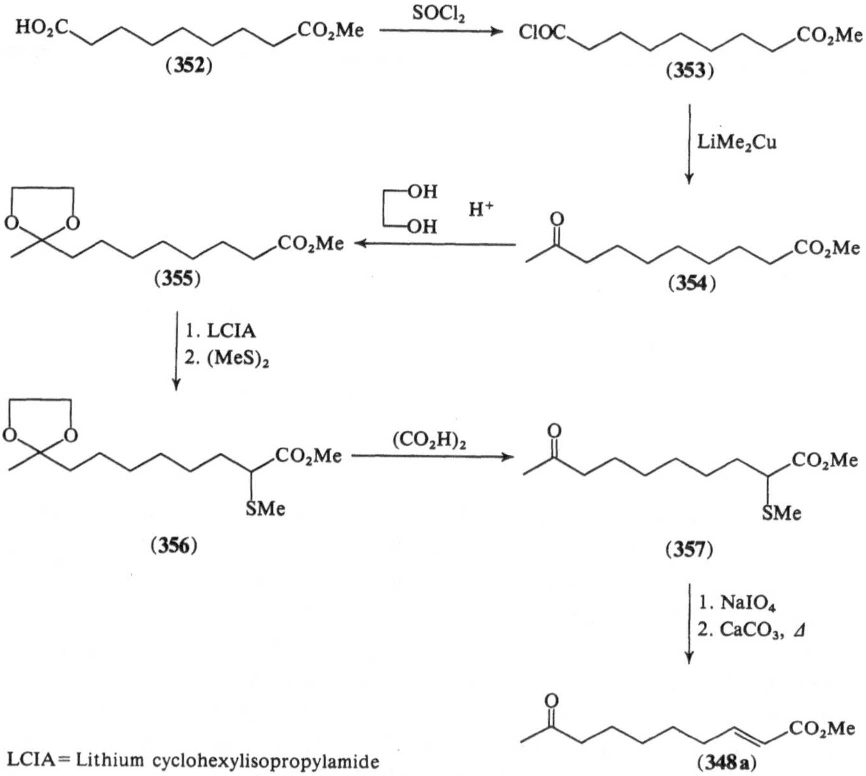

LCIA = Lithium cyclohexylisopropylamide

Scheme 64

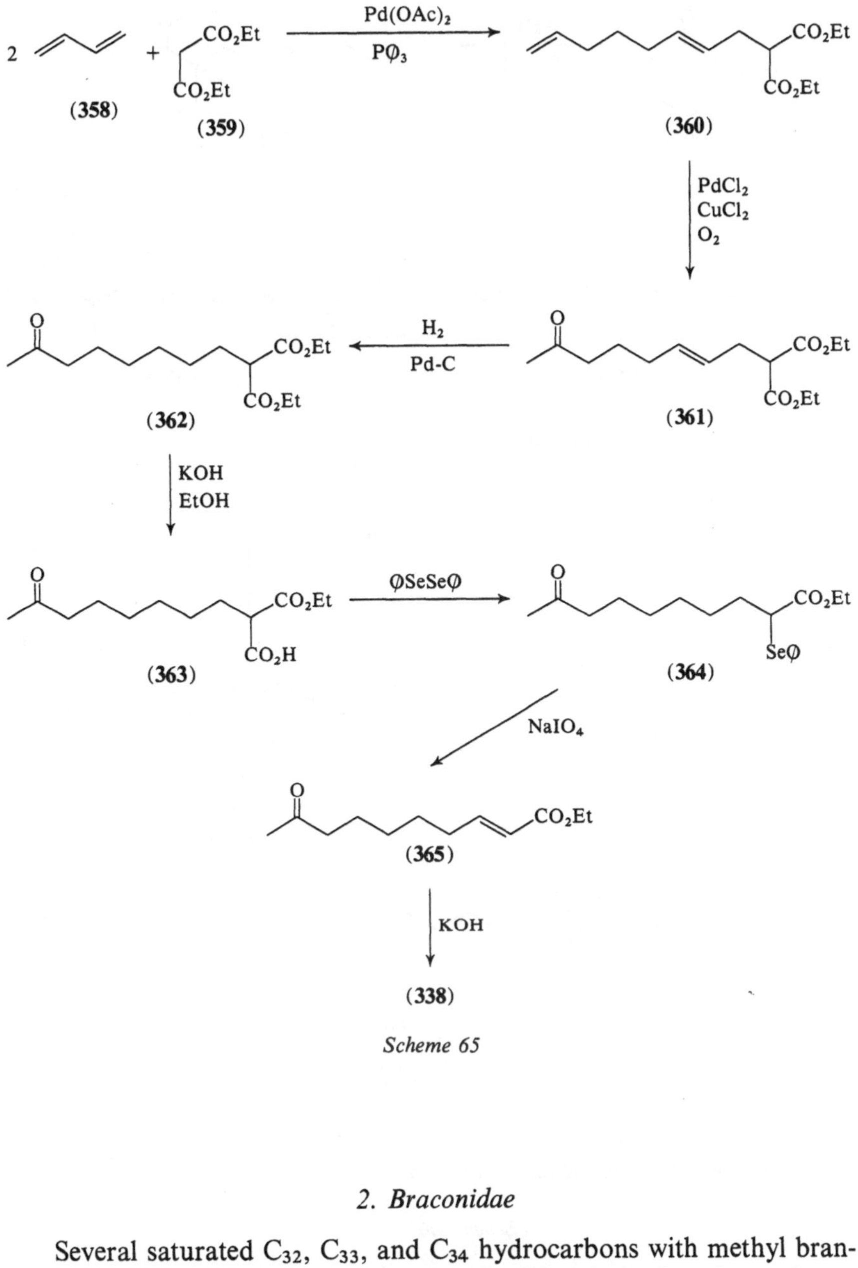

Scheme 65

2. Braconidae

Several saturated C_{32}, C_{33}, and C_{34} hydrocarbons with methyl branches in the C_{11-13} positions have been identified from the tobacco hornworm *(Heliothis virescens)* (Lepidoptera: Noctuidae) that elicit a host-seeking response of its parasitoid, *Cardiochiles nigriceps*. These kairomones have been synthesized by VINSON *et al.* (*170*) using appropriate methyl ketones and Wittig reagents followed by catalytic reduction.

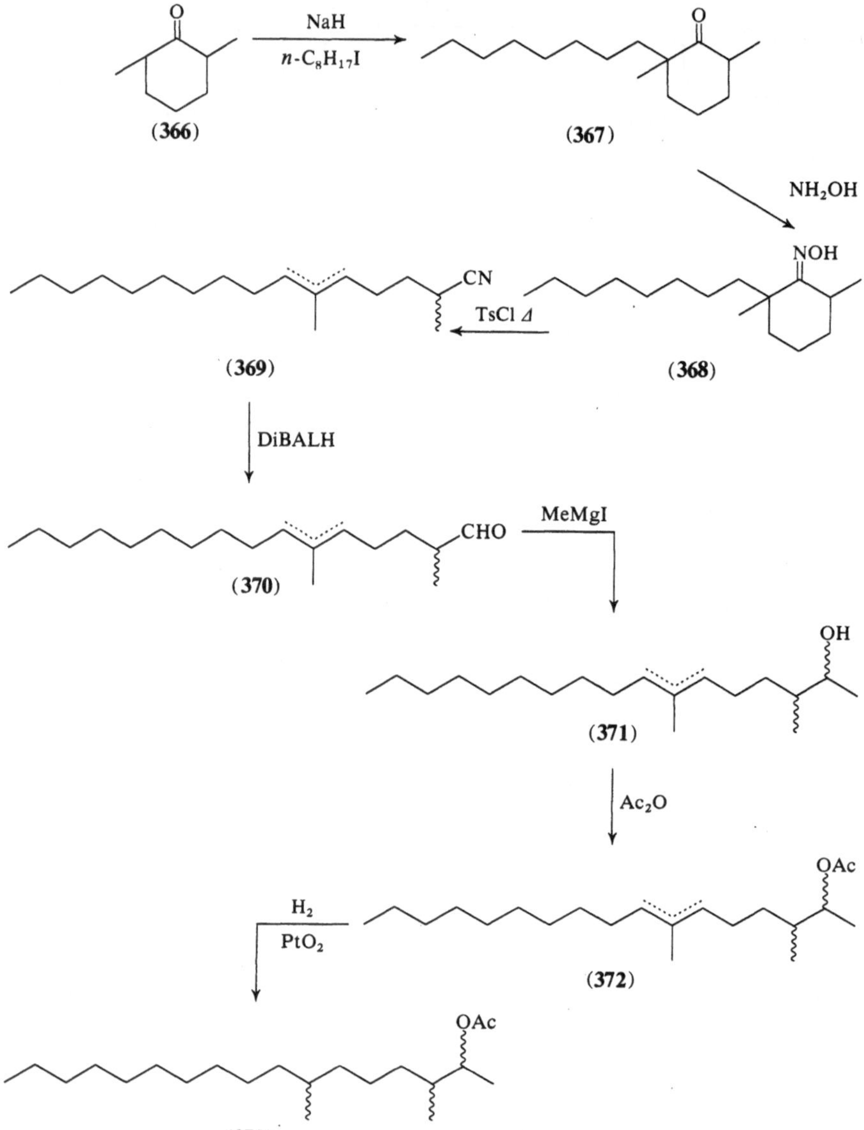

DiBALH = Di-*t*-butoxyaluminium hydride

Scheme 66

3. Diprionidae

The sex pheromone (**373**) (Scheme 66) of the pine sawfly *(Neodiprion lecontei)* was synthesized by Kocienski and Ansell (*171*) as a mixture of epimers starting from 2,6-dimethylcyclohexanone. Alkylation and sub-

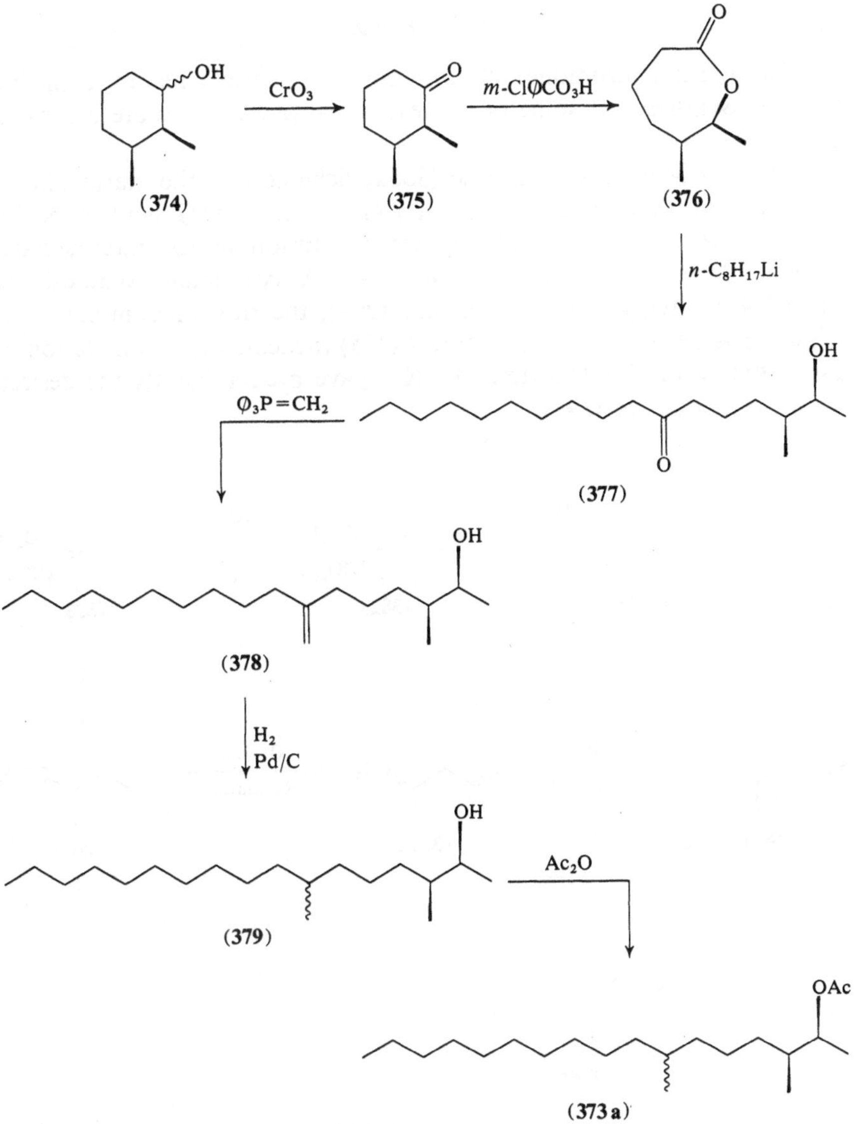

Scheme 67

sequent Beckmann rearrangement of oxime (368) yielded an acyclic un-saturated nitrile (369), which was transformed to the desired acetate. MAGNUSSON (172) utilized pure *trans*-2,3-dimethylcyclohexanone in his synthesis of the *erythro*-isomer of the sex pheromone (Scheme 67). Baeyer-Villiger oxidation and alkylation gave the acyclic hydroxy ketone (377), which subsequently afforded (373a).

4. Formicidae

The alarm pheromones of ants are often simple readily available ketones. Syntheses of some of the less trivial pheromones are described below.

(S)-(+)-4-Methyl-3-heptanone (**386a**, Scheme 68), the alarm pheromone of the Texas leaf-cutting ant *(Atta texana)* was synthesized by Riley et al. (*173, 174*) and required resolution of the intermediate, racemic 2-methyl-4-pentenoic acid (**384**). A convenient synthesis of methyl 4-methylpyrrole-2-carboxylate (**392**), the trail pheromone of *A. texana*, has been reported by Sonnet (*175*) (Scheme 69). Formylation of ester (**390**) with Cl_2CHOMe and $AlCl_3$ gave predominantly the desired 4-formyl derivative (**391**).

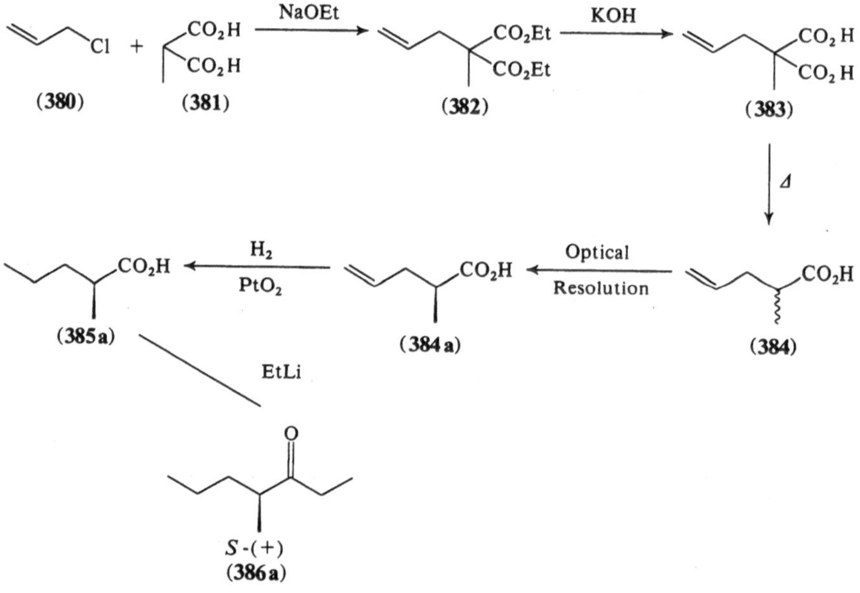

Scheme 68

Caste-specific compounds produced by some male carpenter ants *(Camponotus* spp.) include 2,4-dimethyl-2-hexenoic acid (**396**). Brand et al. (*176*) and Kocienski et al. (*177*) prepared (**396**) as a mixture of *E*- and *Z*-isomers (Schemes 70 and 71), whereas Katzenellenbogen and Utawanit (*178*) obtained the pure *E*-isomer (**396a**) by stereoselective dehydration of β-hydroxy ester (**402**) *via* a β-alanoxy enolate (Scheme 72).

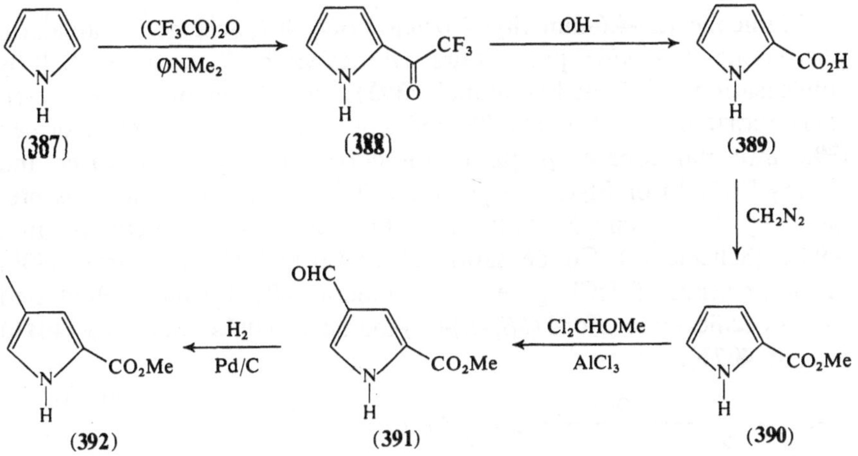

Scheme 69

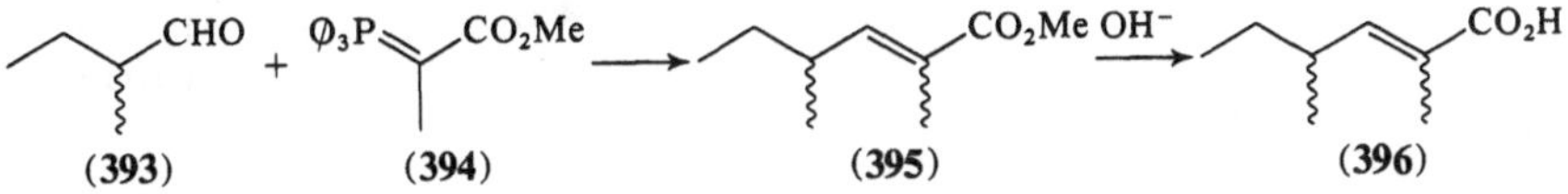

Scheme 70

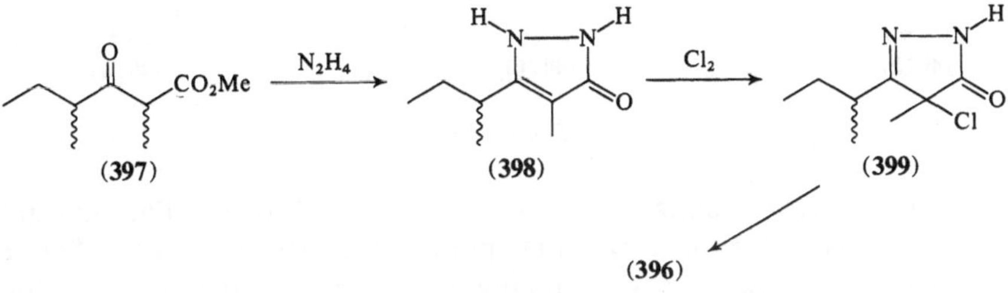

Scheme 71

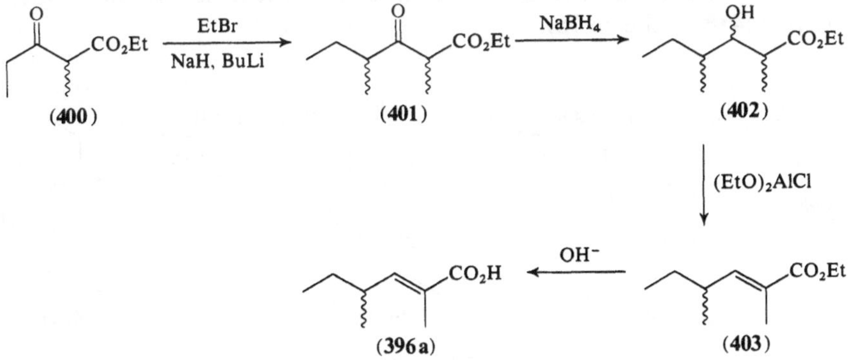

Scheme 72

Manicone, (E)-4,6-dimethyl-4-octen-3-one (**407**, Scheme 73), an alarm pheromone of *Manica* spp. was first synthesized by FALES *at al.* (*179*) by condensation of 2-methylbutanal (**393**) and 3-pentanone. KATZEN-ELLENBOGEN and UTAWANIT (*178*) and KOCIENSKI *et al.* (*177*) transformed (**396**) into manicone by preparing the corresponding acid chloride and adding Et_2CuLi or Et_2Cd, respectively. The S-(+)-enantiomer was prepared by BANNO and MUKAIYAMA (*180*) from (S)-(−)-2-methylbutanol (**404a**) (Scheme 73). Condensation of (**393a**) and silyl enol ether (**405**) in the presence of $TiCl_4$ gave diastereomeric (**406a**), which dehydrated stereospecifically to afford (6S,4E)-(+)-manicone (**407a**). Similarly (**404b**) yielded (**407b**).

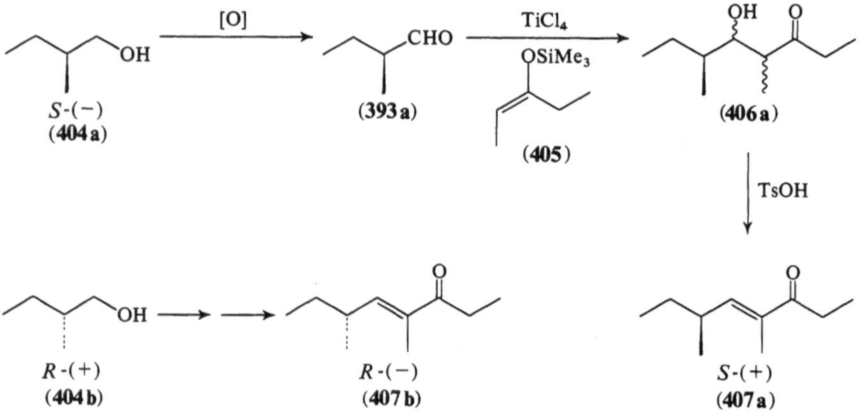

Scheme 73

The octahydroindolizine trail pheromone (**412**) of the Pharaoh ant (*Monomorium pharaonis*) was first prepared by RITTER *et al.* (*181*) as a mixture of stereoisomers, beginning with a termolecular condensation (Scheme 74). OLIVER and SONNET (*182*) then unambiguously synthesized the four stereoisomers from 2,6-lutidine (**413**) (Scheme 75) and 2-butyl-pyrrole (**418**) (Scheme 76).

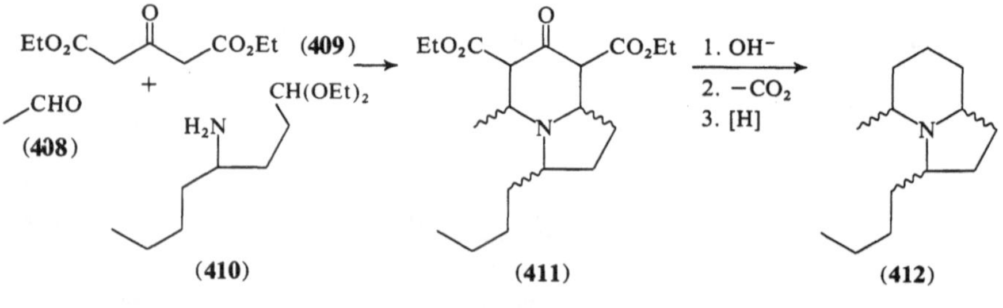

Scheme 74

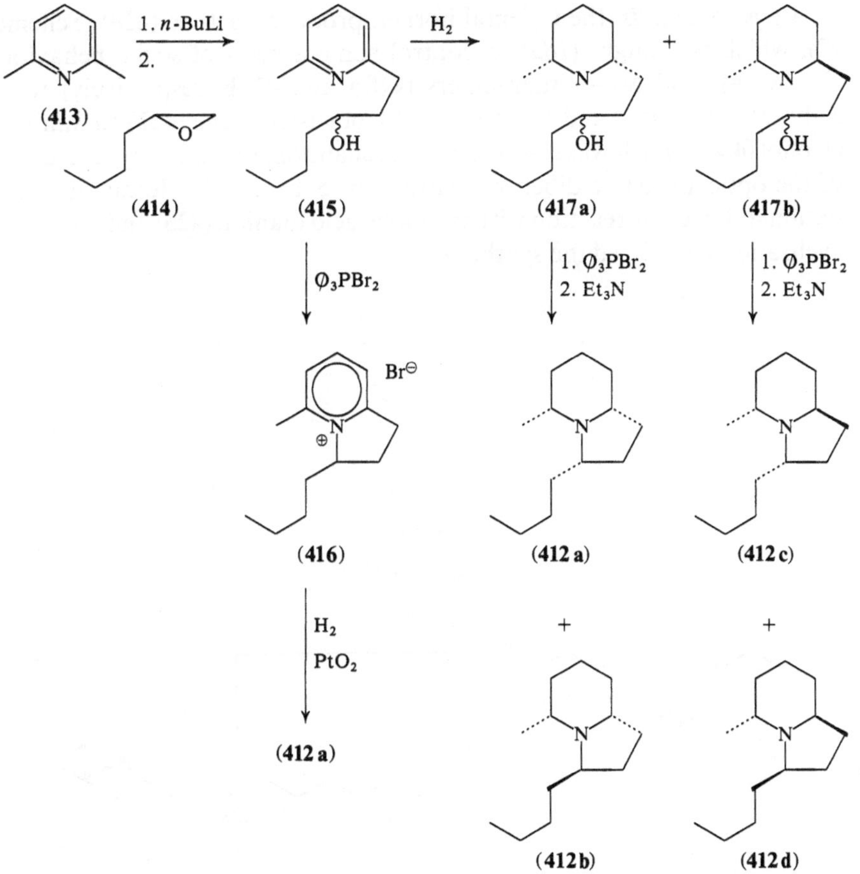

Scheme 75

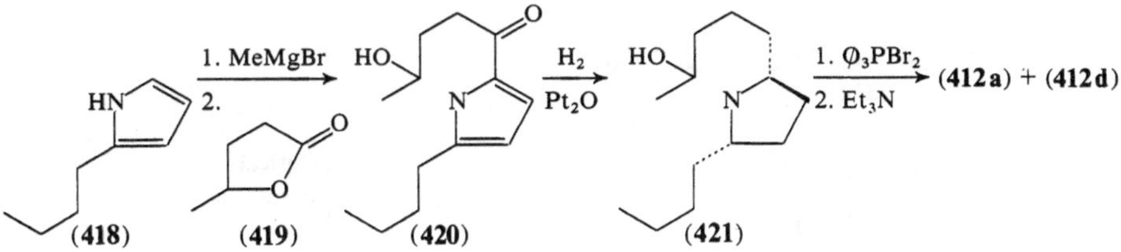

Scheme 76

Vespa orientalis, the Oriental hornet, produces lactone (**428**) (Scheme 77), which is thought (*183*) to control some aspects of social behavior. The *R*-(+)- and *S*-(−)-enantiomers (**428a** and **428b,** respectively) were elaborated by Coke and Richon (*184*) from 1-tridecene (**422**) *via* amino-alcohol (**424**), which was resolved into its enantiomeric forms by resolution of the optically active dibenzoyl tartarates (Scheme 77). Hofmann elimi-nation followed by reaction with propiolic acid dianion (**425**) and reductive cyclization completed the synthesis.

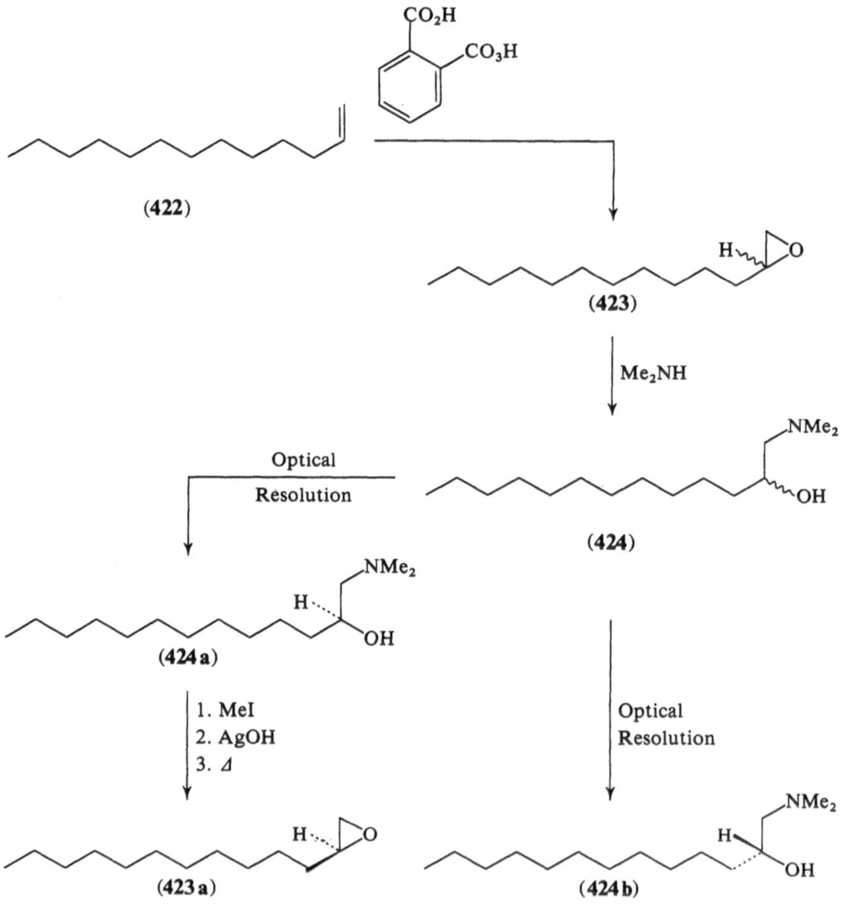

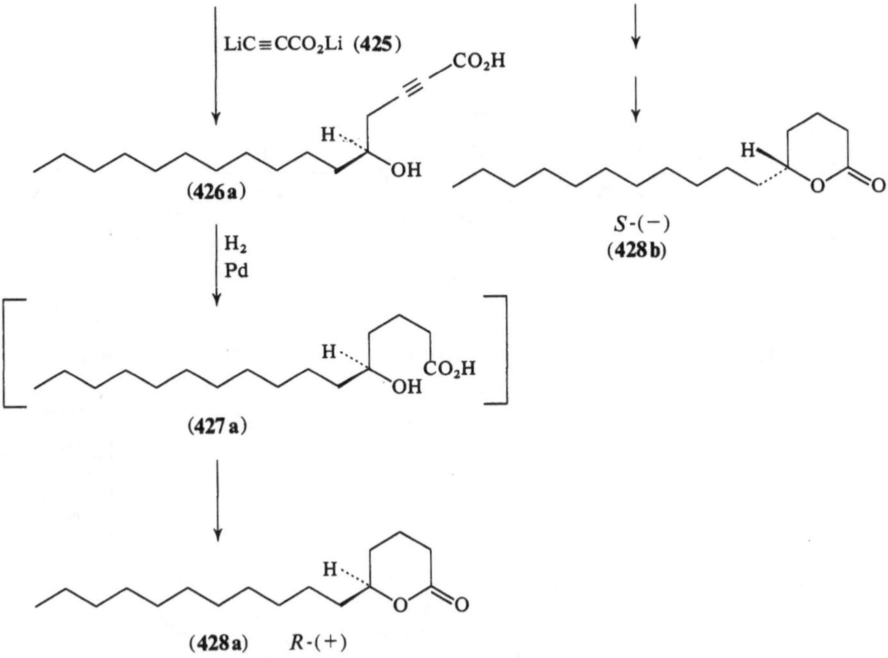

Scheme 77

E. Isoptera

Two syntheses of the trail pheromone, neocembrene (**435**), of *Nasu-titermes* spp. termites have been reported. KODAMA *et al.* (*185*) (Scheme 78) prepared the allylic phenyl thioether (**430**) from *trans,trans*-geranyl-linalool (**429**). Terminal epoxidation of (**430**) followed by intromolecular cyclization, desulfurization, and dehydration led to (**435**). KITAHARA *et al.* (*186*) cyclized *trans*-geranylgeranic acid chloride (**436**) with SnCl₄ to afford chloroketone (**437**) (Scheme 79). Dehydrohalogenation of (**437**) and subsequent reduction of ketone (**438**) *via* acetate (**441**) gave neo-cembrene (**435**).

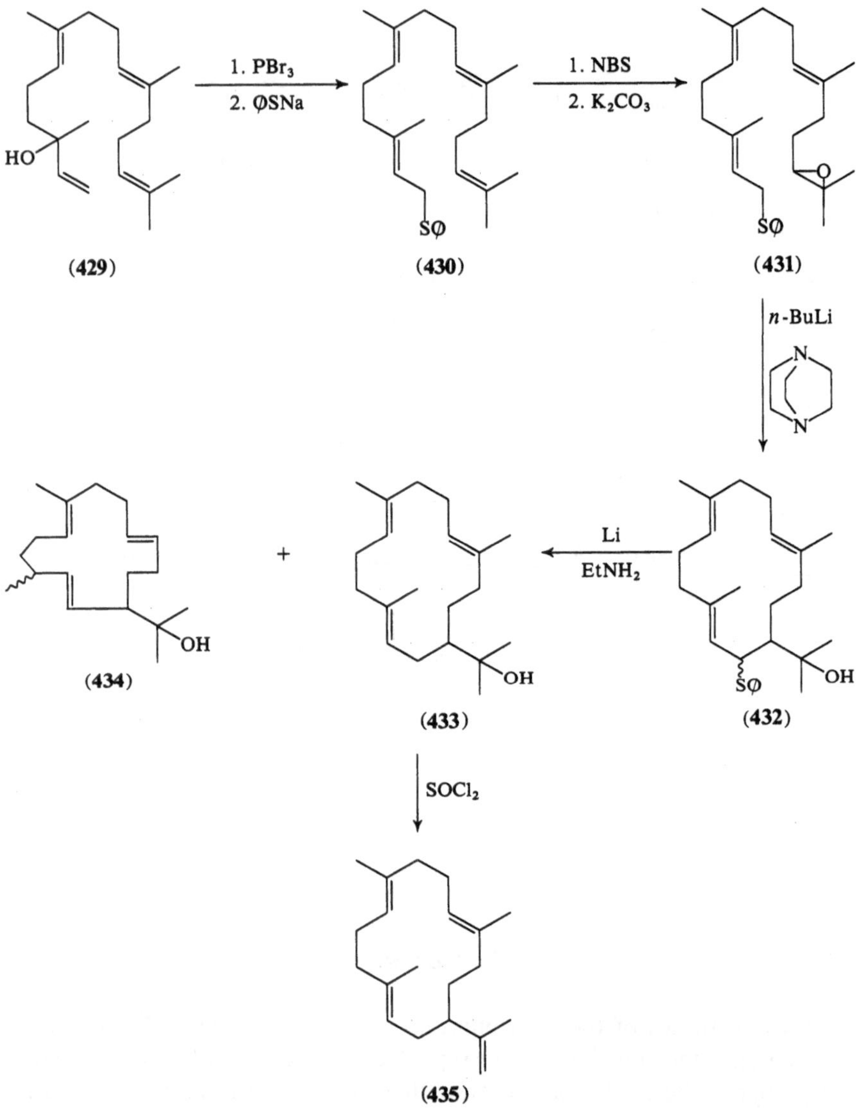

(429)   (430)   (431)

(434)   +   (433)   (432)

(435)

NBS = N-Bromosuccinimide

Scheme 78

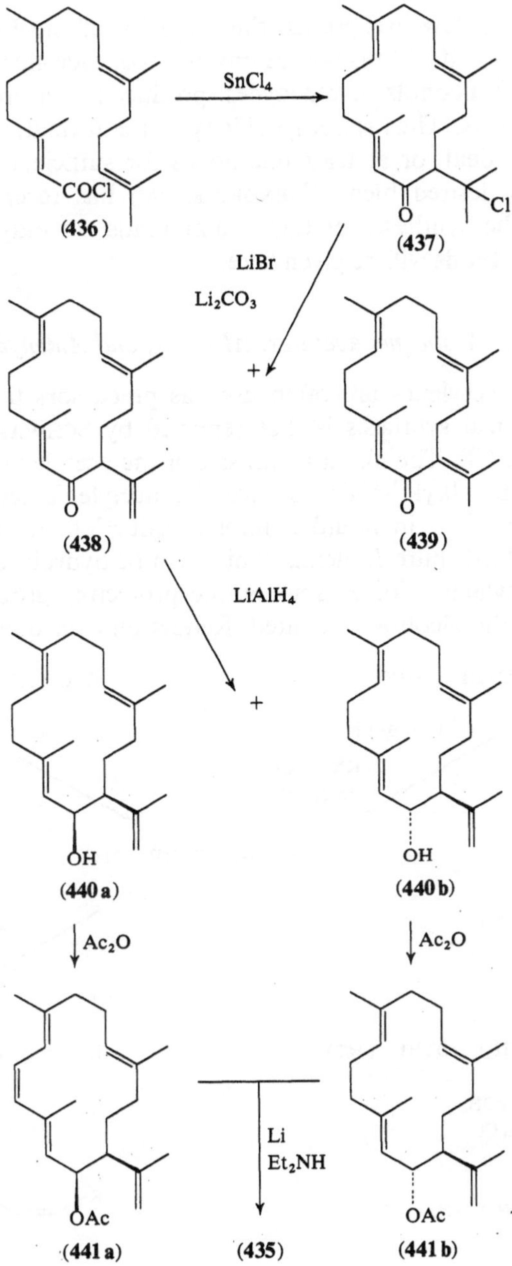
SnCl4
COCl
(436)
Cl
O
(437)
LiBr
Li2CO3
+
O
(438)
O
(439)
LiAlH4
+
OH
(440 a)
OH
(440 b)
Ac2O
Ac2O
OAc
(441 a)
Li
Et2NH
(435)
OAc
(441 b)

Scheme 79

F. Lepidoptera

With only a few exceptions, the majority of the Lepidopteran sex pheromones identified thus far are unsaturated, even-numbered, straight-chain acetates, alcohols or aldehydes, produced as a precise mixture of E- and Z-isomers. Thus stereospecificity in the formation of the desired isomer is the ideal, or at least one hopes for sufficient stereoselectivity to attain the desired blend. HENDRICK (90) has recently reviewed in great detail the syntheses of these compounds, so only a summary of the various methods will be given here.

1. Monoene Acetates, Alcohols, and Aldehydes

Terminal acetylenes are often used as precursors to the E- and Z-olefins. A typical synthesis is that reported by SCHWARZ and WATERS (187) (Scheme 80). The alkali metal salt of the acetylene is coupled with an appropriate alkyl halide to give the acetylenic tetrahydropyranyl ether (444). Sodium in liquid ammonia reduction of the ether stereospecifically affords pure E-olefins, which can be hydrolyzed to an alcohol and then acetylated. For Z-isomers, the protecting group must first be removed and the alcohol acetylated. Reduction over quinoline-poisoned

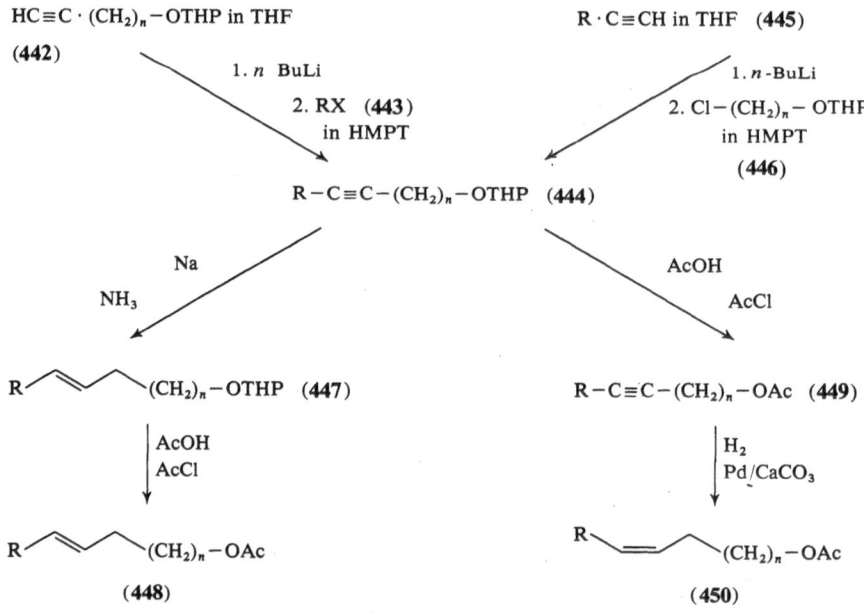

HMPT = Hexamethylphosphortriamide

Scheme 80

Lindlar catalyst stereoselectively gives the Z-isomer contaminated with up to 5% of the E-isomer. Reduction of acetylenes to yield $> 98\%$ Z-olefins has been accomplished by hydroboration with hindered reagents such as disiamylborane (*188*) or 9-BBN (*189*) followed by hydrolysis with acetic acid.

Wittig reactions between aliphatic aldehydes and phosphonium ylids generated with sodium bis(trimethylsilyl)amide in tetrahydrofuran ("salt free") have been used to prepare alkenes with a $Z:E$ ratio of 98:2, respectively (*190*).

HAYASHI and MIDORIKA (*191*) used a [3,3]-sigmatropic rearrangement of an allylic dithiocarbamate to stereospecifically generate E-olefins (Scheme 81). Reduction of alkylated allylic phosphonates with lithium aluminium hydride by KONDO et al. (*192*) exclusively gave E-alkenes (Scheme 82).

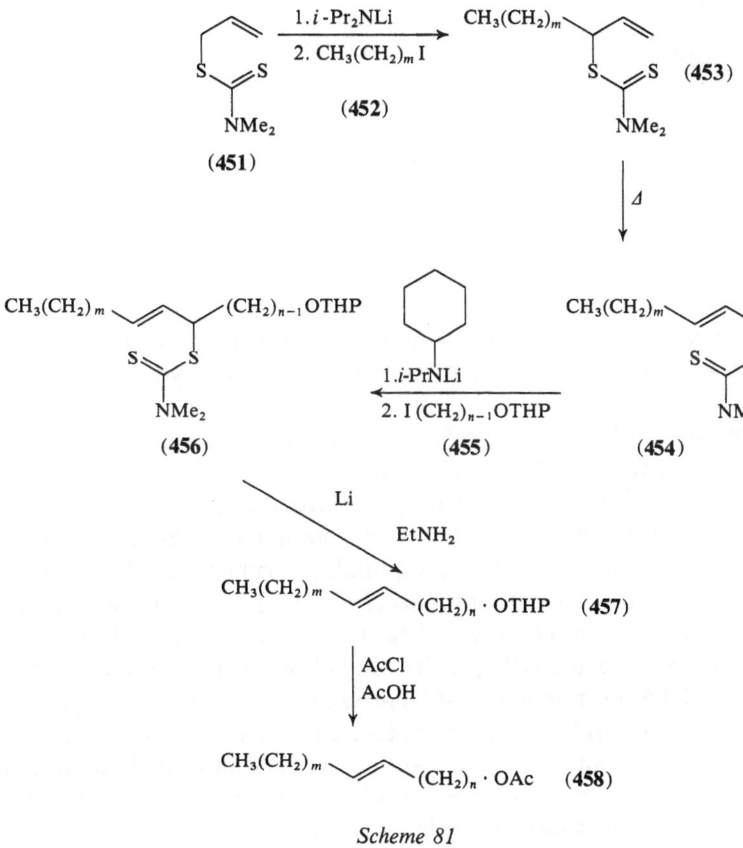

$$1.\ i\text{-}Pr_2NLi$$
$$2.\ CH_3(CH_2)_m I$$

(452) (451) (453)

Δ

$CH_3(CH_2)_m$ ... $(CH_2)_{n-1}OTHP$

(456)

$$1.\ i\text{-}PrNLi$$
$$2.\ I(CH_2)_{n-1}OTHP$$

(455) (454)

Li

$EtNH_2$

$CH_3(CH_2)_m$ ... $(CH_2)_n \cdot OTHP$ (457)

AcCl
AcOH

$CH_3(CH_2)_m$ ... $(CH_2)_n \cdot OAc$ (458)

Scheme 81

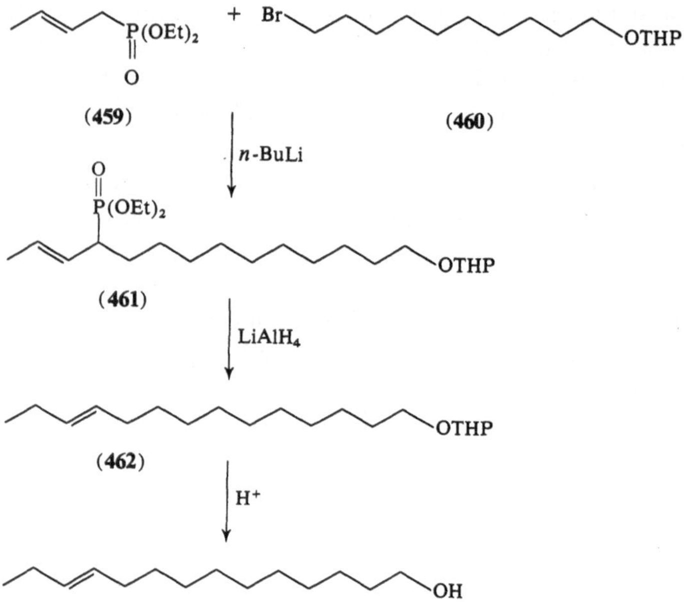

Scheme 82

Solid phase syntheses of alkynols and alkenols have been achieved by LEZNOFF *et al.* (*189, 193*) using polymer-bound symmetrical alkane diols.

MORI *et al.* (*194*) used the 2-alkylated cyclohexane-1,3-dione (**466**) as the source for the straight chain alkynoic acid (Scheme 83). Reduction of alkynol (**468**) with P-2 nickel catalyst afforded (*Z*)-alkenol (**469**).

Olefinic aldehydes have been synthesized by a variety of methods including oxidation of the corresponding primary alcohols with the chromium trioxide-pyridine complex (*195—197*) or N-chlorosuccinimide-dimethyl sulfide complex (*198*), heating a primary alken-1-yl mesylate with dimethylsulfoxide (*199*), or by alkylation of the lithium salt of 5,6-dihydro-2,4,4,6-tetramethyl-1,3-(4H)-oxazine with an alkynyl iodide followed by sodium borohydride reduction and acid hydrolysis (*200*).

Separation of mixtures of *E*- and *Z*-isomers and nearly quantitative recovery of the individual components has been accomplished by selective formation of *E* olefin-urea complexes (*201*).

KOH

(464)   (465)   (466)

1. KOH/(CH₂OH)₂

2. N₂H₄

3. Δ

$$\text{1. KOH/(CH}_2\text{OH)}_2$$
$$\text{2. N}_2\text{H}_4$$
$$\text{3. }\Delta$$

COOH

(467)

LiAlH₄

OH

(468)

NaBH₄
Ni(OAc)₂
(CH₂NH₂)₂

OH

(469)

Ac₂O

OAc

(470)

Scheme 83

2. Conjugated Dienes

A number of conjugated dienes have been identified as sex pheromones. Among these are (8E,10E)-8,10-dodecadien-1-ol (**475**, Scheme 84)
from the codling moth *(Laspeyresia pomonella)*, (7E,9Z)-7,9-dodecadien-
1-yl acetate (**490**, Scheme 86) from the European grapevine moth *(Lobesia
botrana)*, (E)- and (Z)-9,11-dodecadien-1-yl acetates (**518a** and **518b**,
respectively, see Scheme 90) from the red bollworm moth *(Diparopsis
castanea)*, and (10E,12Z)-10,12-hexadecadien-1-ol (**498**, Scheme 87) from
the silkworm moth *(Bombyx mori)*.

Reactions between an α,β-unsaturated aldehyde and a saturated Wittig
reagent *(71, 202—209)* or a saturated aldehyde and an olefinic *(202,
210—212)* or acetylenic *(213)* Wittig reagent have been utilized for

elaborating the conjugated dienyl system. By varying the solvent systems and reaction conditions, some control over E/Z ratios can be exercised. However, pure E- or Z-isomers are rarely obtained and some purification is necessary. Diels-Alder adducts of E,E dienes with tetracyanoethylene (*204, 205, 211*) or sulfur dioxide (*206*), urea complexes (*213*), or recrystallization (*214*) have been used to separate mixtures. Equilibration of isomeric diene mixtures to predominantly E,E dienes has been accomplished by heating without solvent in the presence of benzenethiol (*215, 216*) or in lower yield by photolysis in the presence of I_2 (*210, 217*).

Vig et al. (*218*) coupled an unsaturated phosphonate with a saturated aldehyde, to obtain an E,E diene of undisclosed purity.

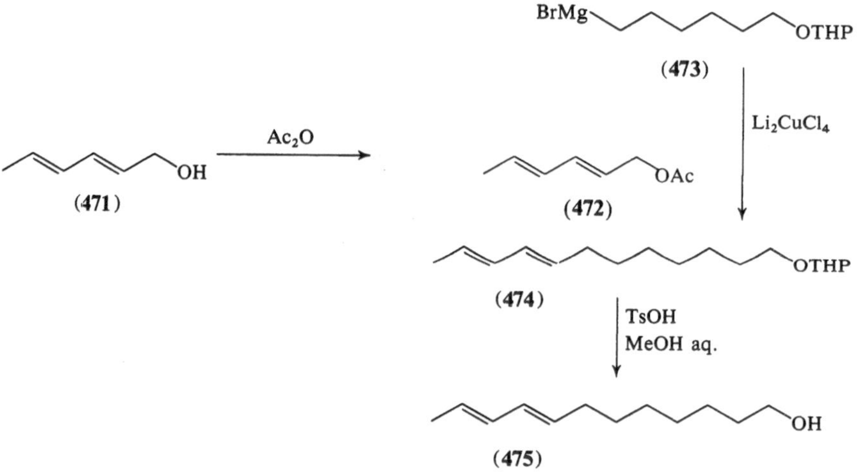

Scheme 84

In several syntheses, a fragment containing the diene system was first prepared and then elongated *via* a Grignard or alkali metal coupling reaction. Sorbyl alcohol (**471**) has been used by Henrick et al. (*219*) (Scheme 84) and others (*220—222*) as an allylic diene precursor, which may be subject to some rearrangement during the coupling reaction. Descoins and Henrick (*214*) used the Julia method to prepare the homoallylic bromide (**479**) (*ca.* 80% *E,E*) en route to (**475**) Scheme 85).

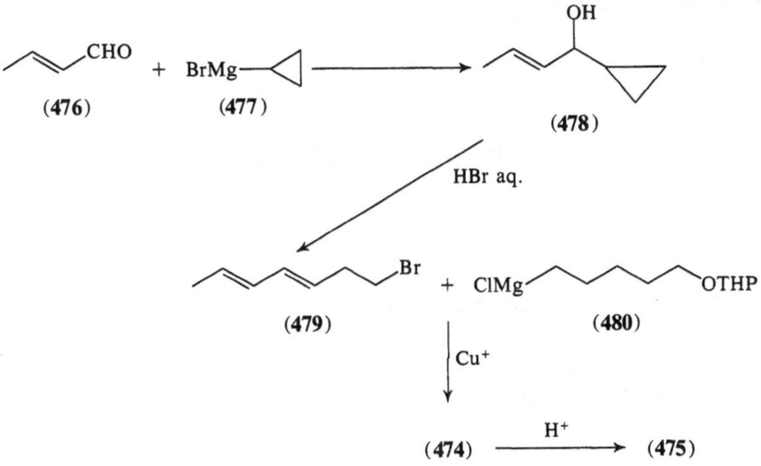

Scheme 85

Allylic nonconjugated enynols have been rearranged (*206, 217, 223*) to prepare mixtures of *E* and *Z* conjugated enynes, which can be separated prior to further reaction, or as in one case (*223*) treated with the alkyl lithium reagent (**486**) in the presence of dilithium tetrachloro-cuprate to give an *E*-enyne (Scheme 86). Conjugated enynes have also been prepared by dehydration of α-ynols (*202, 213, 217, 224*), by Wittig reaction (*213*), *via* vinyl copper intermediates (*225*) (Scheme 87), and from alkenyl boranes (see below). The acetylene can be reduced to a *Z*-olefin with Lindlar catalyst (*213, 217, 226*) or preferably *via* hydro-boration-protonolysis (*223, 224, 227*), or to an *E*-olefin with lithium aluminium hydride (*225*).

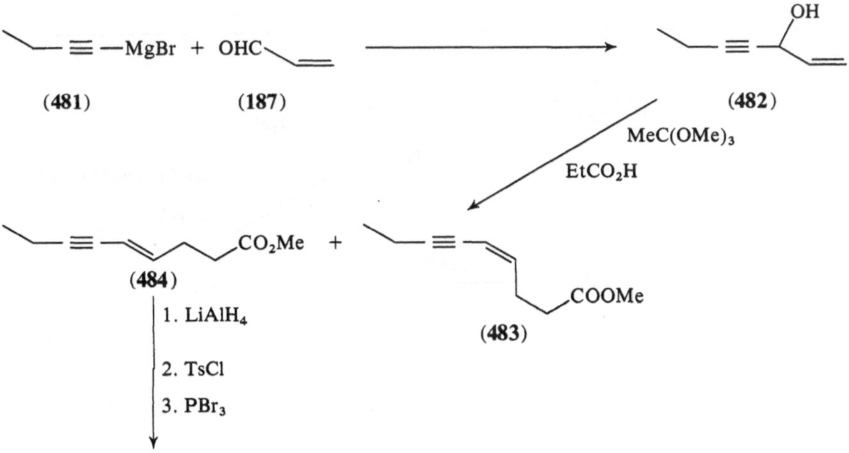

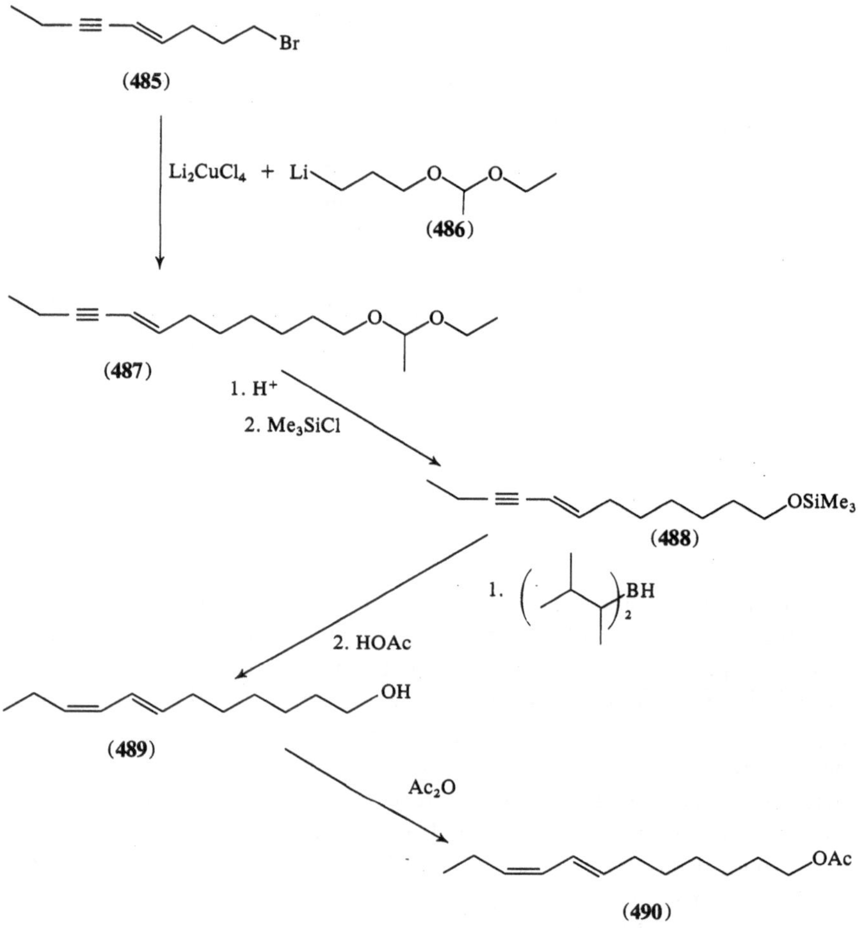
Br
(485)
Li2CuCl4 + Li
O
O
(486)
O
O
(487)
1. H+
2. Me3SiCl
OSiMe3
(488)
1.
BH
2
2. HOAc
OH
(489)
Ac2O
OAc
(490)

Scheme 86

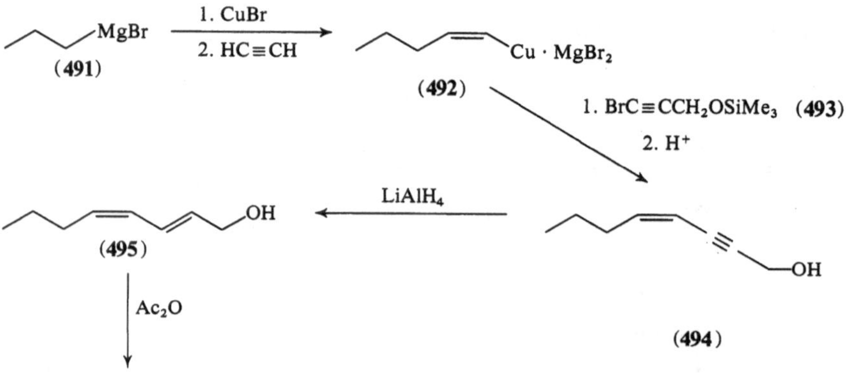
MgBr
(491)
1. CuBr
2. HC≡CH
Cu · MgBr2
(492)
1. BrC≡CCH2OSiMe3 (493)
2. H+
LiAlH4
OH
(495)
OH
(494)
Ac2O

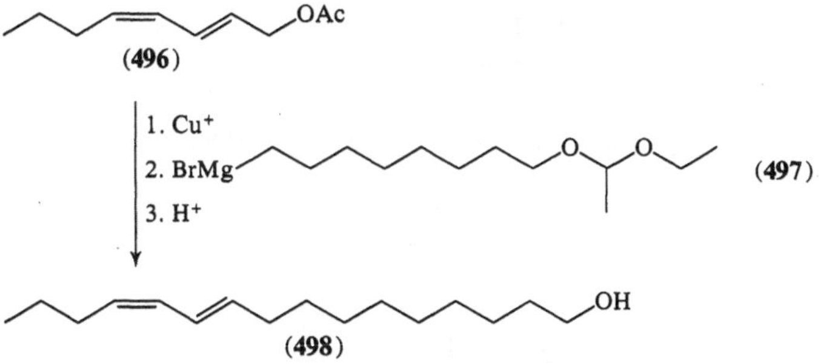

(496)

1. Cu⁺
2. BrMg
3. H⁺

(497)

(498)

Scheme 87

NEGISHI and coworkers have developed a method for preparation of conjugated *E*-enynes *via* alkenylboranes. In one synthesis (*227*), acetylene (**501**) was treated with disiamylborane and then with the lithium salt of acetylene (**503**) to afford complex (**504**), which after treatment with iodine and sodium hydroxide stereoselectively afforded (**505**) (Scheme 88). `Syntheses of other conjugated *E,Z* dienes have been achieved by this method (*228, 229*). By use of thexylborane and 1-bromo-1-alkynes, conjugated *E,E* dienes can be readily prepared (*230*) (Scheme 89).

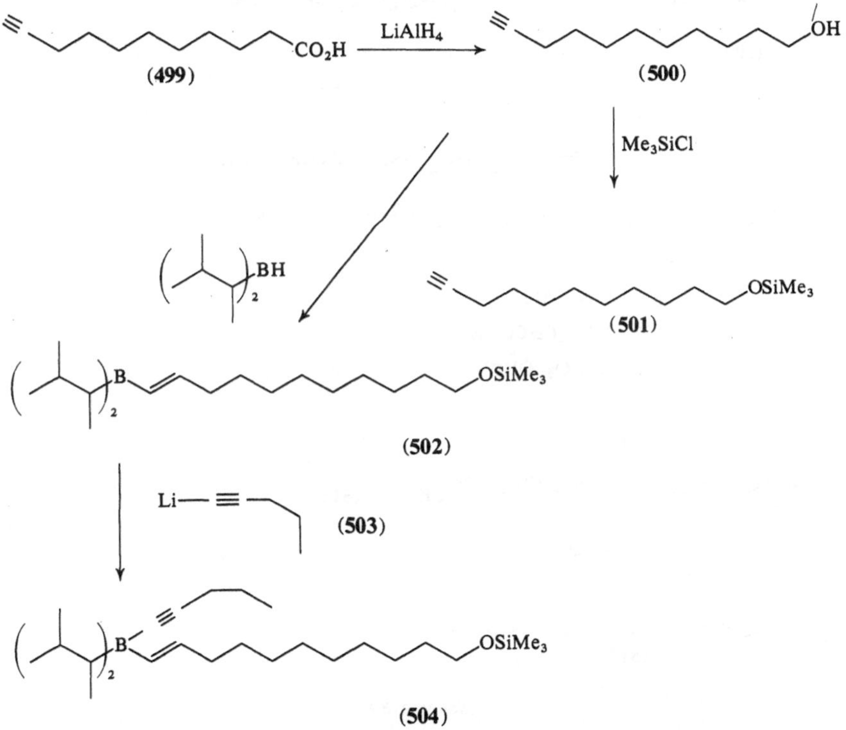

(499) CO_2H — LiAlH₄ → **(500)** OH

Me₃SiCl

BH

OSiMe₃

(501)

OSiMe₃

(502)

Li—≡— **(503)**

OSiMe₃

(504)

Scheme 88

Scheme 89

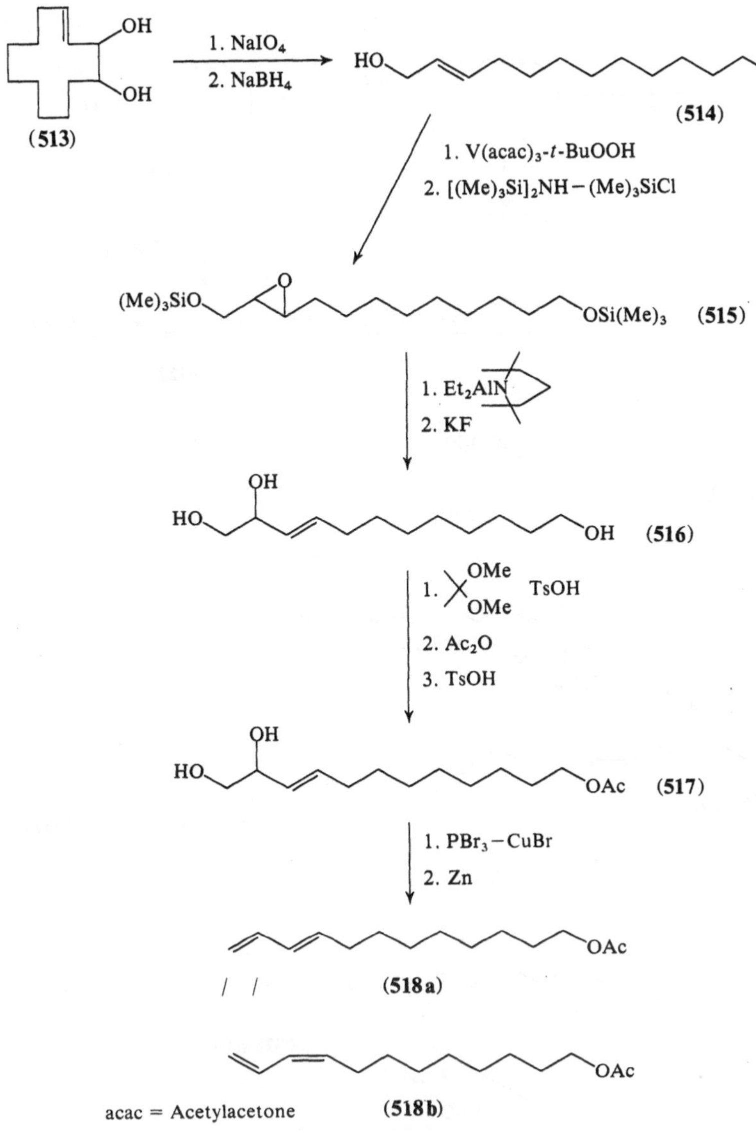

Scheme 90

TANAKA et al. (231) prepared (518a) from allylic alcohol (514) via the epoxy silyl ether (515), which was ring opened with diethylaluminum 2,2,6,6-tetramethylpiperidide to afford the 3-ene-1,2-diol (516) Scheme 90).

Scheme 91

3. Nonconjugated Dienes

Nonconjugated dienes have been identified as sex pheromones from a variety of species, including the pink bollworm moth *(Pectinophora gossypiella)*. The initially assigned structure was that of the branched diene (5) (see p. 15). Numerous syntheses of this molecule have been reported and are reviewed by KATZENELLENBOGEN *(89)*. The pheromone is now known *(79)* to be a 1 : 1 mixture of (7Z,11Z)- and (7Z,11E)-7,11-hexadecadien-1-yl acetate (529a and 529b, respectively, see Scheme 91).

The methods employed in the synthesis of the individual components have tended to follow those for monoenes, namely *via* acetylenic intermediates *(232—236)*, by allylic Grignard coupling *(232, 237)*, and by Wittig reaction *(234, 238)*. SONNET *(239)* isomerized the 7Z,11Z- and 7Z,11E-isomers (as their tetrahydropyranyl ethers) to the 7E,11E- and 7E,11Z-isomers, respectively, by treatment of the corresponding bis-epoxides with lithium diphenylphosphide and then methyl iodide.

A more convenient synthesis is that of ANDERSON and HENRICK *(240)* (Scheme 91). By starting with cyclooctadiene (519), the C-7 bond was fixed as Z, and the C-11 olefin was introduced as a 1 : 1 mixture of E : Z by careful choice of conditions for the Wittig reaction. This synthesis ultimately gave (529a) and (529b) in at least 99% purity.

4. Trienes

Larvae of *Cossus cossus* secrete from the mandibular glands a mixture of 5,13-tetradecadien-1-yl acetate, 3,5,13-tetradecatrien-1-yl acetate and small amounts of the corresponding alcohols *(241)*. The function of these secretions is unknown. GARANTI *et al.* *(242)* synthesized the possible isomers of the acetates *via* acetylenic intermediates and Wittig reactions.

5. Epoxide

The gypsy moth *(Lymantria dispar* [formely *Porthetria dispar])* uses epoxide (536) (disparlure, Scheme 92) as its sex pheromone. Essentially all syntheses have proceeded *via* Z-olefin (535), which is the presumed precursor used by the insect itself. This olefin has been prepared by Wittig reaction *(57, 243, 244)*, with the greatest stereoselectivity achieved by BESTMANN *et al.* *(244)* (Scheme 92), by coupling bromide (532) with the lithium salt of 1-dodecyne and reducing the resulting acetylene over Lindlar catalyst *(245—248)*; and *via* an organosilane intermediate *(249)* (Scheme 93). CHAN and CHANG *(250)* prepared a 1 : 1 mixture of the E- and Z-isomers by condensation of undecanal (534) with the triphenyl-silyl lithium salt (543) (Scheme 94). The olefin mixture has also been prepared in low yield by a transition metal-catalyzed olefin cross-metathesis reaction *(251)*.

Scheme 92

Scheme 93

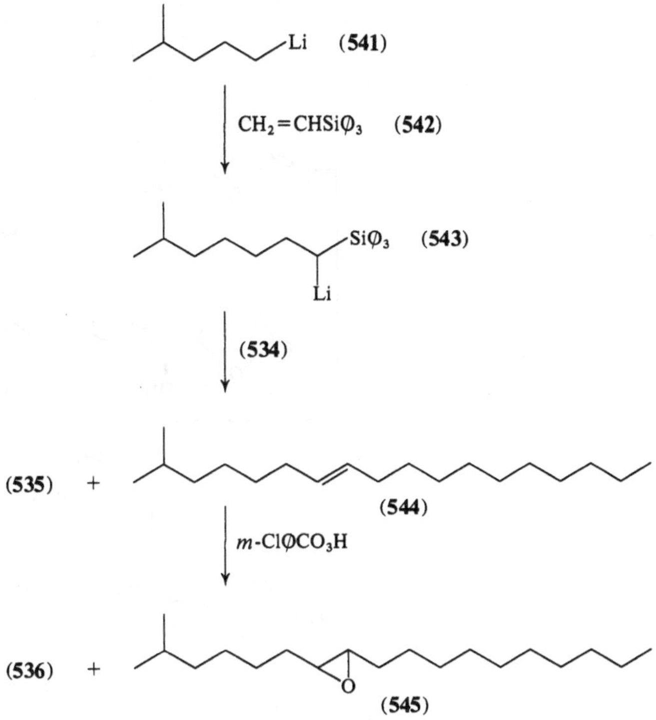

$\diagup\!\diagdown\!\diagup\!\diagdown$Li   (541)

CH$_2$=CHSiØ$_3$ (542)

$\diagup\!\diagdown\!\diagup\!\diagdown\!\diagdown$SiØ$_3$ (543)

Li

(534)

(535)   +

(544)

m-ClØCO$_3$H

(536)   +

O

(545)

Scheme 94

(S)-(+)-Glutamic acid (140b) served as the starting material for the synthesis of (7R,8S)-(+)-disparlure (536a) by IWAKI *et al.* (*252*) (Scheme 95). The final product was *ca.* 94% optically pure. MORI *et al.* (*253*) obtained (536a) in >98% optical purity in their synthesis, which began with L-(+)-tartaric acid (557) (Scheme 96). A shorter stereospecific synthesis of (536a) (>99% pure) has been reported recently by FARNUM *et al.* (*254*) from L-(−)-menthyl-p-toluene sulfinate (566) (Scheme 97).

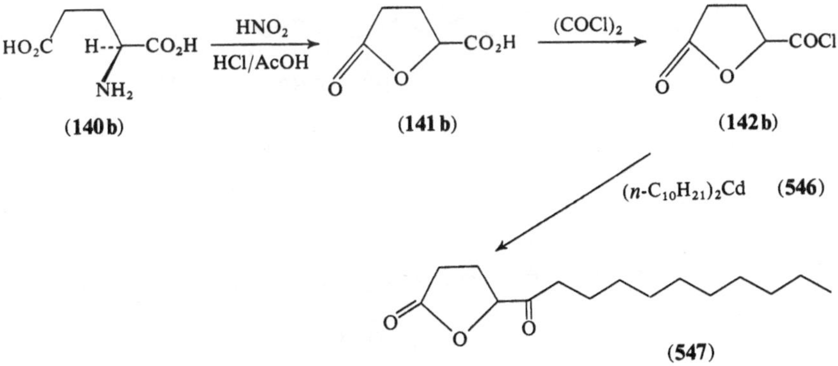

HO$_2$C   H--$\diagup$--CO$_2$H $\xrightarrow[\text{HCl/AcOH}]{\text{HNO}_2}$ -CO$_2$H $\xrightarrow{\text{(COCl)}_2}$ -COCl

NH$_2$

(140b)   (141b)   (142b)

O   O   O   O

$(n\text{-}C_{10}H_{21})_2Cd$ (546)

O   O

O   O

(547)

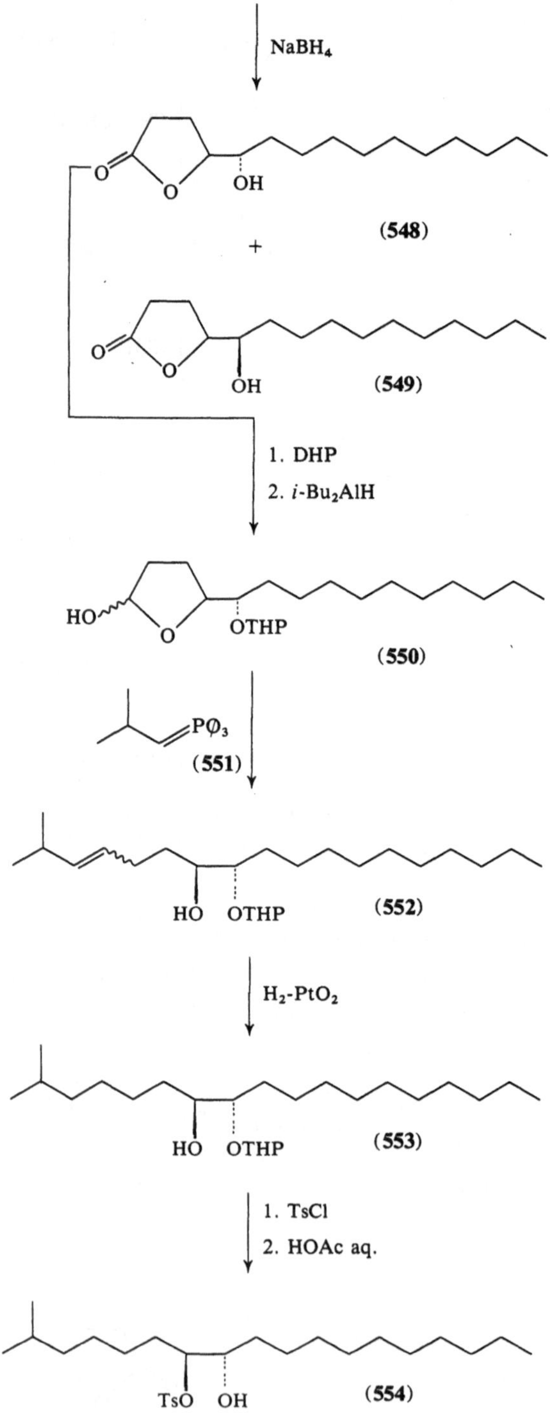

NaBH₄

(548)

+

(549)

1. DHP
2. i-Bu₂AlH

(550)

PØ₃

(551)

(552)

H₂-PtO₂

(553)

1. TsCl
2. HOAc aq.

(554)

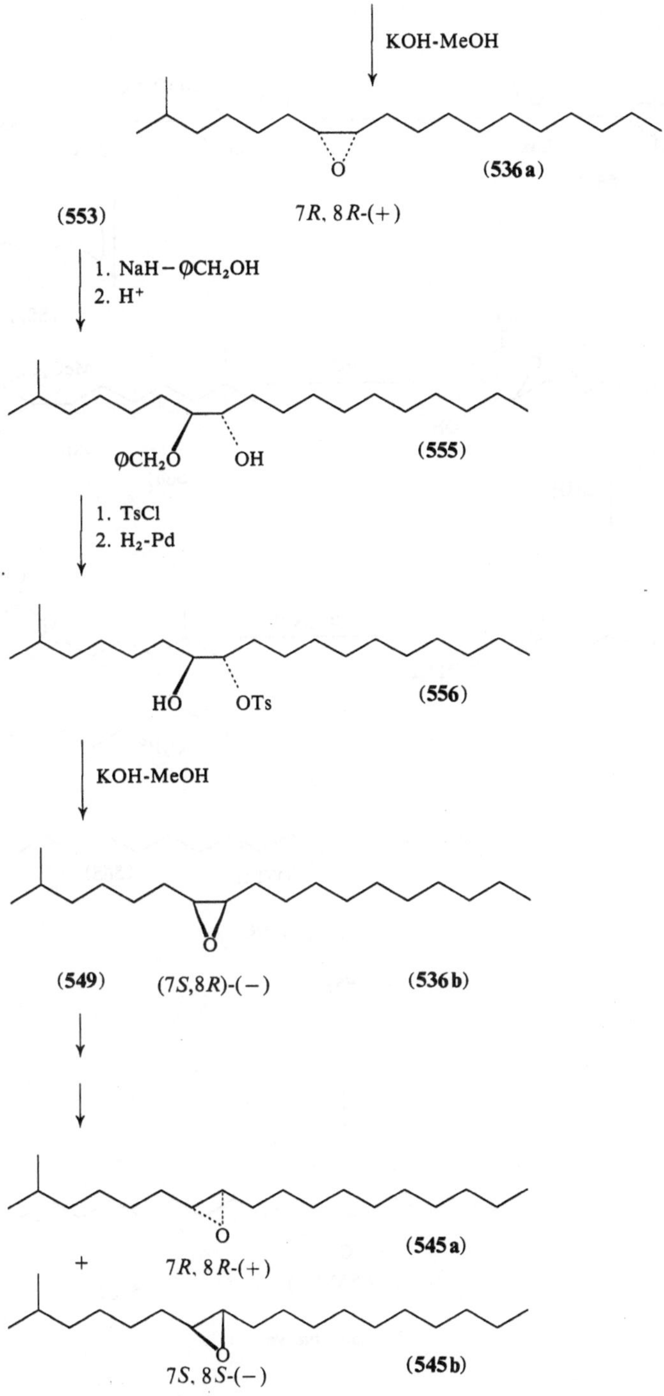

KOH-MeOH

(536 a)

(553)   7R, 8R-(+)

1. NaH−ΦCH₂OH
2. H⁺

ΦCH₂O   OH   (555)

1. TsCl
2. H₂-Pd

HO   OTs   (556)

KOH-MeOH

(549)   (7S,8R)-(−)   (536 b)

(545 a)
+   7R, 8R-(+)

7S, 8S-(−)   (545 b)

Scheme 95

6*

Scheme 96

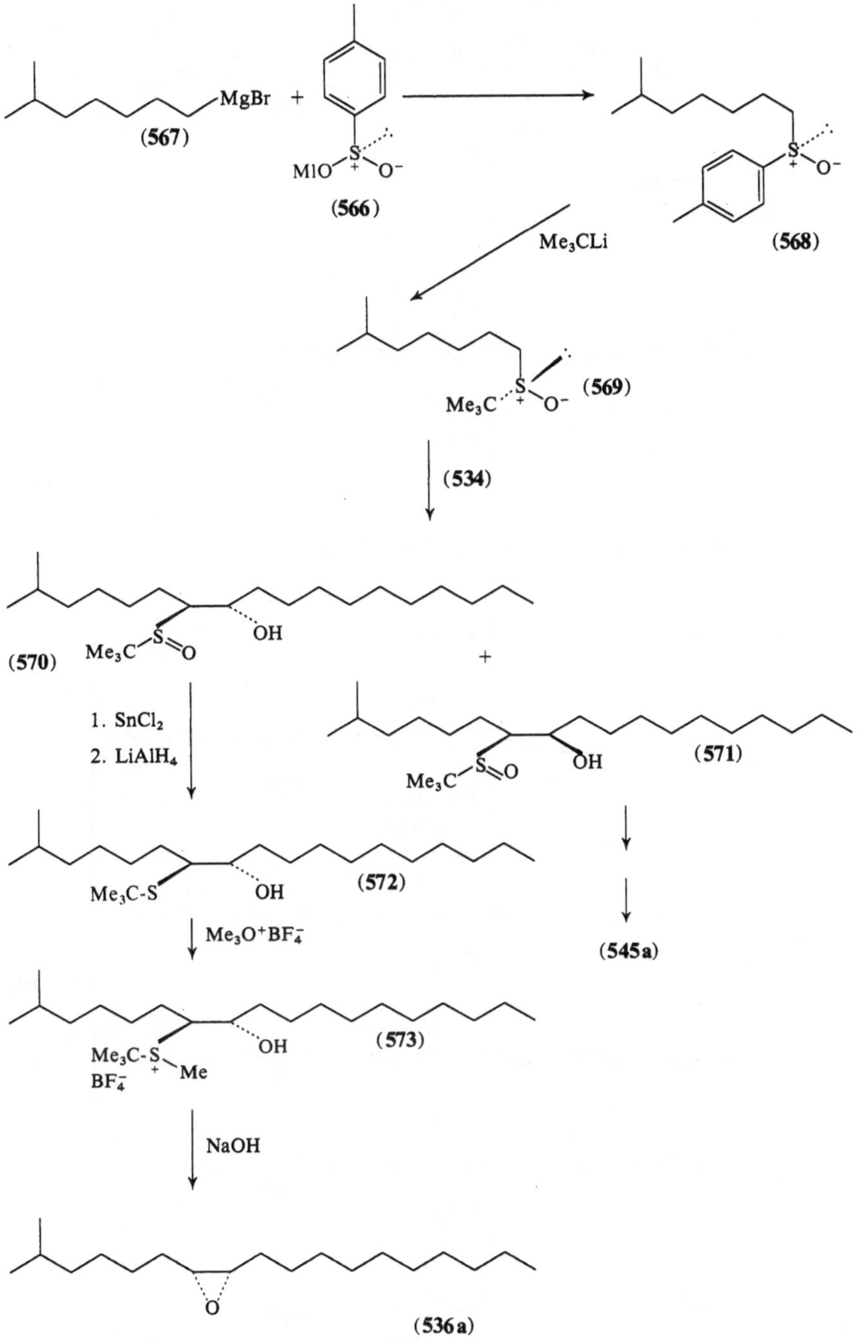

Scheme 97

MI is *L*-(−)-Menthyl

Scheme 98

6. Ketone

Another atypical Lepidopteran sex pheromone is the Z unsaturated ketone (580, Scheme 98) produced by the Douglas-fir tussock moth (*Orgyia pseudotsugata*). The acetylenic ketone (577), produced *via* a dithiane intermediate, was used by SMITH *et al.* (255, 256) to prepare both the Z- and E-isomers (580 and 582, respectively) (Scheme 98). KOCIENSKI and CERNIGLIARO (257) and MORI *et al.* (194) used an Eschenmoser cleavage of an alkyl substituted α,β-epoxy ketone to obtain straight chain acetylenic ketones that were transformed into (580) (Schemes 99 and 100, respectively). Acetylenic alcohol (578) has also been prepared by Grignard reactions (258, 259).

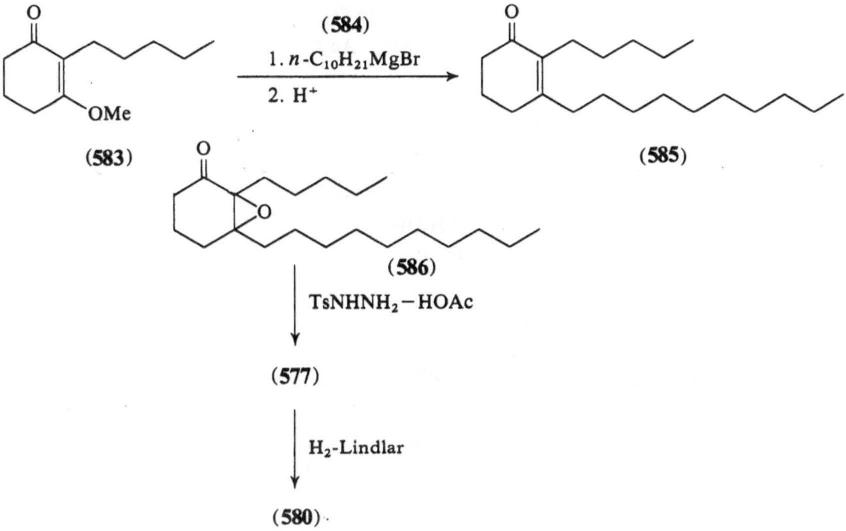

Scheme 99

G. Orthoptera

The German cockroach (*Blattella germanica*) uses the doubly branched methyl ketones (603, Scheme 101) and (615, Scheme 102) as sex pheromones. A number of syntheses of (603) have proceeded *via* bromide (600) or an analog unsaturated at the methyl branch. The branched methyl ketone moiety was introduced by alkylation of ethyl (260) or benzyl 2-methylacetoacetate (261) or of the lithium salt of (2-oxobutylidene)triphenyl phosphorane (262). ROSENBLUM *et al.* (263) coupled

Scheme 100

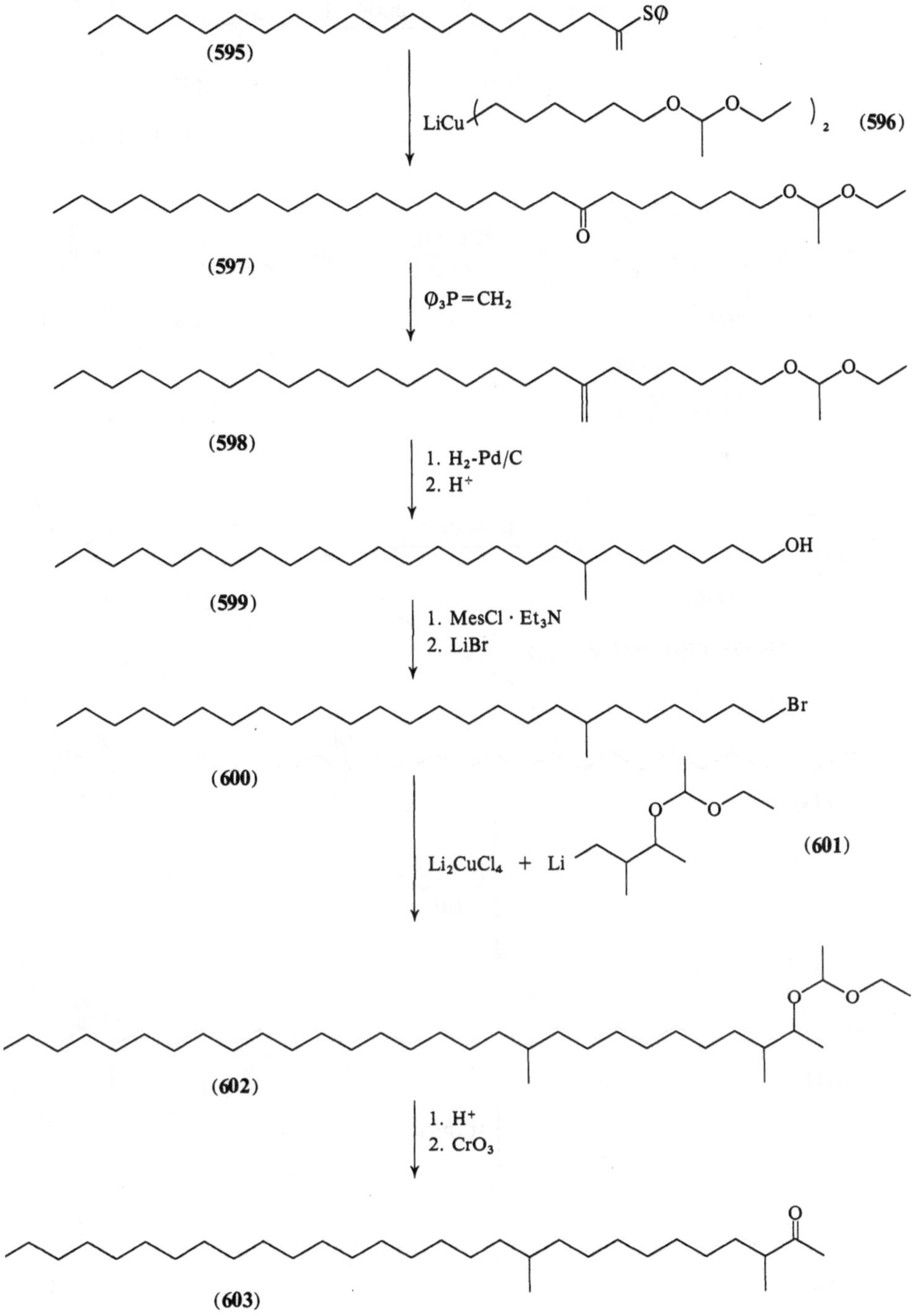

(595)

LiCu$\left(\begin{array}{c} \end{array}\right)_2$ (596)

(597)

$\varphi_3P=CH_2$

(598)

1. H$_2$-Pd/C
2. H$^+$

(599)

1. MesCl · Et$_3$N
2. LiBr

(600)

Li$_2$CuCl$_4$ + Li   (601)

(602)

1. H$^+$
2. CrO$_3$

(603)

Scheme 101

Scheme 102

bromide (**600**) with lithium salt (**601**) (Scheme 101). To obtain (**615**), BURGSTAHLER *et al.* (*261*) used (**604**) as an alkylating agent and proceeded as above, whereas NISHIDA *et al.* (*264*) elaborated the methyl branched portion of the molecule (**611**) by a series of acetoacetate ester condensations and completed the carbon skeleton *via* a Grignard reaction (Scheme 102).

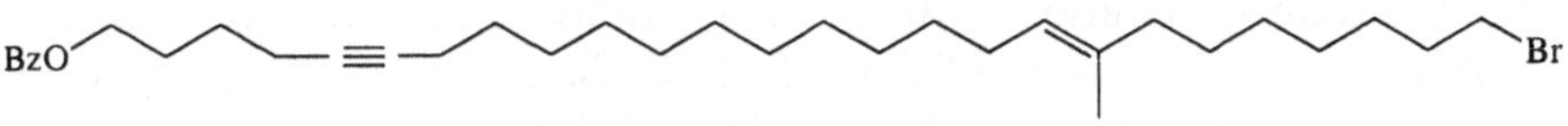

(**604**)

IV. Stereobiology

A. Geometric Isomers

It is now well accepted that pheromone blends are of general occurrence. Four recent reviews (*42, 43, 265, 266*) make this point clear in their thorough treatment of behavior-modifying chemicals of Coleoptera, Hymenoptera, Lepidoptera, and Diptera. For comprehensive lists of the ratios of *E*- and *Z*-isomers in lepidopteran pheromone blends as well as of other necessary components, the reader should consult the articles by SILVERSTEIN and YOUNG (*265*), TAMAKI (*42*), ROELOFS (*43*), ROELOFS and CARDÉ (*21*) and the annotated compendium of MAYER and MCLAUGHLIN (*41*). We intend to concentrate on some of the most recent findings concerning the ratios of *E*- and *Z*-isomers required for pheromone activity.

KLUN *et al.* (*267*), in studies at Ankeny, Iowa, on the European corn borer, found that males were only weakly attracted to highly purified (*Z*)-11-tetradecenyl acetate while the red-banded leafroller was not attracted at all to this compound. However, if small amounts of the *E*-isomer were added to the *Z*-isomer, both species were strongly attracted, yet neither species showed any response to the *E*-isomer alone. It was primarily after this study that it was realized that many lepidopteran sex pheromones are specific blends of the E- and *Z*-isomers of long chain acetate esters. The results on the Iowa European corn borer have recently been confirmed by more sophisticated separation techniques for chemically complex mixtures (*268*) and a number of specific ratios of the non-pheromone, (*Z*)- and (*E*)-11-tridecenyl acetate, have been studied in field trials on the red-banded leafroller (*269*).

Lepidopteran sex pheromones are typically esters, alcohols, aldehydes, a hydrocarbon and an epoxy-hydrocarbon (*270*). TAMAKI *et al.* (*259*) recently added a ketone to this list; females of the peach fruit moth, *Carposina niponensis*, contain (Z)-7-eicosen-11-one (**530**) as a major component and (Z)-7-nonadecen-11-one as a minor component in their pheromone glands. A 20 : 1 ratio of the major to minor component proved most effective in field trials.

Abdominal tips of virgin female Asiatic leafroller moths, *Archippus breviplicanus* contained (E)-11-tetradecenyl acetate and (Z)-11-tetradecenyl acetate in a ratio of 70 : 30 (*271*). Individually these compounds were inactive, but a ratio of E to Z of 70 : 30 competed favorably with virgin females in field experiments. In addition, tetradecyl acetate was detected in extracts of virgin females and a slight synergistic effect of this compound in field trapping experiments was noticed. Dodecyl acetate showed no synergistic effect. The smaller tea tortrix (STT) and the summer-fruit tortrix (SFT) occur sympatrically in Japan (*272*) and their mating times are not significantly different under laboratory conditions (*273*). The sex pheromones of the two species have been identified as mixtures of (Z)-9- and (Z)-11-tetradecenyl acetate (*274—277*), yet field trapping experiments with virgin females and mate-choice experiments in the laboratory clearly indicate a significant sexual isolation (*278*). The ratio of Z9—14 : Ac to Z11—14 : Ac is 63 : 37 in STT and 82 : 18 in SFT (*273*) and the ability of males to discriminate between these ratios to a slight extent was demonstrated (*273*). In addition, there was no difference in the release rate of the pheromonal mixtures between STT and SFT females (*273*). On the basis of other data it was concluded that reproductive isolation could not be based solely on the difference in the mating time of the two species (*273*). Yet orientation of SFT males to virgin females was apparently species-specific. The presence of an additional factor involved in the attractiveness of females of both species is indicated (*273*).

The male oriental fruit moth, *Grapholitha molesta,* is maximally attracted to a ratio of (E)- to (Z)-8-dodecenyl acetate of 7 : 93 (*279—281*). Lower or higher ratios give reduced trap catches (*281*). As certain species are known to utilize variations in their pheromone blend to advantage (*282—284*) and as variation may be genetically controlled, Cardé et al. (*285*) investigated variation in response of males of *G. molesta* to three distinct Z : E ratios. A color-coded fluorescent dye was placed with each blend in nonsticky field traps so that attracted males were appropriately tagged. These traps were then replaced by sticky traps containing the three Z : E ratios in a randomized block design. The results indicated that in this insect the attraction to slight changes in the blend represented a normal distribution about the optimum mixture as males were reattracted to all blends in the same relative proportions as previous field tests (*281*) on these same blends.

The problem of whether intraspecies pheromone polymorphism is genetically controlled at the biosynthetic level, or at the preception level, or neither, is basic to our understanding of its significance (see section on Biosystematics and Speciation). If we are to use pheromone blends in large scale field experiments, we need some understanding of whether we could be reducing one type of behavioral class or encouraging another; i.e., would the trapping of large numbers of a population affect the genotype that will remain?

The most attractive ratio of (Z,Z)- and (Z,E)-7,11-hexadecadienyl acetate (**529a** and **529b**; Scheme 91) for the pink boll worm, *Pectinophora gossypiella*, was claimed to be 1:1 (*79, 233*). ROTHSCHILD (*286*) found that three species, *P. gossypiella*, *P. endema* and *P. scutigera*, were maximally attracted to ratios of 1:1, 1:0.5 and 1:0.1 Z,Z, to Z,E and indicated that these ratios may provide a mechanism for reproductive isolation. FLINT et al. (*287*) have recently shown that the most attractive ZZ to ZE ratio for the pink boll worm varies with the time of year. In the early season a 2:1 ratio was considerably more attractive than a 1:1 ratio, and, as the season progressed, the response maximum was obtained with a much wider range of ratios. Various ratios of ZZ to ZE isomers have been claimed to be most effective (*287*).

EITER (*83*) has recently presented some critical comments on insect attractants and the work on the pink bollworm in particular. As sex attraction activity is brought about by so little material in the Lepidoptera and as it is difficult to obtain really pure configurations of olefins and polyolefins, EITER (*83*) questions the validity of some of the identifications. However, the two-component pheromone described in the preceding paragraph was independently identified by two groups (*79, 233*), and extensive field tests with the synthetic material have been very successful (Section VII D).

The sex attractant produced by the female clover cutworm, *Scotogramma trifolii*, is a mixture of (Z)-11-hexadecenyl acetate and (Z)-11-hexadecenol (*288*). A mixture of these two compounds containing about 10% Z11-16:OH has proved to be most attractive (*289, 290*) while 20—25% Z11-16: OH decreases the catch of males appreciably (*290*). Adult males of the glassy cutworm, *Crymodes devastator*, are attracted by traps baited with equal parts of (Z)-11-hexadecenyl acetate, (Z)-11-hexadecenal and (Z)-7-dodecenyl acetate (*291*). No other species were trapped by this mixture. The addition of small quantities of either (Z)-11-hexadecenol or (Z)-9-tetradecenyl acetate completely inhibited attractancy in the field. As a result of field tests, UNDERHILL et al. (*292*) have recently found that alkenals may be as important in lepidopteran sex attractants as the acetates and alcohols.

The sex pheromone components of the diamondback moth, *Plutella xylostella*, a pest of cruciferous crops, has been identified as (Z)-11-hexadecenal and (Z)-11-hexadecenyl acetate (*293*). More recent data show that (Z)-11-hexadecenol is a synergist of these two compounds in field trapping experiments (*294*).

The sex pheromone of the threelined leafroller, *Pandemis limitata*, is a 91 : 9 ratio of (Z)-11-tetradecenyl acetate and (Z)-9-tetradecenyl acetate (*295*). Field tests in Washington using a 94 : 6 ratio also caught large numbers of a sibling species *P. pyrusana*. Roelofs *et al.* (*296*) have shown that this ratio (94 : 6) of Z11-14 : Ac to Z9-14 : Ac is the actual ratio of these isomers in the pheromone gland of females of *P. pyrusana*. These two species therefore have very similar ratios of their sex pheromone isomers. The individual isomers are not active.

Males of the variegated leafroller moth, *Platynota flavedana*, are attracted in greatest numbers in the field to a ratio of 84 : 16 of (E)-11-tetradecen-1-ol to (Z)-11-tetradecen-1-ol (*297*). Female tip extracts contain these two alcohols in a 9 : 1 ratio (E : Z) as well as the corresponding acetates but the acetates do not appear to be used as pheromones. Trap catches of the alfalfa looper, *Autographa californica*, are greatly increased by the addition of (Z)-7-dodecen-1-ol formate to (Z)-7-dodecen-1-ol acetate (*298*), its proposed sex pheromone (*299*). It is not known whether the formate ester is an actual component of the natural pheromone.

Four species of *Trogoderma* beetles use methyl-branched alkenals as their major sex pheromones (*300*). Aeration of *T. granarium* and trapping of the volatiles on Porapak Q gave a 92 : 8 ratio of (Z)- to (E)-14-methyl-8-hexadecenal. The Z-isomer was the major component obtained from *T. inclusum* and *T. variabile* and the E-isomer was obtained from *T. glabrum*. In laboratory bioassays males could discriminate between the geometric isomers.

Females of the black carpet beetle, *Attagenus megatoma*, produce the sex pheromone (E,Z)-3,5-tetradecadienoic acid (*301*). An important related species, *A. elongatulus*, has also been shown to produce a sex pheromone (*302*) and a major attracting component has been identified as (Z,Z)-3,5-tetradecadienoic acid (*303*).

Several aphid apecies of economic importance use (E)-β-farnesene as an alarm pheromone (*304—307*). Bowers *et al.* (*308*) have studied the activity of various farnesene and nor-farnesene analogs in an attempt to define the structural requirements necessary for alarm activity. They concluded that the following properties were important: 1) there should be an E-configurational double bond in the central position of the molecule, 2) the double bond in the terminal isoprene unit acts as an activator, 3) a π-bond of 1.34 to 1.39 Å must be located ten carbon units from the end of the terminal isoprene unit, and 4) this π-bond must have free rotation about a single bond.

Bombykol, (E,Z)-10,12-hexadecadienol, is usually considered to be the only female sex pheromone of *Bombyx mori*. However, it has recently been established that the abdominal gland of the virgin female contains both bombykol and the corresponding aldehyde, (E,Z)-10,12-hexadecadienal (bombykal) (*308a*). Bombykal inhibits behavioral responses of the male moth to bombykol but further investigation is necessary to elucidate its biological significance,

B. Enantiomers

Chirality is often associated with compounds of biological origin. Many identified pheromones can have enantiomeric forms and it appears, from what follows, that insects often will biosynthesize and utilize as a pheromone only one enantiomer or a specific ratio of enantiomers. It is only in the last few years that stereospecific syntheses of some pheromones have provided pure enantiomers in sufficient quantities for laboratory and field studies. A number of behavioral experiments have now shown that insects can often distinguish between enantiomers.

Employing measurements from single olfactory receptor cells, KAFKA et al. (*309*) showed that the honeybee and the migratory locust respond differently to each of the enantiomers of 4-methylhexanoic acid. Worker honeybees could be trained in the laboratory to associate certain enantiomers of carvone and 2-octanol with the availability of food (*310*). The female spruce budworm is stimulated to oviposit by (S)-(+)-α-pinene but not by the R-(−)-enantiomer (*311*). However, none of these compounds is a pheromone.

The principal alarm pheromone of *Atta texana* is (S)-(+)-4-methyl-3-heptanone (**386a**, Scheme 68) (*173*). Determination of threshold levels of both enantiomers showed that the naturally occurring (+) enantiomer was about 100 times more active than the (−) enantiomer (*173, 312*). It was also established that the (−) enantiomer did not inhibit response to the (+) enantiomer. However, the (−) enantiomer contained 1.3% of the (+) enantiomer, so it is not possible to determine whether the (−) form has some slight activity of its own or not. It was concluded that the receptors on this ant must be chiral and respond maximally to the naturally occurring enantiomer. *Pogonomyrmex barbatus*, which also uses 4-methyl-3-heptanone as an alarm pheromone (*313*), responds to the S-(+) enantiomer up to 10 times more than the R-(−) enantiomer (*314*). It is not known which enantiomer is synthesized by this ant, but we would propose, as a general rule, that an organism responds optimally to its own enantiomer or ratio of enantiomers.

Gnathotrichus sulcatus produces 6-methyl-5-hepten-2-ol (sulcatol) (**229**) as its aggregating pheromone, as a 65 : 35 mixture of S-(+) and R-(−) enantiomers (*315*). Borden et al. (*316*) found that this ambrosia beetle responded maximally to a racemic mixture (50 : 50) of the two enantiomers. In fact, the response to the racemic mixture was significantly greater than it was to a 65 : 35 ratio. More recent work, however, suggests that *G. sulcatus* responds to some extent to S-(+)-sulcatol alone, that it does not respond to the S-(−)-isomer, and that the range of active ratios is much wider than previously indicated (*317*). In addition, Borden and coworkers (*317*) have found that *G. retusus* responds to (*S*)-(+)-sulcatol in an upwind laboratory bioassay and that the response appears to be inhibited to some extent by the S-(−)-enantiomer. They speculate that speciation may depend in part on the enantiomeric compositions of the pheromone.

The response of the European elm bark beetle to isomers of its pheromone blend was determined with two separate laboratory bioassays and field tests (*93, 318*). A combination of 3 compounds that act synergistically was found: (−)-4-methyl-3-heptanol (**294a**), (−)-α-multistriatin (**303a**), and (−)-α-cubebene (**283**) (for formulas, see pp. 48, 49) (*93*). The absolute configuration of the (−)-α-multistriatin is 1*S*,2*R*,4*S*,5*R* (*49*) and that of the (−)-4-methyl-3-heptanol is 3*S*,4*S* (*147*). Both the *A. texana* alarm pheromone, (*S*)-(+)-methyl-3-heptanone, and the alcohol above share 4*S* stereochemistry. Of interest is the fact that the reduction of decalones by microbial enzymes is usually stereoselective, with optically active alcohols of *S*-configurations being obtained (*319*).

Wood et al. (*320*) studied the flight response of both sexes of *D. brevicomis* to a combination of *exo*-brevicomin, frontalin and myrcene. Mori synthesized the enantiomers of frontalin (*133*, Scheme 43) and *exo*-brevicomin (*122*, Scheme 34); Stewart et al. (*46*) established that the natural enantiomers were (1*S*,5*R*)-(−)-frontalin (**195**) in males and (1*R*,5*S*,7*R*)-(+)-*exo*-brevicomin (**150**) in females. Wood et al. (*320*) claim that these two enantiomers are the most active forms of these compounds. Males of *D. frontalis* in a pedestrian type bioassay, are more responsive to (−)-frontalin than to (+)-frontalin, while females are not responsive to either enantiomer (*321*). The natural ratio of the frontalin enantiomers in *D. frontalis* females is 85(−) : 15(+) (*46*) . Data on field tests using optically pure isomers of brevicomin, frontalin and ipsenol indicate that activity is associated with only one enantiomer of each compound (*322*).

Ips calligraphus responds to ipsdienol and *cis*-verbenol (*323*). (*S*)-*cis*-Verbenol (**234**) is the active enantiomer and the response is not inhibited by (*R*)-*cis*-verbenol when present at a 1 : 1 ratio but is inhibited when the *R*-enantiomer is present at a 10-fold higher amount (*324*). *I. typographus* also responds to (*S*)-*cis*-verbenol (*325*). The lack of information on the natural ratio makes it difficult to interpret these results, but we would presume that the R-enantiomer is not dominant.

Racemic disparlure, *cis*-7,8-epoxy-2-methyloctadecane (**536**, p. 80), is not as active in laboratory bioassays and electro-antennogram recordings as the 7R,8S-(+) (**536a**) enantiomer (*252*). The 7S,8R-(−) (**536b**) enantiomer is virtually inactive. Dosage-response effects of "inactive" enantiomers were therefore conducted by VITÉ *et al.* (*324*) on the gypsy moth. The response to low concentrations (10^{-3} to 10^{-5} dilution) of (+)-disparlure was not affected drastically by the addition ef equal or lower concentrations of (−)-disparlure. However, (−)-disparlure in higher concentrations than the (+) antipode lowered the response of moths drastically. It was concluded that the "inactive" enantiomers appear to require a higher concentration in order to saturate the receptor sites for inhibitory effects on the pheromone to appear. A similar result has been obtained by PLIMMER *et al.* (*326*), but again, the natural enantiomer composition is not known.

BIRCH and coworkers (*327*) have shown in laboratory bioassays that female *Ips pini* from the western USA respond well to (−)-ipsdienol (**266a**, Scheme 52) but do not respond to either the (+)-isomer (**266b**) or a racemic mixture. This contrasts with the eastern *I. pini* which responds to the (+)-isomer and a racemic mixture but does not respond to (−)-ipsdienol (*328*). The mechanism for the intraspecific variation in pheromone systems of *I. pini* described by LANIER *et al.* (*329*) seems partially resolved. PLIMMER *et al.* (*45*) found that the natural enantiomer composition of ipsdienol for *I. pini* from Idaho to be 100% (−). The European fir engraver, *Pityokteines curvidens*, aggregates only in response to the S-(−)-isomer of ipsenol, and the R-(+)-isomer is inactive (*330*).

The sex attractant of the female Japanese beetle, *Popillia japonica*, which attracts males, is (Z)-5-(1-decenyl)dihydro-2(3H)-furanone (*73*). The pure synthetic R,Z-isomer (**144**, Scheme 27) was competitive with females, and male response was strongly inhibited by small amounts of the S,Z-isomer. The enantiomeric composition of the natural product has not been determined. The female also produces minor amounts of both the E-isomer and the saturated analog but the roles of these compounds has not been established. Females of the California red scale, *Aonidiella aurantii*, emit a pheromone that attracts males (*331*). Two compounds have recently been identified and synthesized: 3-methyl-6-isopropenyl-9-decen-1-yl acetate (**332a**) and (Z)-3-methyl-6-isopropenyl-3,9-decadien-1-yl acetate (**333**, Scheme 50) (*75*). The four enantiomers of the latter compound were synthesized and only the R,Z-isomer proved to be attractive to male red scale.

The use of the sesquiterpene, (E)-β-farnesene, as an aphid alarm pheromone is well established (*304*—*307*), and the spotted alfalfa aphid,

Therioaphis maculata, has been found to use another sesquiterpene hydrocarbon, (–)-germacrene A (*332*).

The enantiomeric composition of several insect pheromones has been determined by employing a chiral derivative and chiral lanthanide shift reagents (*45*). The amount of material required was of the order of 5—500 µg of substrate. As the enantiomeric composition of chiral pheromones is of fundamental importance to their efficacy and appears to play a role in speciation, one can expect its determination to become a standard procedure.

C. Chemorecognition

This discussion deals largely with the chemorecognition of a stimulus molecule (i. e., a releaser pheromone) by a receptor site. It also deals with how enantiomers may be distinguished. The interactive forces between a releaser pheromone molecule and a receptor site, or their binding, are considered to play a major role in chemorecognition.

The evidence that most insects use their antennae to locate the opposite sex has been reviewed by Jacobson (*11*) and the nature of the olfactory receptors and chemosensory processes has been described by a number of workers (*333—338a*). The concept of receptors was first introduced by Ehrlich (*339*) when he wrote that drugs do not act unless they bind. The best evidence for the existence of selectively sensitive receptors is the difference in the behavioral response to many stereoisomers. Therefore, in the study of the action of pheromones at the molecular level, it is necessary to consider both their conformation and their configuration. The importance of each and their influence on insect behavior has been pointed out.

It is generally accepted that a receptor is an elastic, three-dimensional entity, largely proteinaceous, with a specific region (receptor site) that can interact with the substrate and thereby generate a stimulus. An appropriate interaction results in a biological response. The interaction of the pheromone molecule and the receptor site is likely to involve weak forces such as ionic and dipole-dipole interactions, hydrogen bonding, and van der Waal's forces. The contribution of van der Waal's forces between a relatively non-polar volatile substrate and a non-polar receptor site is likely to be significant.

The binding of pheromone molecules by antennal proteins has been observed (*340, 341*) and transport of a pheromone molecule is considered by some as an integral part of the overall sequence leading to perception (*334, 342*). In addition, antennae of males and females of *Trichoplusia ni* contain an enzyme capable of hydrolyzing its pheromone, (*Z*)-7-dodecenyl acetate, and certain other isomers (*342*). The pheromone was generally degraded less than any of the other isomers and it was suggested that the decreased enzymatic hydrolysis of the pheromone could be the result of nonenzymatic binding to receptors or the absence of a cofactor. In microbial chemoreceptors responsible for chemotaxis, it appears that the binding proteins also serve as part of the transport system for these compounds (*343—345*). However, WRIGHT (*346*) does not consider the binding of an odorant molecule at a receptor site or its transport to be necessary for the information present in the molecule to be translated into nerve impulses, thereby transmitting a signal that elicits a behavioral response. The odorant molecule need only approach sufficiently closely to transfer energy before going on its way unchanged and free to act again and again.

A number of examples have been mentioned which illustrate that the receptor site can differentiate between enantiomeric forms of a pheromone. It is common knowledge that many enzymes are specific for only one of a pair of enantiomers. Also, many differentiate between enantiotopic groups of a single substrate molecule (*347*). If binding of a pheromone to a receptor takes place, it is probably analogous to the formation of an enzyme-substrate complex. The combination of each member of an enantiomeric pair with receptors of a given chirality results in the formation of two complexes that are physically and chemically distinct diastereomeric combinations. Therefore the differentiation between enantiomers may be dependent upon the creation of diastereomeric relationships. Should two enantiomers have equal activity, then the pheromone-receptor interaction may not involve the chiral centers of the enantiomers.

The concept that an enantiomeric pair forms a diastereomeric pair when bound, even if only transiently, to a chiral receptor, fits in with some of the results mentioned previously in which it is found that activity is brought about by one of a pair of enantiomers, and appears to be a plausible explanation of how an insect can discriminate between enantiomers. The initial olfactory process for each enantiomer is different.

Most identified pheromones are blends. In many cases, the response to the major pheromonal component is either synergized or inhibited by other compounds. Both phenomena may be accounted for by each of the components having specific receptor sites; inhibition could also be accounted for by competition for the same receptor site. If the receptor molecule were allosteric, then both inhibition and synergism could be accounted for by the presence of negative and positive modulators. The possibility that a synergistic effect could be due to an allosteric receptor protein has been suggested (*316*). The results reported by AIHARA and SHIBUYA (*347a*) may be pertinent. A better understanding

7*

may result if experiments similar to those in enzyme kinetics employing inhibitors and activators can be devised.

We can safely assume that the receptor site is chiral. It is also clear that there cannot be separate receptor sites for each and every odorous substance. Therefore, many receptor sites must be able to react with many, but not all, substances. A receptor site can distinguish both between a pair of enantiomers and between closely related achiral molecules, but it seems unreasonable to assume that there be two separate mechanisms for the chemorecognition of chiral and achiral molecules.

It has been demonstrated experimentally that when an achiral molecule approaches a chiral one, chirality may be induced in the previously achiral molecule (348—351). This molecular twisting occurs before the two molecules are close enough together to be bound by the formation of a donor-acceptor complex or by hydrogen bonding (348—351) and has been demonstrated by observing the induced circular dichroism of the n-π^* transition of various achiral substances in chiral solvents (348—352). This effect of the twisting of an achiral molecule into a helical form (i. e. into a chiral one), as revealed by the circular dichroism spectrum, has been termed "dispersion-induced circular dichroism" because the twisting occurs at a distance of several van der Waals radii from the chiral molecule (349, 352). At closer distances, (i. e. when binding occurs), "association-induced optical activity" will exist as decribed by Schipper (353). These results emphasize that the solution conformation of a compound may be very different from its conformation at the active site.

To compare the extent to which chirality is induced in a given achiral molecule, Hayward (351) has measured the dichroic and isotropic molar extinction coefficients of the substance when dissolved in two different optically active solvents, and has plotted the ratio of these extinction coefficients in one solvent against the ratio in the other. This was done for a number of aliphatic ketones (351) which previously and independently had been tested as ant alarm pheromones (354). The points for a series of 2-alkanones fell approximately along a straight line and the points for the 3- and 4-alkanones fell along another straight line. The lines intersected in a region of the plot, and those alkanones that exhibit pheromone activity were found in or near the intersection. More recent studies have established the symmetry rules connecting sign and magnitude of the dispersion-induced circular dichroism with the molecular structure of the solutes (355).

It is for these reasons that Hayward (351) considers that the primary physical process in the olfactory detection of either a chiral or an achiral molecule by a chiral receptor involves some measure of induced chirality or modification of existing chirality, and that it is, in fact, this induction or modification. In short, he regards the correlation between biological activity and the nature and amount of the dispersion-induced circular dichroism of a series of compounds to be compelling evidence that induced optical activity occurs at an early stage in the process of chemorecognition, that it is intimately related to recognition-specificity, and that it is important in the process of olfaction as well as in other biochemical processes such as antibody-antigen reactions, drug action, allergies, etc. (355). If the phenomenon of dispersion-induced circular dichroism is firmly established as being important in certain biochemical processes, it certainly will have predictive value.

WRIGHT (356) has proposed a vibrational theory of olfaction that has not hitherto attempted to explain differences in odor of enantiomers. However, in the light of the results discussed above, he now considers that if two enantiomers, on approaching a chiral receptor, are twisted, they will no longer be mirror images and will therefore have different vibrational spectra and may therefore have different odors (346).

PAYNE and DICKENS (357) have described a technique to elucidate the specificity of the receptor system of the southern pine beetle, D. frontalis. The technique employs the differential adaptation of the antennal olfactory receptors to various test compounds; either the single unit recording technique or the electroantennogram (EAG) technique is used. It is designed to determine whether different compounds are recognized by the same receptor site and is based on the exposure of the antennal preparation to one compound until the site is completely adapted, followed by exposure to a test compound. It is claimed that failure to show a response to a test compound after adaptation to another indicates that all the chemorecognition sites for the test compound are occupied by the first compound. As it had been shown that the receptors for both bicyclic ketals and host terpenes respond with equal intensites (358), DICKENS and PAYNE (321) calculated the percent of the acceptors (receptor sites) capable of interacting with the various compounds tested.

However, the components of the pheromone of a butterfly are known to give different amplitude EAG's (359). GRANT (360) found that pheromones from different species elicit EAG's of similar amplitude in a common recipient. EAG's of similar amplitude were obtained when the antenna of the armyworm, Pseudaletia unipunctata, was stimulated with the male's known pheromone components, benzaldehyde, benzyl alcohol and benzoic acid, or with the related aromatic compounds, 2-phenylethanol and benzyl acetate (361). The antennal response of male European corn borer moths, Ostrinia nubilalis, to their two pheromone components, (Z)- and (E)-11-tetradecenyl acetates has recently shown that repetitive stimulation with one of the isomers results in adaptation which affects the amplitude of response in subsequent tests to both compounds (362).

It was suggested that the receptors for the two isomers were either identical or highly interactive.

The differential adaptation technique was used by Dickens and Payne (*321*) to establish the extent to which the known pheromone components of *D. frontalis* and host tree compounds interact with specific receptors. It was concluded that the bicyclic ketal, frontalin (**195**), could react with all the acceptors, and the oxygen-containing pheromones occupied a larger percentage of the acceptors than the host tree monoterpenes. In the female beetle antennal preparation, the attractant (frontalin), the inhibitors (*endo*-brevicomin (**151**) and verbenone (**233**)), and the synergists (*trans*-verbenol (**232**), 3-carene and α-pinene), formed three distinct groups. The inhibitors interacted to a greater extent with the receptor sites than did the synergists. However, Dickens and Payne (*321*) claim that the previous suggestion that there are at least two classes of receptors for the bicyclic ketals and host tree terpenes (*358, 363*) is incorrect. It has also been pointed out that adaptation of receptor cells is a complex phenomenon and may have many origins (*337*).

Adaptation of the antennal preparation to *endo*-brevicomin (**151**) blocked response to *exo*-brevicomin (**150**), indicating that *endo*-brevicomin reacted with a greater percentage of the frontalin acceptors of both sexes of *D. frontalis* than did *exo*-brevicomin (*321*). The reverse order of presentation of these two compounds did elicit a response, and it was suggested that this difference may be due to steric limitations imposed by the *exo*-ethyl group (*321*). Molecular models illustrate quite clearly that there is a distinct difference between the *exo*- and *endo*-brevicomins with respect to the face of the ring containing the two oxygen atoms.

In the eastern spruce budworm, the differential adaptation of a receptor site by a pheromone and an inhibitor acting on a common receptor site probably indicates a different affinity of each of these molecules for the common receptor site (*337*). It seems reasonable to assume that the oxygen atoms of the bicyclic ketals will be involved in the interaction at the receptor site. For this reason, quantum calculations were carried out on all three of these bicyclic ketals in order to determine the relative negative charge on the oxygen atoms (Caputo and Brand,

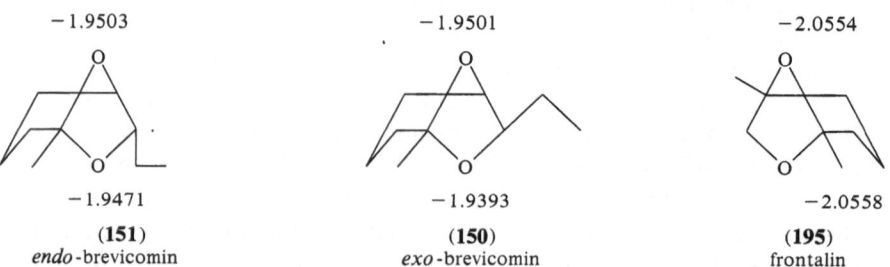

−1.9503	−1.9501	−2.0554
−1.9471	−1.9393	−2.0558
(**151**)	(**150**)	(**195**)
endo-brevicomin	*exo*-brevicomin	frontalin

unpublished data). Calculations were done using the extended Hückel method without charge iteration (QCPE Program Nr. 48). The following values were obtained: (see structures on p. 102).

Similar calculations of the charges on the oxygen atoms of multi-striatin gave values very close to those of *endo-* and *exo-*brevicomin. The substantially greater negative charge on the oxygen atoms of frontalin is not apparent from a comparison of the structures of these compounds.

If we assume that electrostatic forces play a significant role in the interaction of these bicyclic ketals at the receptor site, then the affinity of the receptor site for frontalin may be greater than that for either *endo-* or *exo-*brevicomin. Steric limitations could then account for differences in behavioral activity between the brevicomins. The affinity of the bicyclic ketals, with their two oxygen atoms, for the same receptor site may be considerably greater than that of host terpenes and *trans-*verbenol which only have one π-bond system. Quantum calculations, if done on a sufficient number of molecules of known biological activities, may provide us with a deeper insight into the mechanism of interaction and may possibly have predictive value.

PAYNE and FINN (*364*) have recently studied the pheromone receptor system of females of the greater wax moth, *Galleria mellonella*. The male produces *n-*nonanal and *n-*undecanal in its wing glands. Using the differential adaptation technique on female antennal preparations, the authors concluded that *n-*nonanal reacts with both acceptors (receptor sites), whereas *n-*undecanal reacts with only one acceptor (receptor site).

Contact chemoreceptive sex recognition in the male cricket, *Teleogryllus commodus,* has recently been described (*365*). No behavioral or EAG responses can be obtained from air carrying the odor of males or females. However, touching the antennae of sexually receptive males with antennae of either males or females elicits an agressive song or a courtship song respectively. Because treatment of the test antennae with chloroform eliminated this reaction it was concluded that chemotactile substances are present on the antennae and determine sexual recognition (*365*).

For a better understanding of aspects of insect chemoreception, it may be wise to consider those studies that have led to a model of bacterial chemoreception (*366,* and references therein). Such studies may well provide a sound basis for similar studies on insect olfaction.

V. Biosynthesis

A. Exposure of Bark Beetles to Pheromone Precursors

Many compounds have been isolated and identified from various bark beetle species and claims made as to their effects on behavior. A few of the claims are conflicting and some others are not well proved. It is therefore difficult to state categorically exactly which compounds do what and to whom. Also, the behavioral chemicals to which most bark beetles respond appear to be a complex blend of components and the actual behavioral contribution and meaning of each component part remain essentially unknown. However, it is known that the pheromonal blends, so far identified, to which beetles in the genera *Ips, Dendroctonus* and *Scolytus* respond, have a number of similarities.

Very little definitive work has been conducted on the biosynthesis of insect pheromones, and this is an area that could well be studied by the more biochemically minded. However, a number of investigations to determine the origin of many of the behavioral chemicals of bark beetles have been conducted. Most of these studies are a little different from the usual labelling experiments and we intend to concentrate the first part of this discussion on these studies.

In their natural habitat adult beetles of *Ips* and *Dendroctonus* either ingest, or are in intimate contact with, numerous host plant terpenes, three of the major ones being α- and β-pinene and myrcene. The roles of these and other host-tree monoterpene hydrocarbons have been summarized by Borden (*367*) and their function is generally considered to be synergistic in the overall aggregating pheromone complex. These monoterpene hydrocarbons are found in the insect, and their origin is considered to be the host tree, rather than *de novo* synthesis by the insect itself. However, certain termites (*368, 369*) and ants (*370—372*) can synthesize α- and β-pinene, as well as other monoterpene hydrocarbons.

It is the biosynthesis of the oxygenated monoterpenes that influence bark beetle behavior that provides us with an interesting and largely unresolved problem. The main compounds identified in this category are *trans*-verbenol (**232**) (*373*), *cis*-verbenol (**234**), 2-methyl-6-methylene-7-octen-4-ol (ipsenol) (**240**), and 2-methyl-6-methylene-2,7-octadien-4-ol (ipsdienol) (**266**) (see Chart 1) (*136*). The reaction for the biosynthesis of all these compounds seems to be the allylic hydroxylation of α-pinene and mycrene, both of which occur in abundant amounts in host pine trees.

Let us consider *cis*- and *trans*-verbenol (**234** and **232**). Borden (*367*) has summarized a number of species in which they occur, together with their probable functions. In general, these two compounds are found in the hindgut and the frass and are considered to affect the behavioral response of adult beetles. To date, the precise site and mechanism of their production remain unknown. VITÉ et al. (374) analyzed the hindguts of 12 *Ips* species for the presence of *cis*- and *trans*-verbenol; they concluded that the major pheromones in the hindgut and frass arose by two distinct mechanisms. The synthesis of the verbenols did not require feeding; rather, they were formed on exposure of the insect to oleoresin, whereas ipsdienol production required prior feeding. All stages of bark beetles that attack pine trees will be in an atmosphere rich in monoterpenes, and the indication that exposure to, and contact with, oleoresin leads to the synthesis of the verbenols resulted in a series of experiments by the group at the Boyce Thompson Institute, in which beetles were exposed to the saturated vapors of various monoterpene hydrocarbons.

HUGHES (375) exposed both sexes of *D. ponderosae* to an atmosphere saturated with α-pinene, or with oleoresin obtained from an alternate host, *Pinus lambertiana*. Analysis of the volatiles present in the hindguts of treated beetles showed an increase in *trans*-verbenol content after exposure. HUGHES (376) then extended these experiments to include *D. frontalis, D. brevicomis, D. valens* and *D. pseudotsugae*. The results indicated an increase in the amount of *trans*-verbenol present in hindguts of all species after exposure to α-pinene. HUGHES (376) proposed that the production of certain substances may be under neural and/or hormonal control as first suggested by the work of BORDEN et al. (377). A more thorough study on the exposure of *D. frontalis* to α- and β-pinene was conducted by RENWICK et al. (378), and additional volatile components produced on exposure were identified. The major components in hindguts of males were *cis*- and *trans*-verbenol, 4-methyl-2-pentanol, pinocarvone (**616**), and *trans*-pinocarveol (**617**), while those in females were 4-methyl-2-pentanol and *trans*-pinocarveol.

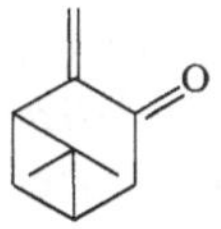

(**616**) 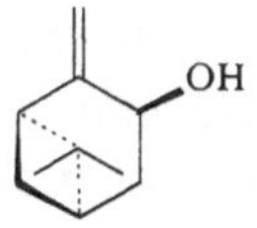   (**617**)

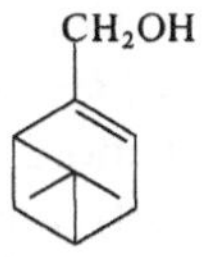

(**618**)

Renwick et al. (379) obtained interesting results on the biosynthesis of the geometric isomers and the enantiomers of the verbenols. All the previous experiments of the Boyce Thompson Institute group showed that α-pinene was converted to cis- and trans-verbenol and myrtenol (618) by all the species they studied. However, the α-pinene used in the experiments was always (±)-α-pinene. Since cis-verbenol is a component of the pheromone complex of I. paraconfusus (136), this species was chosen for experiments in which adult beetles were exposed to (+)- or (−)-α-pinene. Chromatograms of extracts of hindguts of adults of both sexes treated with (+)-α-pinene showed two prominent peaks corresponding to trans-verbenol and myrtenol. In contrast, both sexes produced predominantly cis-verbenol and myrtenol after exposure to (−)-α-pinene. The optical rotation of each purified product was measured in methanol. When (+)-α-pinene was used, (+)-trans-verbenol and (+)-myrtenol were obtained. When (−)-α-pinene was used, (+)-cis-verbenol and (−)-myrtenol were obtained. It was concluded that variations in the optical rotation of the α-pinene in trees under attack would strongly influence the ratio of cis- and trans-verbenol in this species. The implications of this finding are significant and may be important in other bark beetle species as within tree (380), and between tree (381), variations in monoterpene composition are known to occur.

Some confusion has arisen over the relationship between the sign of optical rotation and the absolute configuration, especially with regard to cis-verbenol. This problem has been resolved by Mori et al. (48). As mentioned above, (−)-α-pinene is converted to (+)-cis-verbenol by I. paraconfusus (136). In this case the optical rotation of the cis-verbenol was measured in methanol. Earlier syntheses (51, 52) of cis-verbenol from (−)-verbenone gave a product that was levorotatory in chloroform (52) and dextrorotatory in acetone (51). Both of these results are correct and both groups had in fact obtained (1S,4S,5S)-cis-verbenol (234). The sign of rotation of S-cis-verbenol changes between these two solvents (48). Also, the results of Renwick et al. (136) are quite in order but some of the structures in their publication are incorrect. For example, their structure for (+)-trans-verbenol is in fact (−)-trans-verbenol. However, their structure of (+)-cis-verbenol is correct (see 48). Furthermore, no inversion of the bicyclic terpenoid ring system occurs during biological oxidation. Mori urged that the sign of optical rotation be used in conjunction with the R and S system of configurational nomenclature whenever possible to aboid confusion.

Renwick and Hughes (382) exposed D. frontalis adults to 1-methyl-1-cyclohexene, a non-terpenoid hydrocarbon, and identified 3-methyl-2-cyclohexen-1-ol (seudenol) and 3-methyl-2-cyclohexen-1-one (MCH) among other products. Both seudenol (383) and MCH (384) are identified pheromones of the Douglas-fir beetle, *D. pseudotsugae*. While seudenol and MCH were synthesized from 1-methyl-1-cyclohexene (*382*), this compound is not a known component of the host tree of the Douglas-fir beetle. However, based on these and the previously discussed results, a general allylic oxidation and rearrangement mechanism was suggested in bark beetles.

It should be emphasized that all of these bark beetles are in intimate contact with many monoterpene hydrocarbons and that the normal detoxification mechanism for these substances is their oxidation. Many studies have confirmed the presence of numerous oxidation products of α- and β-pinene and myrcene in bark beetle hindguts. The presence of these oxidation products in the various species, albeit in differing proportions, should be considered biochemically inevitable. What must be stressed is that only certain of these compounds need be used as behavior-modifying chemicals by any one species, even though other related substances are present. The mechanism of perception by a species should be considered more discriminating than the mechanisms of biochemical oxidation. The release of, and the response to, a number of bark beetle pheromones, including MCH, by stridulation have been observed and discussed by RUDINSKY and coworkers (*385—390*). Results obtained by PITMAN and VITÉ (*391*) on the production of MCH by the Douglas-fir beetle led to a conclusion concerning the producing sex different from that of RUDINSKY and coworkers (*385, 386*).

HUGHES (*392*) showed that various *Ips* species produced ipsenol and/or ipsdienol on exposure to myrcene. *I. grandicollis* and *I. calligraphus* both required feeding before metabolizing myrcene, whereas *I. avulsus* and *I. paraconfusus* produced some products without prior feeding. Furthermore, the results indicated that ipsenol is produced by the reduction of ipsdienol obtained by the hydroxylation of myrcene. The biosynthesis of these pheromones appears to be under some form of control that is influenced by feeding in some species (*392*). BAKKE (*393*), studying *I. typographus*, found that ipsdienol and ipsenol were not detected in every male initiating galleries, and therefore presumably feeding, nor were they detected in fed males exposed to myrcene. CHARARAS (*394*), in experiments on the primary and secondary attraction in certain *Ips* species, concluded that pheromone production by males is influenced by nutritional factors such as carbohydrates of the host sap and the host plant terpenes.

HUGHES and RENWICK (*395*) have recently conducted additional exposure experiments with *I. paraconfusus* males. In this species it was found that no appreciable synthesis of *cis*-verbenol occurred in the absence of exogenous (−)-α-pinene. Therefore, unlike *D. frontalis* (*376*), newly emerged *I. paraconfusus* males do not appear to have an endogenous precursor of *cis*-verbenol.

Further experiments of this type by Renwick *et al.* (*396*) established that *D. brevicomis* and *D. frontalis* oxidize camphene (**619**) to 6-hydroxy-camphene (camphenol, **620**). Both sexes of *D. brevicomis* transformed myrcene into 2-methyl-6-methylene-2,7-octadien-1-ol (myrcenol, ´621), whereas ipsdienol was a major product in males only.

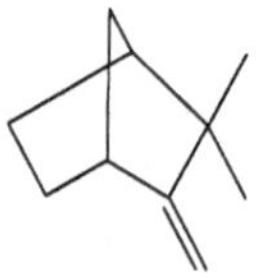

(**619**) 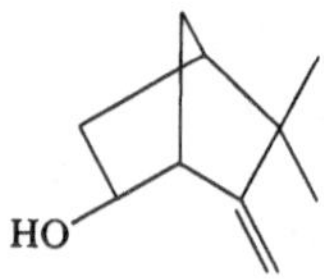   (**620**)

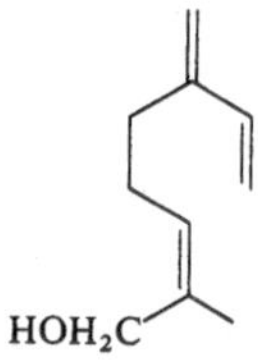

(**621**)

Hughes (*397*) has attempted to elucidate the origin of α-pinene oxidation products by means of exposure experiments. Exposure of both larvae and emergent adults of *D. frontalis* to α-pinene resulted in their producing *trans*-verbenol, whereas exposure of pupae did not. In addition, adult males less than a week old did not contain appreciable quantities of verbenone while males a week old or more did, i. e. after the adult maturation period. These results have been substantiated by Bridges (*398*). In summary, exposure of *D. frontalis* larvae to α-pinene results in the production of *trans*-verbenol, exposure of pupae does not, and exposure of emergent adults does. However, while pupae exposed to α-pinene do not produce *trans*-verbenol, this compound appears later in the adults that develop from the pupae. To continue with the puzzle, callow adults obtained from pupae removed from the bark do not contain *trans*-verbenol and verbenone, but these compounds can be detected after a week or so *without* any contact of these insects with α-pinene (*397*). Where did the compounds come from? Hughes (*397*) suggested that the pupae conjugate some form of the terpene molecule with an unknown compound. Attempts by Brand (unpublished data) to obtain *trans*-verbenol or ver-benone from pupae by mild hydrolysis of homogenates have proved unsuccessful. Taskinen (*399*) has shown that cyclic allylic alcohols such as *trans*-verbenol can react with ethanol under acid catalysis to form ethyl ethers. This question remains unsolved and should be considered of prime importance for the understanding of the biosynthesis of these oxygenated monoterpenes in these beetles.

B. Hormonal Influence on Pheromone Production

BORDEN et al. (377) showed that topical application of 10,11-epoxy-farnesenic acid methyl ester to male *I. confusus* (*I. paraconfusus* Lan.) improved the bioassay response to the hindgut region. In studies on the control of pheromone production in *I. paraconfusus*, the main findings of HUGHES and RENWICK (395) focus on neural and hormonal control of the synthesis of ipsdienol and ipsenol from myrcene. When myrcene is presented in the vapor form, feeding stimulates its metabolism to ipsdienol and ipsenol. This stimulation was also brought about by distension of the gut with air and by topical treatment with juvenile hormone (JH). Implantation of corpora allata into, or JH treatment of, decapitated males did not stimulate synthesis of the pheromones on exposure to myrcene. However, implanation of the corpora allata and corpora cardiaca together, or the corpora cardiaca alone, did stimulate synthesis. The following sequence for the neural-hormonal control of pheromone synthesis from myrcene was suggested (395). "1) Distension of the gut by feeding removes neural inhibition at the corpora allata, resulting in the release of JH; 2) JH acts through the brain neurosecretory cells and/or the corpora cardiaca to stimulate the production and/or release of brain hormone (BH); 3) BH stimulates the synthesis of enzyme(s); 4) the enzyme(s) converts myrcene or a myrcene-derived intermediate into ipsdienol, which is then reduced to ipsenol." GERKEN and HUGHES (400) have also reported that exposure of certain bark beetles to synthetic juvenile hormone analogs stimulates the biosynthesis of pheromones and therefore may have practical value in their isolation and identification.

HUGHES (397) confirmed that the gut of adult emergent *D. frontalis* males contained a considerable amount of verbenone. However, verbenone was essentially undetectable in either black adults enlarging the pupal chamber or in younger stages. In the *I. paraconfusus* study (395), corpora allata and attached corpora cardiaca from feeding males were implanted into callow males. Exposure of these males to myrcene did not result in the production of ipsdienol and ipsenol. Many insects cease the production of JH during the pupal and early adult stages, and, if this is the case in these bark beetles, it may provide part of the explanation for these findings. However, the implantation experiment of glands into callow adults suggests that more than a mere lack of JH and BH is occurring at this young stage as enzyme synthesis is apparently not stimulated by these two factors. In addition, it is suggested that feeding removes neural inhibition at the corpora allata, resulting in the release of JH (*395*). Although synthesis of JH occurs in the corpora allata, its storage in appreciable quantities in this organ is thought not to occur. Cyclic changes in the amount of hormone secreted from the corpora allata of cockroaches control some aspects of reproductive behavior including pheromone production (*401—408*).

Hughes (*409*) has summed up his views on the significance and logic of the various exposure experiments as follows. "Chemically, scolytid pheromones identified to date may be placed into three categories: 1) bicyclic ketals, 2) terpene alcohols and a corresponding ketone, or 3) a group of simple cyclic or acyclic alcohols. Nothing is known about the synthetic mechanism for the production of the bicyclic ketals or the simple non-terpenoid alcohols except that, at least in the cases studied, their synthesis can be stimulated by juvenile hormone and/or host compounds with no requirement for exogenously supplied pheromone precursor. The precise synthetic pathways for the terpene-derived pheromones are also unknown, but a considerable amount has been published on their precursors and the control of production."

"Bark beetles invading conifers contact externally and/or ingest large quantities of monoterpenes known to be toxic to these insects. Alcohols and ketones of these terpenes have been identified from the hindgut contents and fecal material of all species examined, and more recent work has shown that these alcohols and ketones are produced by the metabolism of the host-derived terpenes. It is not clear at this time to what degree their production depends on microorganisms and to what degree it depends on the insect's own enzymatic system(s), but the mechanism is general for oxidation at allylic positions."

"The terpene alcohols and ketones identified to date are common to virtually all of the species studied, although the conditions under which they are found may vary from species to species and there may be both quantitative and, to a lesser degree, qualitative differences; sex-related differences in their occurrence have also been noted. Generally, these compounds appear to be innocuous with respect to the distance orientation behavior of the bark beetles, and their production depends primarily on the penetration of the terpenes through the integument or absorption by the intestine. However, a few of these compounds do function as aggregation pheromones or deterrents in certain species; the production of most of these terpene-derived pheromones is also passive, but active control of the mechanism for synthesizing pheromones from myrcene has been demonstrated in some *Ips* species."

"Other studies have shown that the absolute configuration of the precursor determines the activity of the pheromone both by fixing the absolute configuration and, in one case, the geometrical configuration of the pheromone. It has also been demonstrated that adults can produce the pheromones from terpenes acquired in the pupal and teneral adult stages, suggesting a conjugated intermediate that is utilized by the adult at the time of maturation."

"The studies on the synthesis of the terpene-derived pheromones have led to the hypothesis that these compounds are waste products from the detoxification of host terpenes that, as a consequence of the timing and conditions of their production and release, have secondarily been utilized as chemical messengers. The nature of these oxidations, the large quantities of products formed, and the formation of the same products by other insects such as house flies when exposed to the terpenes led to the suggestion that the mixed-function oxidase system in the insect may well be involved."

We have covered these experiments on the biosynthesis of certain bark beetle pheromones by exposure to precursors and JH in some detail. These studies do not present us with a clear understanding of all the factors involved at this stage, but they do make us aware of the complexity involved in the regulation and control of pheromone synthesis in these economically important insects.

C. Possible Role of Mixed-Function Oxidases

The most likely reaction for the biosynthesis of the verbenols, as well as of ipsdienol and ipsenol, is allylic hydroxylation of α-pinene and myrcene. A reaction of this type could be carried out by microsomal mixed-function oxidases, a group of enzymes important in the degradation of insecticides and drugs. These enzymes are present in insect gut tissue, e. g. American cockroach (410), gypsy moth (411), and honeybee (412). These findings are significant in the biosynthesis of the bark beetle pheromones, particularly since the recent findings of BRATTSTEN et al. (413) in which mixed-function oxidases were induced in the midgut tissue of southern armyworm larvae by a variety of secondary plant substances, the most potent inducers being (+)-α-pinene and myrcene. However, only the enzyme activity was measured, and the oxidation products of these monoterpenes were not determined. The results of BAKER (414) which support a secretagogue mechanism for the control of digestive enzyme synthesis in insects, are also pertinent.

In the case of bark beetles invading a host tree, one would expect from the preceding that the ability to produce enzymes capable of oxidizing the monoterpenes would be rapidly induced. If this were the case then many experiments exposing various bark beetle species to α-pinene and myrcene are more understandable.

D. Possible Involvement of Microorganisms in Pheromone Synthesis

The natural habitat of most insects in their various stages of development dictates that they will come into intimate contact with numerous microorganisms. For example, many insects live in holes in the ground, in cavities in trees, in decaying logs, in leaf litter and compost, in sewers, and numerous other places all of which have an abundance of microorganisms. As might be expected, this constant exposure to microorganisms over the ages has led to many and varied associations between insects and microorganisms. Some associations may lead to disease while others may be of a beneficial and symbiotic nature (415).

Of the large number of insect pheromones identified in recent years, some are also known to be substances produced by microorganisms. For example, 3-octanone and 3-octanol produced in the mandibular glands of many ant species (416) are also produced by *Aspergillus flavus* (417). The typical defensive substance of ants in the subfamily Dolichoderinae, 6-methyl-5-hepten-2-one (418), recently also found in the ant genus *Formica* (419) and in a staphylinid beetle (420), is produced by *Endoconidiophora coerulescens* and *E. virescens* (421, 422) and a mycangial symbiont of *D. frontalis* (423). The monoterpene hydrocarbons, *dl*-limonene and α-pinene, produced by termites (368—369) and ants (370—372) also occur in algae (424), and the oxygenated monoterpenes neral and geranial which are widely distributed in the Hymenoptera (425) are also synthesized by *Ceratocystis variospora* (426).

The possibility that microorganisms associated with insects may synthesize insect pheromones has been demonstrated in a number of cases. The production of phenol in the colleterial gland of the New Zealand grass grub beetle is claimed to be due to a bacterium (427), and the termite trail pheromone, (Z,Z,E)-3,6,8-dodecatrien-1-ol (428), occurs in greater amounts in a wood-rotting fungus eaten by these insects than in the termites themselves. Recent attempts have been made to resolve the question of the origin of the trail pheromones of certain termites (429). While the authors found some degree of species specificity of termite trail pheromones, due possibly to secondary components, the biogenetic origin of the trienol was shown to be due to neither the termites alone nor fungal sources alone. Investigations are continuing to resolve these points (429). Matsumura et al. (430) have recently studied the production of (Z,Z,E)-3,6,8-dodecatrien-1-ol by the brown rot fungus, *Gleophyllum trabeum* (*Lenzites trabea*), on various carbon sources.

Larvae of the seedcorn maggot, *Hylemya platura*, damage certain crop species by feeding on the cotyledons and plumules. It has been assumed that the stimulus for oviposition was provided by the germinating seeds. However, Eckenrode et al. (431) found that significantly fewer eggs were laid on microbe-free squash seeds than on regular non-sterile seeds. The most effective microbial elicitors of oviposition were a *Pseudomonas* sp. and the yeast, *Torulopsis aeria*. Ethanol is a common end product of anaerobic fermentation by yeasts. It has been identified in extracts of wood and bark which were attractive to the ambrosia beetle, *Trypodendron lineatum* (*432*) and it also has a synergistic effect on the aggregation pheromones of *D. pseudotsugae* (*433*). The response of males of the ambrosia beetle, *Platypus flavicornis,* is increased by the addition of ethanol (*434*).

In the case of certain bark beetles a number of their identified pheromones are derivatives of α-pinene (*136, 373, 435*), and it is probably correct to assume that these derivatives arise by enzymatic action on the α-pinene of the host plant. There are two possibilities for the origin of the enzymes concerned. Firstly, they could be secreted by the gut tissue of the insect and effect the transformation of α-pinene in this region. The induction of mixed-function oxidases in insect gut tissue by host plant monoterpenes has been mentioned. Secondly, they could be produced by microorganisms present either within the gut or in the host plant tissue itself or in the frass after excretion. It is the second possibility that concerns us now.

It is well known that many microorganisms are able to oxidize both nonactivated carbon-hydrogen bonds and allylic systems to produce alcohols (*436*). The fungus, *Aspergillus niger*, was shown to convert *dl*-α-pinene into (+)-*cis*-verbenol and (+)-verbenone (*437, 438*). Both of these α-pinene derivatives occur in the gut of various bark beetles and are known to be part of the pheromone system of these insects (*136, 373, 435*). The lower alimentary tract of insects usually contains a large number of microorganisms and it is therefore quite plausible that microorganisms associated with bark beetles could be responsible for the oxidation of some of the α-pinene present in the phloem of the host plant thereby forming behaviorally active substances. With this point in mind BRAND *et al.* (*439*) isolated, under aerobic conditions, various microorganisms from the gut of adult male and female *I. paraconfusus* and determined their ability to transform α-pinene into *cis*- (**234**) and *trans*-verbenol (**232**). One organism, identified as *Bacillus cereus,* was found which produced these two compounds, together with *trans*-pinocarveol (**611**) and myrtenol (**612**) in low yield. BRAND *et al.* (*439*) stated "while our preliminary results do not prove conclusively that *B. cereus* actually synthesizes the verbenols from α-pinene in the hindgut, our data clearly indicate that this is a distinct possibility and substantiate the hypothesis that microorganisms may play a significant role in the synthesis of certain pheromones occurring in the frass of these bark beetles." CHARARAS (*394*) has concluded from feeding experiments involving broad spectrum antibiotics that bacterial conversion of ingested monoterpenes is possible but not essential for phero-mone production in certain *Ips* species.

From these results we must conclude that certain gut-associated pheromone components may be produced by the enzyme systems of the insect itself and/or by microorganisms. If this is the case, then the resulting blend occurring in the frass will be the result of the dynamic balance between those substances that are most readily produced by both microbes and insect and those that are most readily metabolized by both microbes and insect. Furthermore, the types of microbes present in the gut and the nature of the enzymes secreted by the gut tissue will be affected by the diet of the insect. Therefore, the occurrence of frass-associated pheromones should probably be considered the result of a delicate equilibrium involving a number of biological species.

The female of *D. frontalis* has a mycangium, which usually contains two fungi, *C. minor* var. *barrasii* (SJB-133) and a Basidiomycete (SJB-122), and two yeasts, *Hansenula holstii* and *Pichia pinus* (*440, 441*). These mycangial microorganisms are introduced into the phloem of host pines upon attack by female beetles and result in rapid invasion of the gallery system and surrounding phloem. Chemical transformation experiments showed that the one mycangial fungus, SJB-133, could quantitatively convert either *cis-* or *trans*-verbenol into verbenone (*442*). The *trans*-verbenol used in these experiments had an enantiomeric ratio of (+) to (−) of 60 to 40 (Brand, unpublished data). It was suggested by Brand *et al.* (*442*) that SJB-133 growing in the phloem could produce verbenone from *trans*-verbenol *in situ*, and, if this is the case, then a microorganism external to the beetle would be responsible for part of the production of at least one of its behavioral chemicals.

An increase in the verbenone concentration released from a success-fully colonized tree has been proposed as an important factor in inhibiting further attacks on the tree by both *D. frontalis* and *D. brevicomis* (*443*). It is therefore possible that SJB-133, which is important in the beetle's nutritional regime (*440*), could also play a significant role in regulating response to the plant host. More recently *endo*-brevicomin has been implicated in the shifting of attack from one tree to another (*444*).

Inquiries into the possible production of behavioral chemicals of *D. frontalis* by microorganisms associated with it have been carried further. As mentioned above, two fungi and two yeasts are associated with the mycangium of the female (*440, 441*). The production of various volatile substances, other than ethanol, by actively fermenting yeasts is well established (*445*). Brand *et al.* (*446*) grew three yeasts obtained from *D. frontalis,* namely *H. holstii, P. pinus,* and *P. bovis,* on Sabouraud's dextrose broth, and identified isoamyl alcohol, 2-phenylethanol, isoamyl acetate and 2-phenylethyl acetate as the main volatile substances (other than ethanol) produced. The presence of 2-phenylethanol in the hind guts of emergent *D. brevicomis* males, and feeding *Ips paraconfusus* males has recently been reported (*447*). In field bioassays, the response of *D. brevicomis* was not affected by the addition of 2-phenylethanol to its known attractant, whereas the response of *I. paraconfusus* to male infested log sections was greatly enhanced by the addition of 2-phenylethanol (*447*).

The behavioral activity of the compounds isolated from the yeasts was tested in laboratory bioassays on pedestrian male and female *D. frontalis* (*446, 448*). In this bioassay procedure, a standard attractant mixture of frontalin : *trans*-verbenol : loblolly turpentine (1 : 1 : 12), referred to subsequently as the triplicate standard, was used. None of the yeast metabolites exhibited any activity alone. The two acetate esters were found to enhance the attractiveness, mainly of males, to the triplicate standard, especially at low concentrations of triplicate standard and ester. 2-Phenylethanol decreased the response of females to the triplicate standard. More recent results on the inhibitory effect of 2-phenylethanol on the response of females has shown that a concentration of triplicate standard that gives a response of 50—60% can be substantially decreased by the addition of 2-phenylethanol at concentrations up to 10^5 times lower than that of the triplicate standard (BRAND, unpublished data).

If it is assumed that these yeast volatiles are produced in the tree under active attack, it is very likely that they would be perceived by attacking beetles and could influence their behavior. *Scolytus multistriatus* often initiates attacks on bark directly over sapwood streaks, which are indicative of invasion of *Ceratocystis ulmi* (LANIER, *328*). The beetles' preference for attack at this site may be directed by odorants released through lenticels of the bark as it has been shown that *S. quadrispinosus* regularly initiates attack at the lenticels (*449*). Evidence has been obtained which indicates that host finding by the braconid, *Biosteres (Opius) longicaudatus,* involves attraction to specific fermentation products of fungi coming from rotting fruit, and not to compounds produced by host larvae (*450*).

We consider that thorough investigations on the origin of bark beetle pheromone blends will establish microbial systems as an important source. However, the real credibility of the examples cited remains to be firmly established; future results will modify or strengthen this idea. We hope that some of our statements will provoke discussion, criticism and experimentations among our colleagues.

E. Biosynthesis of the Bicyclic Ketals

The bicyclic ketals, frontalin (**195**) (*130*), *exo*-brevicomin (**150**), *endo*-brevicomin (**151**) (*62, 117*), and multistriatin (**303**), all shown in Chart 1,

p. 8, have been identified from various bark beetles and appear to be of major importance in their pheromonal blends. No one has commented in any detail on the biosynthetic origin of these compounds. Let us consider frontalin. The immediate precursor of this bicyclic ketal may be either 6,7-dihydroxy-6-methylheptan-2-one (**211,** Scheme 43) or 6,7-epoxy-6-methylheptan-2-one (**205,** Scheme 42), both of which would have originated in 6-methyl-6-hepten-2-one (**204**). This latter compound is an isomer of 6-methyl-5-hepten-2-one. This methyl ketone is produced by one of the mycangial fungi (SJB-122) of female *D. frontalis* (*423*) but we do not wish to suggest at this time that the precursor of frontalin is produced by a mycangial fungus. However, it is food for thought. 6-Methyl-5-hepten-2-one is also produced by ants (*418, 419*) and certain other microorganisms (*421, 422*).

In a similar manner, the alicyclic precursor of multistriatin may be 4,6-dimethyl-7-octen-3-one (**301,** Scheme 55). Similar compounds have been obtained in alarm and defensive secretions of certain insects. Fales *et al.* (*451*) found that the mandibular gland secretion of the ant *Manica mutica* contained mainly 4,6-dimethyl-4-octen-3-one. Meinwald *et al.* (*452*) identified 4,6-dimethyl-6-octen-3-one in the defensive secretion of the daddy long legs, *Leiobunum vittatum*.

Gore *et al.* (*453*) recently analyzed extracts of emergent and boring females of *S. multistriatus* for 4,6-dimethyl-7-octen-3-one (**301**) and 4,6-dimethyl-6,7-epoxyoctan-3-one (**302**) but were unable to detect either compound. While these compounds are the most likely precursors of multistriatin, the possibility of a large pool of these intermediates is ruled out. Both the ant (*451*) and the daddy long legs (*452*) secretions also contained 4-methyl-3-heptanone (**386a**). This finding is pertinent as 4-methyl-3-heptanol (**294**) is a component of the pheromone complex of *S. multistriatus* (*146*). We must surely conclude that the biosynthesis of these various compounds is intimately related but no studies have been conducted in this area.

F. Boll Weevil Sex Attractant

Chemical studies on the sex attractant of the boll weevil led to the isolation (*454*) and identification (*38*) of four terpenoid compounds. Tumlinson *et al.* (*455*) suggested a hypothetical biosynthetic scheme in which all four compounds could be derived from a geraniol-like compound. Hardee (*456*) showed that male boll weevils required feeding for the synthesis of the attractant substances. Cotton squares proved to be the best diet, but pheromone production was demonstrated on a variety of diets. Mitlin and Hedin (*457*), using ^{14}C tracers, obtained evidence

that the biosynthesis of the pheromone compounds may be *de novo*.
Adult males were injected with either [1-^{14}C] acetate, [2-^{14}C] acetate,
[2-^{14}C] mevalonic acid or [U-^{14}C] glucose and the feces were steam distilled.
Approximately 0.02% of the label was incorporated into the volatile
fraction. While all four components accounted for only 39% of the volatile
fraction they contained 57—80% of the radioactivity. It is assumed
therefore that in spite of the boll weevil being an obligate insect of cotton,
it does not seem to require any specific component in the cotton for the
synthesis of the four terpenoid sex attractants.

HEDIN (458) has recently summarized factors that influence the biosynthesis of the boll weevil pheromone complex. Total synthesis was at
a maximum during the summer and a minimum during the winter. In this
study HEDIN (458) did two types of experiments. The first series exposed
10-day-old adult weevils to a saturated atmosphere of a number of terpene
hydrocarbons. It was found that only myrcene and limonene produced
detectable amounts of oxygenated substances and that males produced
more than females. No compounds isolated suggested that the weevils are
capable of cyclizing alicyclic terpenes or their pyrophosphates.

The second series of experiments conducted by HEDIN (458) involved
the incubation of whole abdomen homogenates with added (+)-grandisol
[(+)-*cis*-2-isopropenyl-1-methylcyclobutaneethanol, (33)] and (Z)-3,3-
dimethyl-$\Delta^{1,\beta}$-cyclohexaneethanol (110b, Scheme 18). After an overnight incubation at 37°, pentane extracts were analysed by GC-MS.
Boiled homogenates and homogenates to which neither (33) nor (110b)
were added served as controls. From the various identified products it
was suggested that male abdomens possess three major enzymatic
capabilities: (i) oxidation of the alcohols (33) and (110b) to aldehydes
by a dehydrogenase, (ii) dehydration of the alcohols to hydrocarbons by
a hydrase, and (iii) conversion to other alcohols by an isomerase. The
dehydrogenase activity is the most significant as it can explain the oxidation of (110b) to (E)- (111a) and (Z)-3,3-dimethyl-$\Delta^{1,\alpha}$-cyclohexane-
acetaldehyde (111b), two of the identified boll weevil sex attractants
(Scheme 18).

The conclusion drawn by HEDIN (458) from these results is that male
boll weevils possess enzymes that bring about the chemical transformations observed in the incubation studies. We suggest how this conclusion
could be drawn with greater conviction. It has been reported by GUELDNER
et al. (459) that the presence of a high bacterial load in the gut of boll
weevils decreases the amount of pheromone produced, and BRAND *et al.*
(439) have suggested that microorganisms may lead to the production or
modification of certain pheromones in an insect gut. HEDIN (458)
employed non-sterile abdomen homogenates for an overnight incubation
at 37° in a pH 7.0 phosphate buffer. This medium would be suitable for

the growth of many microorganisms, and their enzymatic potential cannot merely be overlooked. Filter sterilization of the homogenates and incubation under aseptic conditions would make these experiments far more convincing.

G. Miscellaneous Labelling Studies

Biosynthetic studies usually imply the use of ^{14}C-labelled compounds, but the majority of studies on insect pheromones have dealt with their identification and synthesis and with field applications; very few have used ^{14}C-labelled compounds.

Gordon et al. (460) established that the defensive aldehydes, hexenal and decenal, are synthesized from [1-^{14}C]-acetate in the green vegetable bug, Nezara viridula var. smaragdula (F) and it has also been established that the ant, Acanthomyops claviger, synthesizes monoterpenes from acetate and mevalonate de novo (461). As mentioned previously, the boll weevil is capable of pheromone synthesis from labelled acetate, mevalonate and glucose (457). The pheromone components of the wax moth, Galleria mellonella, are n-nonanal and n-undecanal (462). Injection of labelled acetate, propionate and oleic acid into moths indicated that these aldehydes arise most readily from oleic acid (463). This insect contains a large amount of oleic acid (29.6—43%) in its total fatty acids, and its diet, is also a rich source (22.2% of the total fatty acids) (464). It was concluded from these studies that odd-numbered straight chain aldehydes are synthesized from many different precursors, both even- and odd-numbered.

The hairpencil secretion of the adult male bertha armyworm, Mamestra configurata, contains 2-phenylethanol (465, 466). Radioactive labelling experiments with male pharate adults indicated that it is synthesized from phenylalanine (466). The proposed biosynthetic pathway entails the irreversible loss of ammonia from phenylalanine to yield trans-cinnamic acid, decarboxylation to styrene, and hydration of the styrene to yield 2-phenylethanol (466). The production of 2-phenylethanol from phenylalanine is well established in microorganisms (467—470).

The ponerine ant, Paltothyreus tarsatus, produces various alkyl sulfides in its mandibular glands (471, 472). With the use of doubly labelled methionine, it has been established that the CH_3-S-group is incorporated intact but the origin of the third sulfur atom in dimethyl trisulfide remains unknown (473, 474).

H. Dietary Origin of Pheromones

The suggestion that the sex pheromone composition of the oak leaf roller was dependent to a large extent on diet was made (*82, 475, 476*) and refuted (*477, 478*). It is now generally accepted that the sex pheromone of this tortricid moth species is a specific blend (67 : 33) of (*E*)-11- and (*Z*)-11-tetradecenyl acetates (*477*). It is also accepted that in this species this ratio is not influenced by diet (*478*). However, the effects of diet on the amount of pheromonal compounds have been suggested for the summer fruit tortrix moth (*479*), the gypsy moth (*480*) and the smaller tea tortrix moth (*481*).

There are cases where the host plant supplies the precursor to compounds exhibiting pheromone activity. For example, adult male danaid and ithomiine butterflies are attracted to plants containing pyrrolizidine alkaloids which they modify into dihydropyrrolizines (*482—486*). These substances then occur in the hairpencil secretions.

The results obtained by RENWICK *et al.* (*379*) on the exposure of adult *I. paraconfusus* beetles to (+)- and (−)-α-pinene should be reiterated. (+)-*trans*-Verbenol and (+)-myrtenol were the major products in the hindgut after exposure to (+)-α-pinene whereas (+)-*cis*-verbenol and (−)-myrtenol were obtained after exposure to (−)-α-pinene. Therefore, variations in the enantiomeric composition of the α-pinene in trees under attack could influence the ratio of *cis*- to *trans*-verbenol occurring in the gut of this species.

The process of melanization in the desert locust, the tropical migratory locust and the brown locust is stimulated by crowded conditions and it was suggested that this process is influenced by an airborne pheromone (*487*). A pheromone, called locustol, was isolated and characterized as 2-methoxy-5-ethylphenol. It was postulated that this compound was formed by the degradation of lignin in ingested food in the crops of larvae (*488*) and a biosynthetic pathway has been proposed (*489*). The question has been raised as to whether microorganisms present in the crop may be responsible for the production of locustol (*490*) and recent evidence strongly suggests that they are (*491*).

The heteropteran, *Eurygaster integriceps,* uses ethyl acrylate and vanillin as components of the sex pheromone of the male (*492*). These compounds induce specific behavioral responses in sexually receptive females and may also act as short range attractants. Both compounds may be metabolic products from the degradation of ingested lignin (*492*).

VI. Chemosystematics and Speciation

The methodology employed in the identification of pheromones is usually rigorous and precise and not subject to personal opinion. However, the same degree of precision and impartiality cannot be applied to the use of chemical data for solution of taxonomic problems. Chemosystematics would best be served by the detailed comparison of the enzymology of the biosynthetic pathways. In addition, comparisons between chemical compounds are best made by those familiar with the many ramifications of metabolic pathways.

Some aspects of the value of pheromones in speciation of Coleoptera have been discussed by Lanier and Burkholder (493). They concluded that in spite of recent advances in the chemistry of pheromone systems of beetles, our understanding of the role of pheromones in speciation is limited, because the data are scattered among various groups of beetles and good data on response specificity are lacking.

The interspecific attraction of males of certain *Trogoderma* species to extracts of female beetles has been studied by Vick et al. (494) and by Levinson and Bar Ilan (495). The extensive cross attraction between some species and not between others suggests the involvement of several active compounds. The degree of phylogenetic relatedness of seven *Trogoderma* species has recently been studied by Greenblatt et al. (496). The most important compound in the volatile fraction of four species is 14-methyl-8-hexadecenal. The active isomers are *E* in *T. glabrum*, *Z* in *T. inclusum* and *T. variabile*, and 92% *Z* : 8% *E* in *T. granarium*. This aldehyde is not found in extracts of macerated females. The next most active compound in these four species is the corresponding alcohol, 14-methyl-8-hexadecen-1-ol, which is present only in trace amounts in the volatile fraction. However, this alcohol has been found in extracts of macerated females of *T. glabrum* (*E*-isomer) (497), *T. inclusum* (*Z*-isomer) (498), and *T. variabile* (*Z*-isomer) (499). The third most active compound is the corresponding ester, methyl 14-methyl-8-hexadecenoate and is found in *T. inclusum* (*Z*-isomer) (498) and *T. glabrum* (*E*-isomer) (497).

Greenblatt et al. (496) concluded that the response of males to calling females is largely due to the aldehyde, and that the response to extracts of macerated females is due largely to the alcohol. Strong interspecific responses between *T. inclusum, T. variabile* and *T. granarium* are a consequence of the presence of the *Z*-isomer of each pheromone component and the different response of *T. glabrum* lies in its use of the *E*-isomer. Neither *T. sternale* nor *T. grassmani* respond to the *Z*- or *E*-aldehyde (496). *T. simplex* males respond strongly to extracts of *T. inclusum, T. granarium* and *T. variabile* females which would indicate that all four species share one or more pheromone components and that *T. simplex* probably emits the *Z*-aldehyde.

On the basis of these and other data (morphological and interbreeding responses) Greenblatt et al. (496) presented a cladogram of the possible relationship between these seven species. *T. sternale* and *T. grassmani* were placed together in one group and *T. simplex* was placed in a group on its own. *T. inclusum* and *T. variabile,* both of which have the Z-isomer of the aldehyde, are considered to be closely related, while *T. granarium* and *T. glabrum* were placed off a fourth main branch (*T. glabrum* has the *E*-isomer of the aldehyde and *T. granarium* has a mixture of *E*- and Z-isomers). Chemical investigations on the pheromones of *T. sternale* and *T. grassmani* are continuing.

Many sympatric tortricine moths possess overlapping chemical communication systems because at least one component of their pheromones is either a 14-carbon chain acetate, alcohol or aldehyde, with unsaturation in the C_{11-12} position. The species recognition of distinctive pheromone blends, together with non-pheromonal reproductive isolating mechanisms such as habitat preferences and differential mating times have been discussed by CARDÉ *et al.* (*500*). This article, together with the literature cited therein, should be read to obtain an informed view of pheromone specificity and its significance in tortricine moths.

(*Z*)-11-Hexadecenyl acetate is an important component of the sex attractant blend of several noctuid moths. Species specificity of blends containing common constituents can be attributed to the presence of different additional coattractant compounds, and each coattractant could be effective for a single species, or the ratio of components may be the deciding factor (*270*). Most laboratory and field bioassays are designed to show attraction. However, STECK *et al.* (*501*) have recently reported some experiments designed to show interspecies inhibition in a group of moths, all of with require (*Z*)-11-hexadecenyl acetate. Their results indicate that dual attractant/inhibitor roles are common occurrences among sex attractant compounds. The generality of this phenomenon is being investigated, as it could be of importance in the use of atmospheric permeation with pheromones for mating disruption in pest species.

The European corn borer (ECB), *Ostrinia nubilalis,* is known to employ different ratios of its sex attractants, (*E*)- and (*Z*)-11-tetradecenyl acetate, in various geographical areas. KLUN (*502*) has recently obtained some particularly interesting and significant results on this pest insect. (*E*)-9-Tetradecenyl acetate has always been observed in GC analyses of heptane surface washes of individual female ovipositors. This suggests that this compound is part of the insect's pheromone signalling system. However, in field and laboratory assays, this compound suppressed male attraction and precopulatory behavior. By the use of mixtures of (*E*)-9-tetradecenyl acetate and *Z*:*E*-isomer blends of 11-tetradecenyl acetate in bioassays, it was suggested that these positional isomers are perceived through separate sensory channels. The production of an "anti-sex substance" by the ECB is enigmatical. As (*E*)-9-tetradecenyl acetate does not deter the redbanded leafroller, a species that also uses a $Z:E$-isomer blend of 11-tetradecenyl acetate as a sex pheromone, it does not serve a role in maintaining pheromonal specificity between these two species.

The likelihood of genetic control of the ratio of pheromone components has occurred to many. Some recent results provide an exciting beginning to what is bound to follow on this topic. Bioassays of hybrids of species in three groups of *Ips* bark beetles showed that hybrid males were intermediate in attractiveness to the parental types (*493*). The possibility that hybrid females were slightly more attractive to males of their own kind than to those of the two parental species was also suggested. However, in these bark beetles it was concluded that the genes controlling pheromone production and reception were not sex linked (*493*). The genetic basis of intraspecific pheromonal polymorphism in the ECB has been investigated more recently by KLUN (*502*). The Iowa strain of the ECB responds maximally to a 97:3 ratio of $Z:E$-isomers of 11-tetradecenyl acetate. The reverse is true for the New York strain. KLUN and coworkers have now shown that the geometric isomer composition of 11-tetradecenyl acetate in the female secretion is controlled by single Mendelian inheritance involving one pair of genes. The female AA genotype secretes an isomer blend of approximately 97:3 ($E:Z$) and the aa genotype secretes a blend of approximately 3:97 ($E:Z$). F_1 hybrid females from crosses between Aa and aa genotypes secrete an isomer mixture that approximates 65:35 ($E:Z$). Of particular significance is the finding that F_1 males from the same cross respond preferentially to the 65:35 $E:Z$-isomer combination rather than to either of the parental mixtures.

Therefore, both the isomeric composition of the sex pheromone secretions in female ECB and the mechanism of isomer-ratio perception in male ECB are genetically regulated. These studies by KLUN and coworkers clearly demonstrate the need to know the genotypic profiles of females and the isomeric ratios of their sex attractants at any location where behavior-modifying chemicals are to be used for the suppression of this species. CARDÉ et al. (*502a*) have recently suggested that the E- and Z-strains of the ECB may be semi- or sibling species.

Many species of ants exhibit a synchronized swarming of male and female alates from many nests thereby ensuring that large populations of reproductives are available to each other at the same time. The coordination of this synchronized swarming in alates of the carpenter ant, *Camponotus herculeanus,* is governed by volatile substances secreted from the mandibular glands of the males, as well as by climatic factors (*503*). It has been reported that the males of a number of *Camponotus* species contain substances in their mandibular glands which do not occur in females or workers (*504, 505*). The quantitatively most important compounds among those species studied were methyl 6-methylsalicylate, methyl anthranilate and mellein (3,4-dihydro-8-hydroxy-3-methylisocoumarin) (504, 505). The possibility that these caste-specific compounds may play a role in the regulation and coordination of swarming was inferred, but not established, and their application as an additional aid to the taxonomy of this large genus of ants was indicated.

The volatile products in the heads of a large number of Nearctic species of *Camponotus* have been studied more recently by DUFFIELD (506). In general, mandibular gland secretions were restricted to males although some species produce the same compounds in both sexes and castes. When a compound was produced by both sexes and castes it appeared to be a non-specific alarm pheromone, but the role of cast-specific compounds was not demonstrated. The results obtained by DUFFIELD (506) led to the preparation of a taxonomic key to the various species studied which is based primarily on chemical characters.

The identification of 6-methyl-5-hepten-2-one as the major detectable volatile substance in the mandibular gland secretions of workers of eight *Formica* species in the subgenera *Neoformica* and *Proformica* (419) indicates that this ketone may have some taxonomic value in this ant genus.

VII. Practical Applications of Pheromones: Status and Projections

A. Plea for Sanity and Integrated Pest Management

The era immediately following World War II saw the intensive use of organopesticides, starting with the archetypal DDT—the agricultural analog of the modern "magic bullet" of medicine, penicillin. The pesticides salesman prescribed his magical potions on some arbitrary schedule and vast amounts were applied worldwide. Extraordinary successes abounded. Crop yields were increased, pest populations declined and millions of lives were saved throughout the world, but the long-term effects of such indiscriminate application of pesticides were neglected. FLINT and VAN DEN BOSCH (507), and DETHIER (508) eloquently describe how preventive pest control practices were discarded and biological studies languished as reports on insecticide testing proliferated. Entomology, for a generation, suffered under the stigma of being the only branch of science dedicated to eradicating what should have been the object of its study. It should be noted, however, that basic studies of insect toxicology and metabolism flourished.

The world is never without prophets and during this era, several spoke; and they—in the usual manner of prophets—spoke in vain. Development of resistance to insecticides was the first theme of the prophets' warnings, and as the inevitable happened, the predictable—and profitable—response was to increase the amount and frequency of application. Rebound of the target pest began to be noted as broad spectrum pesticides wiped out the natural enemies of the pest. In a very short time, new major pests were *created* as minor pest populations, freed of their natural enemies, exploded to devastating levels. Finally the problems of environmental contamination by persistent pesticides were brought home to the public by the appearance in 1962 of "Silent Spring" (*509*).

Since then the excesses of pest management seem to have peaked; starting with the ban on DDT in the United States in 1972, far-reaching, decisions have severely restricted the uses of other hard pesticides. The concept of "Alternatives to Hard Pesticides" was taking hold, and during the 1960s these efforts multiplied and resulted in a number of promising approaches. At the present time, we can list, in addition to sound agricultural practices, a fair number of such alternatives:

1. Biological control through manipulation of predators, parasites, and disease organisms.

2. Genetic selection of resistant plants.

3. Reproductive suppression by radiative or chemical sterilization.

4. Introduction of reproductively incompatible strains.

5. Use of hormones and hormone analogs.

6. Use of behavior modifying chemicals, of which pheromones and pheromone analogs are most important (*507, 510—517*).

Gradually the concept of Integrated Pest Management developed as an integration "of all of the factors impinging on the pest control decision so as to determine when control should be practiced and what would comprise the best method or combination of methods to employ" (*510*). Synthetic chemical pesticides would be relegated to an appropriate role within a holistic program that considers entire ecosystems.

In the following we will deal with the present and potential roles of behavior modifying chemicals (mainly pheromones) as one of the tools available to the manager of an integrated pest management system.

At the outset, we must state our own biases as participants in pheromone research. We subscribe to the statement from "Advancing toward Operational Behavior-Modifying Chemicals" (*518*): "Some observers have evidenced impatience with the various programs that have been directed toward the development of behavior-modifying chemicals (especially the pheromones) for insect pest management. However, it must be recalled that almost all of the practical research directed toward operational control programs has been initiated only during the past 5 years [written in 1975] and by a small number of investigators. Additionally, only limited resources have been made available for this research. Thus, on any reasonable basis of expectation in science, progress should be considered quite remarkable." The lack of basic information on insect behavior and population dynamics of the important insect pests is obvious. The problems of determining reasonable economic thresholds are complicated by expectations—unblemished apples for example—that are themselves unrealistic. The basic problem of devising control situations against which to assess the effects of manipulating insect populations is a formidable one. Nonetheless, assessments of control strategies have been attempted, and as we shall see, with at least partial success. G. E. DATERMAN (private communication, *519*) believes that "for control purposes, the appropriate question is not 'will pheromones be used', but 'how soon'."

The availability of synthetic pheromones has provided a remarkable tool for investigation of insect behavior, ranging from studies of potentials generated in antennal receptors to studies of responses in the field over large distances. Practical applications can be categorized as follows:

1. Trapping to collect species that are otherwise difficult to obtain.

2. Trapping for monitoring and survey; timing of pesticide treatment can be based on results.

3. Luring to areas that are treated with pesticides.

4. Luring to areas that are treated with pathogens, which can then be spread by the infected individuals to the rest of the population.

5. Mass trapping for population suppression.

6. Disruption of communication by permeation of areas of insect population. In most cases, the most important function that can be disrupted is mating, the end result being population suppression.

Parapheromones (chemicals that mimic pheromones) or anti-pheromones (those that block responses), which are not part of the natural communication system, may also be used in some of these categories. A list of pheromones, parapheromones, and anti-pheromones active in the field has been compiled by INSCOE and BEROZA (*40*).

At the present time the first two categories are operational, although more research is needed to relate trap catches to insect populations. Use of pheromones for population suppression by the other techniques is now in the "highly promising" stage for many pests, and is operational for several; we shall consider the promises and problems on a crop-by-crop, insect-by-insect basis. Coverage will be selective and emphasis will be on the more advanced systems and those most familiar to the authors.

Although the U. S. Environmental Protection Agency (EPA) is constrained by law to treat pheromones as pesticides for registration purposes (*520*), it should be obvious that, operationally, there is a vast difference between spraying pounds of a liquid or solid, persistent, broad-spectrum pesticide to kill insects, and releasing a fraction of a gram of a biodegradable, species-specific, natural product in the vapor phase to lure an insect to a trap or to disorient the matefinding process.

Certainly, individuals within the EPA will acknowledge these differences in private conversations and there is some indication that official policy will be modulated; at this writing, EPA is handling pheromone registration on a case-by-case basis. A Task Group convened by the American Institute of Biological Sciences is advising EPA on efficacy test procedures. The EPA response will be critical to the role of behavior modifying chemicals in control programs, because an unrealistic policy will stifle the already marginal interest of industry in these materials. The wary posture of EPA is motivated in part by past industrial practices resulting from lack of knowledge and from the inherent need for short-term benefits; because of this, the innovative industrialist is penalized by unrealistic restrictions (521).

For the present, at least, the impetus for promoting the use of behavior modifying chemicals within the context of integrated pest management must come from government agencies and private foundations. A vast amount of basic behavioral research is needed, but eventually the materials and methodology must be produced by industry under government regulation, and given the nature of the task, very likely under some form of government subsidy. Some cost estimates involved in bringing behavior regulating compounds from chemical synthesis to commercial use were presented by Siddall and Olsen (522), who conclude that not even a research-oriented concern can currently justify development of a pheromone for control of an insect pest by disruption of its communication system.

An interdisciplinary group has examined the economies from the industrial viewpoint of several components of pest insect control: bacteria, viruses, pheromones, conventional pesticides and a few miscellaneous agents. The following table from their report (511) summarizes the relative likelihood of pest suppression programs using pheromones:

	Relative Likelihood*	
Pest	1980	1985
Pink bollworm	3	3
Stored products pests	3	3
Western pine beetle	2	3
Boll weevil	1	2
Codling moth	1	2
Leaf roller complex	1	2
Cabbage looper	1	2
Tussock moth	0	1
Gypsy moth	0	1
Spruce budworm	0	1
European corn borer	0	1
Southern pine beetle	0	1

* Relative likelihood: 0 = negligible
                          1 = low
                          2 = medium
                          3 = high

This report also presents a detailed decision analysis for the manufacture of gossyplure, the pheromone of the pink bollworm. The material was to be used in permeation schemes to disrupt mating of the pink bollworm in the cotton fields of the Southwest. The conclusion was that such a venture "appears modestly attractive". Although some of the assumptions were based on solid research results, many of them were necessarily somewhat arbitrary. Note that the report is already dated by recent progress. For example, mass trapping of the ambrosia beetle *Gnathotrichus sulcatus* is now on a commercial basis (see below).

In the time frame of overall pest control we should realize that the first *field* test involving a correctly identified insect pheromone component was reported by GARY in 1962 (*19*), who showed that flying honey bee drones were attracted to (*E*)-9-oxo-2-decenoic acid (**338**), a component of the mandibular gland of virgin queens. BUTLER and FAIREY in 1964 (*20*) similarly showed that another component, (*E*)-9-hydroxy-2-decenoic acid, was also attractive. However, the honey bee is hardly a pest insect. In 1966, SILVERSTEIN, WOOD, and RODIN (*525*) reported that both sexes of *Ips paraconfusus* (formerly *I. confusus*), the California five-spined Ips, were trapped from a natural population by the 3-component pheromone isolated from the males. Results of further field tests by WOOD *et al.*, reported in 1967 (*526*), demonstrated the synergistic effects of the three components, and a masking or species isolation effect towards a sympatric species, *Ips latidens*. The extensive field studies of the early 1960s on the gypsy moth were carried out with a compound that had been erroneously described as the sex attractant.

As noted earlier, a small group of investigators has, in a very brief period, established behavior modifying chemicals as one of the more promising components of integrated pest management. We shall sample the opinions of several investigators active in developing this methodology. An earlier general review which includes some references from 1973 should be consulted (*527*). WOOD (*528*) recently discussed manipulation of several forest insect pests, and reviews by ROELOFS (*529*) and MITCHELL (*530*) are in press at this writing. A brief discussion of some of the variables in trapping is given by MINKS (*531*). BONESS *et al.* (*532*) have briefly summarized studies carried out by the Bayer AG group on a variety of insect pests in Europe. They describe several successful experiments, but overall results varied widely. BONESS (*533*) points out that pheromones are a potent tool, but do not readily lend themselves to manipulation by industry without involvement of official agencies.

This section will review selected field tests that have been carried out to establish monitoring or control systems with behavior modifying chemicals. Actually, four orders of insects, Lepidoptera, Coleoptera, Hymenoptera, and Diptera, contain most pests for which manipulation by behavior modifying chemicals has been attempted. (A few studies of tick pheromones have been reported.) It is convenient to divide them into pests of forest and shade trees, orchard trees and vines, field crops, stored products, and those that directly afflict humans or animals.

B. Forest and Shade-Tree Insect Pests

1. Coleoptera

Many coleopteran pests of forest and shade trees produce an aggregation pheromone that attracts both sexes of adults, which represent the destructive stage. This is in contrast with many lepidopteran pests whose sex attractant is specific for the adult male and whose larvae are the destructive stage. Thus, manipulation of coleopteran populations, from this point of view, should have a greater impact than manipulation of lepidopteran populations.

a) Western Pine Beetle

The aggregation pheromone of the western pine beetle, *Dendroctonus brevicomis,* which kills stressed or even apparently healthy ponderosa pines, consists of three components, one produced by the male, one by the female, and one by the host tree (*266* and refs. therein). Wood (*528*) and Wood and Bedard (*534*) have summarized the results of large-scale attempts at population suppression based on mass trapping. Wood *et al.* (*515*) discuss integrated pest management of the western pine beetle; Bedard *et al.* (*535*) discuss the role of behavior modifying chemicals in the management of the western pine beetle.

Since the survival of the beetle depends on overwhelming individual trees in the aggregation phase mediated by the pheromone, control strategy depends on reducing the intensity of these attacks. Two approaches have been used. 1. Large sticky traps baited with the pheromone have been distributed throughout the area of infestation to reduce the number of beetles available to attack trees (trap-out method). 2. The forest canopy has been permeated either with the pheromone or with an anti-attractant.

In 1970, an experiment was carried out to suppress the western pine beetle population in plots within a moderately infested area of 65 km² at Bass Lake, California. Large baited sticky traps in a grid pattern captured approximately one million beetles—essentially the total population estimated for the area—and the tree mortality dropped from 283 ±89 before the experiment to 91 ±28 afterwards; tree mortality remained low for the following four years. A second experiment was carried out on a much larger scale in a more heavily infested region at McCloud Flats, California. Although about 7 million beetles were trapped, the total population was not decreased enough to affect tree mortality. However considerable redistribution of the population occurred in response to the pheromone. Results are still being analyzed. The following is a statement from Wood (*515*): "Evidence from these two experiments indicates that western pine beetle populations can be manipulated on a large scale. This strengthens our belief that an effective control strategy can be developed using methods similar to those employed in these experiments. ... Because chemicals used in this method are considered pesticides by the Environmental Protection Agency (EPA) they are subject to the same scrutiny as insect toxicants in the areas of efficacy, toxicology, and production and environmental chemistry to insure that any recommended use will be safe and effective ... The compounds used in these field tests: 1. were nontoxic in our panel of tests; 2. occur in nature; 3. would be applied only in their normal environment; 4. would probably be applied at rates that do not exceed those found in nature and 5. are not introduced into plant, animal, soil, or aquatic systems (the trap-out method employs bait stations). We feel therefore that there is a high likelihood that these compounds can be registered under EPA requirements."

As part of both the Bass Lake and the McCloud Flats population suppression experiments, small survey traps (in contrast with the large suppression traps) were located on a grid pattern throughout the area to monitor the in-flight population through time and space. Studies to interpret and correlate the results are still in progress. Small survey traps for early detection of infestations are on extremely valuable contribution. If relationships between trap catch, insect population, and tree damage can be developed, this information will form an important basis for management decisions.

Permeation with the 3-component pheromone over a 0.81 hectare plot prevented the beetles from being trapped on traps baited with the same pheromone in the center of the treated area (*536*). When verbenone was released from formulations attached to ponderosa pine trees that were also baited with the pheromone, no mass attack occurred on these trees, in contrast to the mass attacks that occurred on trees that were simply baited with the pheromone. Further development of this approach is warranted to develop a method for protecting individual high-value trees.

One further feature that requires additional study is the effect of the pheromones on natural enemies. VITÉ (*537*) cites this effect as a disadvantage of trap-out; he prefers the use of baited trees when timely removal is feasible.

As a general summary statement, the following, taken from WOOD and BEDARD (*534*), will serve: "Attractant pheromones used over large areas have potential for estimating the size and distribution of WPB [western pine beetle] populations, and for manipulating populations as means of determining the interactions of adult WPB populations with its host and natural enemies. Further, these attractans can be used experimentally to redistribute populations so that we can study the dynamics of population behavior. At the same time, we can expect that the outcome of this research will have immediate practical benefits for the use of behavioral chemicals in pest management systems."

Wood (538) further suggests: "It is important to note that only very limited attempts have been made to develop the use of any of the available compounds in pest management programs. In a few cases where attempts have been made, the reasons for failure or limited success have not been made clear. Undoubtedly, the high cost of research following the identification phases and the inherent complexities of reasearch on the dynamics of highly mobile, widely distributed pest populations are the underlying causes for the current status of IBRs [Insect Behavior Regulators] in forest pest management."

b) Gnathotrichus sulcatus

The aggregating pheromone produced by the male ambrosia beetle, *Gnathotrichus sulcatus*, consists of a "single" compound, actually a 65/35 mixture of two enantiomers. In field tests, traps baited with the racemic mixture of this compound, sulcatol (**22 g**), caught large numbers of both sexes of *G. sulcatus* in competition with natural host and beetle odors. Since infestation of freshly sawed lumber is of continuing concern to the forest industries of British Columbia, a survey to determine the population distribution and seasonality of flights within a commercial sawmill at Chemainus, B. C. was undertaken in 1974 (539). The results indicated that "sulcatol could be used as an inexpensive, reliable detection and survey tool that is considerably more accurate than currently used, visual methods". Furthermore, it was suggested that sulcatol and other scolytid pheromones might be used "as a sensitive detection tool at unloading and processing areas" in countries that import logs and unseasoned lumber. Finally, it was proposed that "sulcatol should be tested as a means of intercepting *G. sulcatus* [beetles] before they are able to infest valuable lumber. Unlike programs directed at bark beetles in large tracts of forest ..., a sawmill-based program would challenge a more limited and potentially manageable population. This situation is particularly true for the Chemainus sawmill in which the only source of beetles appears to be infested logs transported to the mill site from distant logging operations."

Suppression experiments were carried out at the Chemainus sawmill in 1975. The conclusions (540) were that "sulcatol baited traps can capture most of a *G. sulcatus* population, that 2-to-4-week old lumber sawed from sapwood is attractive to flying beetles in sawmills, and that highest attack densities can be expected on loads [i.e. sawed lumber] adjacent to sulcatol-baited traps". The recommendation was made that piles of freshly sawed sapwood slabs be placed around the mill site and that "sulcatol-baited traps placed alongside each slab pile would attract and capture most beetles; those not caught could attack the slabbing, ...

which could then be removed and chipped ...". These recommendations have been accepted and the procedure is in commercial operation at the Chemainus sawmill site "as part of their normal quality control operation" (*317*). BORDEN's general assessment is that "with lower beetle populations in limited areas in an uneven aged forest, I do believe that pheromone-based mass trapping, disruption, and tree-baiting techniques will find a useful place in forest pest management". However he feels that efficacy has not yet been demonstrated for general use, but sees immediate application for sawmills and dryland sorting areas for logs on the west coast. In these cases, "resident beetle populations, and the area to be protected, will be relatively small. Moreover, the prior investment in growing, surveying, harvesting, and hauling timber by the companies justifies a considerable additional investment in beetle control to protect a high value product ...".

c) European Elm Bark Beetle

Introduced from Europe half a century ago, the European elm bark beetle, *Scolytus multistriatus* (together with its associated pathogenic fungus) has virtually eliminated the elm as a shade tree throughout most of the United States. Weakened or dying elms emit volatiles that are weakly attractive, but the addition of volatiles produced by boring females greatly increases the rate of attack by both males and females so that the tree is rapidly overwhelmed by the Dutch elm disease fungus. In 1970, efforts were initiated to mass-rear beetles in elm bolts and collect the volatile compounds for isolation and identification. The most effective procedure for collecting volatiles was by aerating the boring females and trapping the volatiles by passing the airstream through Porapak Q. (*541*). The isolation procedure was monitored by laboratory and field bio-assays, and three compounds were identified; two are produced by the female beetle and one by the stressed tree, each synergistic as components of the aggregation pheromone. The compounds were synthesized; a field test in 1974 confirmed that the mixture (multilure **283, 294a, 303a**) was a powerful aggregating agent (*146*). Since that time, baited traps have been used in many locations in attempts to detect, survey, and suppress populations (*542*).

In 1974 and 1975, large sticky traps were deployed in a grid pattern throughout heavily infested areas in Detroit, Michigan. In 1974, the approximately 1,000,000 beetles trapped in one section were estimated to represent only 20% of the population, and no significant impact on tree mortality was noted. In 1975, an improved trap and formulation resulted in trapping nearly 4,000,000 beetles, and in an actual increase in the number of attacked trees in the treatment area. Apparently the grid deployment within a treatment area resulted in significant beetle immigration from adjacent areas.

Tests now in progress are based on a barrier trapping strategy. "Multilure-baited traps are deployed on utility poles or non-elms in one or more rows encircling an area containing high-value elms. The area may be an entire city with several thousand elms, or it may be a relatively small area having only a few (25—100) elms. In theory, the barrier traps will lure beetles out of the area within the barrier, and will intercept beetles flying in from outside of the barrier. The annihilation of these beetles should reduce the number of beetles within the plot that are available to feed on, or breed in, elms, thereby bringing about a corresponding reduction in the incidence of beetle-vectored DED [Dutch elm disease]." (543). Such a large-scale barrier trapping system is in effect at Ft. Collins, Colorado, and at Evanston, Ill. At Ft. Collins, the area involved is 3 × 5 km and contains 4,300 American elms *(Ulmus americana)* and 8,000 Asiatic elms *(U. pumila)*. The plot at Evanston is 346.5 km and contains about 15,000 elms, mainly *U. americana*. The elm population in Ft. Collins is virtually isolated, whereas the Evanston elms are contiguous with other elm forests to the north. Survey traps are deployed in both cities as well as in the control cities of Loveland and Greeley, Colorado.

In both test cities, a large drop in beetle population occurred, presumably as a result of the barrier traps. No significant effect on tree mortality, however, could be found during the single trapping season. The investigators propose to continue the study for a total of five years and they expect to see a decrease in tree mortality over this period. Such factors as "annual variation and lag between inoculation and symptom expression" require a longer test period.

In a complementary series, Lanier and collaborators (329) are conducting a number of barrier trapping tests by encircling 12 small discrete clusters of elms in 8 eastern states. Over the past three years, the "DED (Dutch elm disease) rate is down in every area. Whether or not, we can credit this effect to beetle trapping will depend on the accuracy with which we can determine historical rates in comparable areas where trapping is not done."

Another study was initiated in 1976 by Birch and collaborators (327) to survey and suppress the beetle population in the towns of Lone Pine, Independence, and Big Pine all in Owens Valley, California. These towns are separated from one another by at least 24 km of open, high elevation desert, and each contains a moderate number (300—500) of elms and beetles, but apparently no Dutch elm disease. Traps were deployed on the perimeter of Lone Pine, throughout Independence on a grid pattern and throughout Big Pine in four lines. Traps were also placed between the towns and in outlying areas. A total of about 600,000 beetles were trapped in the three towns. Some of the outlying traps that were 8 or more km from any known elm also caught beetles. Apparently the beetles may disperse over greater distances than previously suspected. Some brood sources in dead branches in trees within the towns were located, but apparently the major source of beetles was elm woodpiles.

Further trapping was carried out during 1977 to correlate trap catches with beetle population and to assess the impact of the 1976 trapping. The first and second counts for 1977 were about 3% and 40% of the corresponding counts in 1976. It is tempting to ascribe these results to the trapping program, but obviously other factors could be involved. The major source on one town for 1977 appeared to be a stand of elms that died from lack of water. Since the disease is not present, it is obviously not possible to relate any population decrease to tree mortality.

BIRCH' assessment (327) for pheromones in general is: "Outlook good (some imminent successes) with integrative use and with increased biological input." On control of the Dutch elm disease by trap-out, he is dubious about the feasibility of attempts on the scale of the Detroit experiment, but feels that pheromones will certainly play a role as part of an integrated strategy applied to small discrete areas.

ARCIERO (544) has been trapping elm bark beetles throughout the Bay area counties in California to study the seasonal variation in population and to estimate how many beetles are carrying the Dutch elm disease fungus.

d) Douglas-Fir Beetle

The Douglas-fir beetle, *Dendroctonus pseudotsugae,* periodically builds up to epidemic levels mainly in windthrown Douglas-fir trees, and several attempts have been made to prevent this build-up with behavior-modifying chemicals in the Idaho-Montana region. The female produces two attractant compounds, which are synergized by host volatiles. However, a compound, 3-methyl-2-cyclohexen-1-one (MCH, **35**), produced mainly by the male, apparently functions to shut down the aggregation, and this "antiaggregative pheromone" was used in 1974 by FURNISS et al. (545) to reduce the level of attack on felled trees.

In 1975, FURNISS et al. (546) carried out a series of tests with several controlled-release formulations of MCH on 11 m × 43 m plots, each containing a single felled Douglas-fir tree. Attack density and brood density were significantly reduced by some of the MCH treatments. A pilot test is planned for 1978 if sufficient windthrown Douglas-fir becomes available during the previous winter. Plants (private communication, 547) call for application of formulated MCH by helicopter on 4-hectare plots. "Evaluation would include measurement of dosage received, counts of frass on the windthrown trees in June, and measurement of number of egg galleries and progeny on bark samples in late summer. Resultant infestation of live trees near the plots may be evaluated also, depending on the presence of untreated windthrow outside the plots." FURNISS points out the difficulties in utilizing aggregative or antiaggregative compounds in procedures for control of bark beetles in forests, but, on balance, he favors the use of anti-aggregative compounds: "The trick is going to be to see what level of attack density is required to obtain a lessening of damage to live trees by the subsequent beetle generation." He summarized as follows: ". . . Antiaggregative pheromones may still prove useful in disrupting attraction to beetles such as the pine engraver, Douglas-fir beetle, and spruce beetle, all of which depend on felled trees for massing numbers necessary to cause significant damage in the forest."

The Douglas-fir beetle has also been attracted with pheromones to "bait trees" that were scheduled for logging (548). No attempt at economic assessment was made, but the method is certainly limited to areas that are readily accessible and are scheduled for prompt logging.

e) Southern Pine Beetle

"Aggregation of the southern pine beetle, *Dendroctonus frontalis* . . . on loblolly pine *(Pinus taeda)* under beetle attack was not disrupted by aerial application of frontalure, which is a mixture of the attractant pheromone frontalin (**195**) and the host terpene α-pinene (**230**). Instead, aerial saturation with the pheromone in a heavily beetle-infested pine forest resulted in a rapid increase in the aggregation of beetles on pine trees undergoing attack" (549). In this experiment, a ten hectare section of pine forest including 1.6 hectares of a *D. frontalis* infestation was treated twice by aircraft with rice seed soaked with frontalure. This formulation released virtually all of the frontalure within 24 hr (45 g/ hectare for the first application and 450 g/hectare for the second).

"Reduced landing of beetles on host trees" was achieved by PAYNE *et al.* (549) with a mixture of *endo-* and *exo-*brevicomin (**150, 151**) from a controlled-release formulation in dispensers positioned throughout the test plot, which consisted of a 3×3 grid of 15 m^2 blocks. A total of 36 dispensers was used, each releasing ca. 1.5 mg/day of each compound (300 mg of each compound/hectare/day over a period of 30 days). Addition of verbenone (**233**) to the brevicomin mixture gave a 74% reduction in beetles landing on treated trees, which also showed a significant decrease in the number of galleries constructed (550).

Earlier studies (1971 and 1972) involved the use of frontalure to bait trees that were killed with cacodylic acid to reduce brood survival (*551, 552*). No further work in this direction has been reported.

f) Miscellaneous Bark Beetles

Spruce beetles *(Dendroctonus rufipennis* Kirby) were attracted to lindane-treated trees baited with frontalin (*553*), but baited trees could not compete with windthrown trees (*554*). FURNISS (*547*) reports the following studies in progress: tests of the Douglas-fir beetle pheromone components against the Eastern larch beetle *(Dendroctonus simplex* Le Conte) in Alaska; use of the antiaggregative compound MCH against the spruce beetle *(D. rufipennis)* in white spruce in Alaska, the anti-aggregative effect of ipsenol against the pine engraver, *Ips pini*, in ponderosa pine in Idaho, and similar studies involving both ipsenol (**240**) and ipsdienol (**266a**). The mating response of the European pine shoot moth, *Rhyacionia buoliana*, was disrupted by a controlled release formulation of the pheromone in small (3 m × 3 m) plots (*555*).

2. Lepidoptera

a) Gypsy moth, *Lymantria dispar* (formerly *Porthetria dispar)*

The gypsy moth, because of its highly visible defoliation of trees in populated areas and its rapidity of spread, has been the target of intensive efforts to eradicate, or at least control, this introduced pest. Use of massive amounts of hard pesticides for this has generated controversy, and alternative measures have also been unsuccessful, albeit less counter-productive. The pheromone was identified and named disparlure in 1970, and a number of promising field tests have since been carried out. In his review of the ecology and control of the gypsy moth, LEONARD (*556*) doubts that the "brush fire response" [present authors' term] to cyclical outbreaks has accumulated basic knowledge of the biology and ecology of the insect in proportion to the time and money spent.

In 1973, CAMERON (*557*) considered the decidedly mixed results of the 1972 trapping and confusant tests with disparlure, and stated: "There is reason for optimism." However, he pointed out the serious lack of behavioral studies and decided that "we do not now have the data which would justify any operational control programs in 1973; we need at least one more research field season for extensive testing." He concluded: "I seriously doubt that we ever will be able to manipulate well-established populations of the gypsy moth solely with disparlure (**536**). Its use is likely to be confined to fringe, newly infested, or isolated areas with very low populations, unless, of course, other agents can reduce outbreak populations to levels at which disparlure may be effective. Under no circumstances can I see eradication of the gypsy moth from North America, and the use of disparlure, in whatever manner, to eradicate local infestations is still open to question. And, I feel that our attempts to establish a barrier zone to confine the insect to the Northeast are still premature. Evidence in hand simply does not support the practicality of this approach. Disparlure is a potential tool for gypsy moth population manipulation. We still have not been able to determine how to utilize its potential effectively."

Cameron and Mastro (558) reported on the results of a 1974 experiment designed to test whether subsequent application of disparlure to an area treated with a pesticide would further reduce the population: "With the technology and formulations available, we could not demonstrate that disparlure further reduces a gypsy moth population sprayed with Sevin 4 oil during the larval stage."

In 1974, Cameron (559) felt that disparlure is operational for survey and detection, but not for "population monitoring and/or prediction, eradication of isolated infestations, the establishment of a barrier zone, or use in a chronic infestation". In 1975, Cameron et al. (560) reported that the olefin precursor (535) of disparlure, which had earlier been proposed by Cardé et al. (561) as a natural inhibitor of the attractant, did not disrupt mate-finding or reduce trap catches to useful levels.

Beroza (562) summarized the results of large scale disruption experiments carried out in 1974: "Overall, the 1974 field results indicate that air permeation with slow-release disparlure microcapsules applied at the rate of 20 g lure/hectare is effective in reducing mating success in low-level infestations of the gypsy moth or in infestations brought to low levels with an insecticide." To date, there is no assessment of the result of disparlure treatment on defoliation over a large area.

The state of knowledge of pheromone-mediated behavior of the gypsy moth is discussed in a provocative paper by Richerson (563) in which he comments that "most of the 'modes of action' of disruption are symptomatic of the 'dogma of immaculate perception'". Obviously Richerson takes issue with several of the neat formalizations of insect behavior—at least as applied to the gypsy moth—and he points out some of the limitations of mating-disruption by pheromones in high populations.

In a recent presentation, Cameron (564) once again summarized the promises and problems of the field work with disparlure: "We do need more time for testing—BUT—we do *not* need any large scale tests in 1978. If this approach is to be pursued, emphasis in 1978 tests should be directed to formulation development and rates of lure release. . . ."

References, pp. 157—190

PLIMMER *et al.* emphasize the need for effective formulations, and discuss disparlure formulations and strategies of applications (*565, 566*). They agree with CAMERON (see above): "We [Agricultural Research Service] have considerable investment in this research and it is continuing. However, it would be a mistake to oversell until we have done much more homework. I would like to be able to define precisely and reproducibly where and when mating disruption will succeed and whether it can be effectively combined with insecticide treatment at the larval state (for Lepidoptera). Our formulations need much development before we can be sure of their economy and reproducibility." PLIMMER also notes (*567*) that even the basic question of responses to the enantiomers of disparlure still needs to be resolved.

BONESS (*568*) used disparlure in field trials against the nun moth, *Lymantria monacha*, which is closely related to the gypsy moth. After treating 5 hectares of pine forest with 5.4 g of microencapsulated disparlure, he distributed 40 small disparlure-baited traps in the central part of the plot. Catches were reduced by 99% in comparison with a control plot.

b) Eastern Spruce Budworm, *Choristaneura fumiferana*

This native defoliator of balsam fir and white spruce ranges from the Yukon through the northern Prairie Provinces into eastern Canada and the northeastern United States. Massive outbreaks have been countered with massive applications of hard pesticides with dubious long-range results. "There is some evidence that extensive aerial spraying operations using conventional pesticides have been successful in keeping trees alive, but there is mounting evidence that preserving the insects' food supply in this way also prolongs the outbreak." Furthermore, gravid females are dispersed in enormous numbers out of densely populated areas (*569*).

The male responds to one attractant pheromone component and is inhibited by two others. Surveys with small attractant traps have been shown to be "extremely effective in catching male spruce budworms in areas where conventional larval sampling has failed to locate any insects. Again, the potential use of the attractant is considerable, but implementation of the system for monitoring low density populations requires calibration of the catches with population density which must await the decline of the current extensive infestations to endemic levels." (*570*). SANDERS (*570*) has summarized the results of a semi-operational attempt at disruption of spruce budworm mating by the synthetic attractant in Ontario in 1977: "Semi-operational trials of the aerial dispersion of the synthetic attractant of the eastern spruce budworm from aircraft have shown conclusively that mating behaviour is profoundly disrupted. In 1977 a release rate of 5 mg/ha/hr resulted in 99.4% reduction in catches in traps baited with virgin female moths. Assessment of the effects of this on population density in the following generation was complicated by late application of the attractant. However, apparently no further eggs were laid following the application. Further trials are planned for 1978 using different formulations and release rates with the assessment focusing on population regulation.

"The potential of the synthetic attractant for population regulation is therefore considerable. The major problems now are technological: inexpensive attractant and effective formulation. So far, formulations have been derived from the spin-off from various industries. There is an urgent need for custom formulations designed specifically for the slow release of synthetic attractants."

c) Other Forest Lepidopteran Pests

Daterman and his colleagues have been concerned with three western forest pests: Douglas-fir tussock moth *(Orgyia pseudotsugata)* and western spruce budworm *(Choristoneura occidentalis)*, both defoliators of Douglas-, grand-, and white firs; and the western pine shoot borer *(Eucosma sonomana),* stunter of ponderosa and Jeffrey pines. Daterman summarizes these studies as follows *(519)*:

"For tussock moth work, we have two major efforts underway; 1. to develop pheromone-baited traps for detecting critical increases in population density (early-warning system for impending outbreaks), and 2. to evaluate the pheromone for control purposes via the mating disruption approach. Our results to date have been promising in both areas. In 1977 we have gone an additional step with the control effort by applying aerial applications to test plots in New Mexico and Oregon. The dispenser system used in 1977 was the Conrel hollow fibers. Evaluation of 1977 tests is still taking place, but the initial returns are promising *(571, 572)*.

"Our efforts on western spruce budworm have thus far been limited to survey applications. Specifically, we have been attempting to correlate moth captures (in pheromone-baited traps) with larval densities and/or defoliation on a series of test plots. We would like to expand this effort in the future, and also explore use of the budworm attractant for control purposes.

"The western pine shoot borer work dates only from 1976. We have a tentative structural identification, and using that material on very small scale $^1/_{10}$ acre) plots, were successful in demonstrating disruption of pheromone communication. We have plans to expand this effort in

1978, and have a very optimistic outlook for achieving damage reduction through control by pheromone application."

C. Orchard and Vineyard Insect Pests

1. Lepidoptera

Apple orchards in the eastern U. S. are massively afflicted by a variety of pests, mainly moths, and in consequence, the orchards are blanketed with insecticides during the entire growing season. Vineyards in the same area are considered to have only two major moth pests. After the pheromones of a number of these pests had been identified, the group at the New York State Agricultural Experiment Station at Geneva adopted the strategy of using small baited traps to monitor the population so that insecticide application for a particular pest could be timed to its arrival. Together with other information, moth catches can be used to predict population densities and egg hatch. Zoecon Corp. has been selling traps and lures for this purpose for the past few years, and several examples habe been reported of appreciably reduced insecticide application in conjunction with monitoring (573—576). Pheromone monitoring is an important component of the farm advisory pest management system set up in Wayne County, N. Y. Savings in pesticide costs are reported to range up to 50% of the costs under the previously used spray regime (577).

Population control by mass trapping in low populations has been demonstrated for the redbanded leafroller (RBLR), *Argyrotaenia velutinana,* in apple orchards and vineyards, and for the grapeberry moth (GBM), *Paralobesia viteana,* in vineyards. However it was concluded that the benefits relative to cost could not justify this means of control, and research was diverted to mating disruption programs.

Microencapsulated and hollow fiber formulations of the pheromone were shown to disrupt mating of both the RBLR and the GBM (578). However attempts to disrupt mating for the complex of moth pests were less successful; improved formulations with better control of release rates of individual components are needed. TUMLINSON et al. (579) and ROELOFS (580) have dealt with the problems involved in manipulating complexes of insect pests. ROELOFS and CARDÉ (21) have analyzed communication systems and their disruption by pheromones and parapheromones.

Throughout the apple and pear orchards of Canada and northwestern U. S., only a few major pests need be contained; population monitoring and control are especially promising. Monitoring, combined with visual inspection for the codling moth, *Laspeyresia pomonella,* is successfully used to time sprays in British Columbia (*581*). Madsen *et al.* (*582*) have reported that, without the use of sprays, mass trapping of the codling moth resulted in population suppression and in the maintenance of fruit damage within acceptable commercial standards. Since the experiment was carried out in an isolated apple orchard, it is hard to predict the general applicability of a trap-out strategy.

An extensive study of the use of survey traps in the management of the codling moth was carried out in the Rhone Valley and in the vicinity of Paris, France, by Audemard and Milaire (*583*). They concluded that critical levels of trap catches in conjunction with temperature readings can be used to determine spray schedules.

Recently Audemard *et al.* (*584*) reported a successful mating disruption test against a natural population of the codling moth: "An attempt at controlling the codling moth *(Laspeyresia pomonella)* was made with the synthetic sex pheromone $8E,10E$-dodecadien-1-ol (**475**) in 1976 in a commercial apple orchard of 0.75 ha in Avignon. The technique of 'male disruption' was being used. The pheromone was being diffused in dispensers made of rubber tubing hung in trees. The quantity so used was 132 g/ha for the four series successively placed in the orchard. But only 0.55 g/ha/day have been effectively diffused, that is to say 84 g for the whole season. The results were very satisfying. The codling moth infestation at harvest was less than 0.5% injured apples. The codling moth population was drastically cut down by three times, thus making unnecessary the control of the first generation in 1977."

The pheromone of another major apple pest in British Columbia, the fruit leaf roller, *Archips argyrospilus*, has been used successfully to monitor activity and estimate populations so that sprays may be timed more effectively (*585*).

Moffitt (*586*), describing recent results on the codling moth in the Pacific Northwest, writes: "We have achieved season-long control of the codling moth on pears with three applications of the sex pheromone during the season." At this point, he is convinced of the technical feasibility of control of the codling moth with pheromones in conjunction with other tools.

Minks (*587*) has developed a system for determining the most effective time of spray against the summerfruit tortrix moth, *Adoxophyes orana,* in Dutch orchards; predictions are made on the basis of survey trap catches and temperature readings. Minks *et al.* (*588*) found that the geometric isomers of the pheromone of this moth blocked the attractant response (antipheromone), but large-scale attempts to reduce population by spraying a microencapsulated formulation of the antipheromone in apple orchards gave "variable results". Reasons for failure are discussed.

References, pp. 157—190

ARNE et al. (589) demonstrated that the mating disruption of the plum fruit moth (Grapholitha funebrana) was maintained throughout the growing season by permeating a plum orchard with the pheromone. Numbers of moths caught on survey traps and damage to fruit were comparable with the results in an orchard that was treated with the usual schedule of insecticides.

The number of sprayings against the apple maggot, Rhagoletis pomonella, in eastern Canada was reduced by coordinating the sprays with catches on baited traps (590).

2. Diptera

Several species of fruit flies have been controlled with a bait consisting of a carrier impregnated with an attractant (parapheromone) and a pesticide, and with attractant baited traps. Since the extensive literature has been recently reviewed by CHAMBERS (591), a brief summation here will suffice.

Food baits for fruit flies go back over 60 years, but an extensive, empirical, screening program for fruit fly attractants was carried out by the USDA during the 1950s. An interesting aspect of this work is that a male response was obtained to a whole series of natural fragrances and synthetic compounds. "... this olfactometer screening backed by field tests of promising candidates produced attractants in a relatively short time that have provided some of the greatest economic benefits yet demonstrated" (591). Attractants are now available for males of the following important pests: oriental fruit fly, mango fly, melon fly, Queensland fruit fly, Mediterranean fruit fly, and the Natal fruit fly. In addition, protein hydrolysate attracts both sexes of most fruit flies. CHAMBERS (591) gives a number of examples of population suppression and a few instances of apparent eradication from isolated areas.

D. Field Crops Insect Pests

1. Coleoptera

a) Cotton boll weevil, Anthonomus grandis

A recent thorough review of attempts to control the cotton boll weevil by means of pesticides (592) makes it unnecessary to go into detail here, but some perspective is presented, since the problem is massive and attempts to control this insect with pesticides have contributed to the difficulties alluded to in Section VII. A. One third of the agricultural pesticides used in the U. S. is dedicated to attempted control of the cotton boll weevil (592).

The pheromone consists of four compounds (**33, 110b, 111a, 111b**), produced by the male, that in combination attract the female and also serve in the spring and fall as an aggregant. The compounds (the mixture is grandlure) have been synthesized and are currently available is several formulations, and several effective traps were devised. A two-year experiment was started in 1971 to determine whether it was feasible to suppress or even eradicate a population from an isolated area. Grandlure baited traps were part of a program that included sound agricultural practices, insecticide treatment in the spring, and release of sterile boll weevil males. Despite the necessity to use supplemental sprays in peripheral area and despite evidence that mated female weevils had migrated into these areas from up to 40 km away, the technical guidance committee report concluded "that the boll weevil could be eliminated as an economic pest". The pheromone traps "improved the performance of the trap crop over unbaited trap crops" (*593*).

Subsequently, trapping tests and disruption tests have been conducted on a smaller scale. Mitchell *et al.* (*594*) reported: "Grandlure-baited in-field traps at the rate of 10 per acre captured 76% of a population of overwintered *Anthonomus grandis* Boheman estimated to number ca. 25 per acre. ... A combination of in-field traps and insecticides captured or killed 100% of the emerging overwintered beetles ... the traps alone captured about 96% of a late emerging population ... These experiments were carried out on a well-isolated cotton planting totaling 108 acres ..." In contrast, Huddleston *et al.* (*595*) reported disappointing results in an attempt to disrupt communication with grandlure.

Knipling (*593*) in 1974 discussed the use of population models to predict the feasibility of "suppression or eradication of the boll weevil through the use of insecticides, sterile insects, and the boll weevil sex pheromone employed alone, employed simultaneously or in sequence". Basic to the trapping strategy is the assumption that "the estimated effect of traps will be considerable only for low level populations". He concludes that "pheromone traps should come close to maintaining subeconomic populations in most boll weevil areas without the need for insecticide application if the populations are once brought down to a level of about 10 boll weevils per acre". "...If the population has been reduced to a level of about two per acre, which seems readily possible by the use of a thorough reproduction-diapause spray program, the in-field traps alone, based on the parameters, should achieve of the order of 98% capture of the females. ..." Two further uses for grandlure, Knipling points out, are to reduce the risk of developing insecticide-resistant strains, and to detect incipient infestations.

Lloyd (*596*) found that "the theoretical study of Knipling was validated by our [Mitchell *et al.*, above] field research. The debate on the feasibility of boll weevil eradication will be resolved by the Trial Boll Weevil Eradication Program which will begin in North Carolina and Virginia in 1978. The use of the boll weevil pheromone, grandlure, will represent a significant technical component in this Trial Program."

2. Lepidoptera

The communication system of each of the following important lepidopteran pests of field crops has been disrupted in field studies by either a pheromone, a parapheromone or both (597): pink bollworm (*Pectinophora gossypiella*), cabbage looper (*Trichoplusia ni*), alfalfa looper (*Autographa californica*), soybean looper (*Pseudoplusia includens*), corn earworm (*Heliothis zea*), tobacco budworm (*H. virescens*), yellow-striped armyworm (*Spodoptera ornithogalli*), fall armyworm (*S. frugiperda*), beet armyworm (*S. exigua*), European corn borer (*Ostrinia nubilalis*), Egyptian cotton leafworm (*S. littoralis*).

With the exception of the corn earworm, fall armyworm, pink bollworm, and Egyptian cotton leafworm programs, these disruption tests were preliminary studies carried out on small plots without optimization of disruptant formulations and with little assessment of crop damage.

The corn earworm and fall armyworm are serious pests on corn crops. MITCHELL et al. (598, 599) have demonstrated disruption of mating for both insects in corn fields by permeation with either the pheromone or a parapheromone. Suppressions reported ranged from 85—97%. Furthermore, they showed that simultaneous mating disruption could be accomplished for both insects with a mixture of the pheromone and parapheromone. Disruption was in the range of 88—97%. A four-year program will be initiated in 1978 to develop the necessary technology for suppressing populations of these insects in corn, so that the large number of insecticide applications currently used on this crop can be reduced.

MITCHELL (597) has discussed the feasibility of using multicomponent formulations for control of several important pests of field crops. HENDRICKS et al. (600) showed that a mixture of looplure (622) and virelure (623, 624) were compatible and caught the cabbage looper (*T. ni*), the soybean looper (*Pseudoplusia includens*), and the tobacco budworm (*H. virescens*) on the same traps. It seems likely that pesticide applications to field crops could be greatly reduced by means of pheromone-baited monitoring traps, and such efforts will undoubtedly become a part of the integrated pest management projects now established by the Extension Service of the USDA.

(622)

(623)

(624)

Because of the rapid spread of the pink bollworm through the cotton fields of the Southwest into California despite massive pesticides applications, and the realization that its establishment in the San Joaquin Valley of California would be a disaster, the impetus for developing control procedures was present. Between 1971 and 1973, Toscano et al. (601) showed that the number of pesticide applications could be reduced by monitoring the arrival of males by means of traps baited with the parapheromone, hexalure. In 1972, McLaughlin et al. (602) reported disruption of the mating response by hexalure on small plots; Shorey et al. (603) repeated the demonstration on larger fields, and also showed that permeation with hexalure throughout the growing season was as effective as treatments with the pesticide, carbaryl. Subsequent large-scale tests were conducted with the pheromone gossyplure, which was much more effective than the parapheromone. "We have now released [middle of May through early September, 1976] economically practical amounts of the pheromone into the air within three cotton fields [5, 6, and 12 hectares] throughout an entire growing season and have suppressed the population of pink bollworm larvae infesting the cotton bolls of those fields to an extent comparable to that achieved in ten comparison fields treated with conventional insecticide applications." The growers were instructed wo use insecticides in both sets of fields as counts of larvae in cotton bolls indicated. This resulted in a ninefold reduction in insecticide usage in the pheromone-treated fields. The technology used in this test was fairly primitive. Hoops of hollow plastic fibers containing the pheromone were attached by hand to the cotton plants in a grid pattern, and were replenished at 3-week intervals to keep the release sources near the tops of the plants (*604*).

In 1976, Brooks and Kitterman (*605*) treated 1,160 hectares of cotton in Arizona and California with a hollow fiber formulation of gossyplure, initially applied with ground equipment designed especially for this application; subsequent applications were made by air. "Efficacy of treatment was assessed by moth captures in monitoring traps, one trap per 10 acres [4 hectares], and twice weekly sampling of bolls for larval infestation. Since untreated control fields were not practical, the effectiveness of pheromone treatment was gauged by comparison of larval infestations in fields which were under conventional treatment regimes with insecticide." Pink bollworm levels were historically severe in one area and moderate to severe in two areas. In the severely infested area, "it is believed that early season suppression was helped by such [phero-mone] treatments". However, a rapid midseason buildup in a few problem fields coupled with a tobacco budworm infestation made it advisable to terminate the pheromone treatments and launch a regular insecticide treatment. The tentative conclusion is that "pheromone treatments postponed the onset of test field infestations by about one month". On a 132 acre farm in a moderately infested area, "boll infestation suppression ... indicates that pheromone treatment is almost directly comparable in effectiveness to insecticide treatment". Prophylactic insecticide treatment was used on several of the fields as a precaution against in-migration late in the season.

In 1977, a similar operation by Conrel Co. was carried out over 20,000 acres of large isolated blocks of 2,000—3,000 acres in Arizona and California. At this writing, the data have not been completely ana-lyzed, but D. W. Swenson states (*606*) that a large reduction in insecticide application was achieved, and that the economics are feasible for this type of operation. He points out however that timing is critical, and that in several fields, insecticide applications were needed to control the tobacco budworm, cotton leaf roller, and lygus bug. In general, the technology seems in place to make suppression by pheromone permeation a feasible component of an integrated pest management system for cotton crops in the southwestern part of the U. S. Further large-scale operations are scheduled for 1978.

Another cotton pest in several parts of the world is the red boll-worm, *Diparopsis castanea*. Its pheromone and an inhibitor have been shown to disrupt mating, but no crop damage assessment was made (*607*).

On the basis of their success in trapping *S. littoralis* at a distance of 3 km from cotton and alfalfa fields in Israel, Neumark et al. (*608*) proposed that "safety belts of baited traps be established around cultivat-ed fields".

3. Hymenoptera

Several species of leaf-cutting ants do massive damage to a number of field crops in tropical countries. The trail pheromones of three species, *Atta texana, A. cephalotes,* and *A. sexdens rubropilosa* have been used in attempts to improve the pickup of poison bait. This was moderately successful for baits formulated with an inert substrate, but did not improve pickup in the field for baits formulated with such food materials as citrus pulp, grain, and bagasse fortified with molasses (*609*).

E. Stored Products Insect Pests

Stored food products are high-value items that have been planted, weeded, watered, fertilized, harvested, processed, transported, and stored in facilities, which also represent an investment. Since insect pests cause massive economic losses, strenuous efforts for detection and control of these pests are warranted and have been actively undertaken. Most of these pests are coleopteran or lepidopteran, and a number of pheromones have been identified in both orders (see reviews by BURK-HOLDER (*610*) and LEVINSON (*611*)).

1. Coleoptera

Practical application of pheromone traps for detection and control is furthest advanced for several species of *Trogoderma* and *Attagenus,* which include some of the more notorious pests. In fact, small detection traps containing the pheromone and an insecticide are now in use in a number of storage facilities, and on ships for quarantine purposes. The very high levels of response make early detection in a sparse population much more feasible than previous procedures; the use of numerous small traps allows for pinpointing the sources of the infestation. Current studies are aimed at improved trap and dispenser design.

Although mating disruption of *Attagenus megatoma* by pheromones has been demonstrated, permeation of a storage facility is not considered a good tactic, principally because residues absorbed on the stored products or containers may attract insects after the material has left the warehouse.

Two concepts for population suppression with baited traps are under investigation. The first is simply the pheromone-insecticide trapping system. The second concept (*612*) involves the use of a pheromone trap treated with spores of a pathogenic protozoon. The males, which are attracted to the trap, are contaminated with the spores and transmit them to the rest of the population. Under simulated warehouse conditions, "seminatural" populations of *T. glabrum* were effectively suppressed.

These systems may be very effective in mixed populations of *Trogoderma* because many species share some of the same pheromone components. One of the major targets is the khapra beetle, *Trogoderma granarium*, a notorious pest in warmer regions throughout Europe, Asia and Africa. Only a tight quarantine prevents its entry into the U. S. Ships and port facilities are being monitored with pheromone-baited traps.

2. Lepidoptera

Several members of the family Phycitidae share the same pheromone, and field studies involving several of these moths have been carried out. REICHMUTH *et al. (613)* showed the utility of pheromone traps for early detection of the tobacco moth, *Ephestia elutella,* and the Indian meal moth, *Plodia interpunctella.* SOWER *et al. (614, 615)* showed that mating frequency of the Indian meal moth and the almond moth, *E. cautella,* was effectively reduced at low population densities. WHEATLEY *(107)* and HAINES *(617)* also reported that mating frequency of the almond moth was reduced, and that detection and survey traps would be feasible and very useful for this pest and several related species.

F. Pests That Directly Afflict Humans or Animals

Pheromones have been identified and field-tested in only a few of the host of pests that directly afflict man or animals.

Muscalure (**318**), a weakly attractant pheromone of the common housefly, has been registered with the EPA (in fact, it is the only registered pheromone to date) and is available in a bait (Zoecon Corp.) containing sugar, a pesticide and the attractant. Both sexes are attracted in field use (animal barns), and enhanced catches of up to sevenfold over control traps have been reported *(618, 619)*.

The sex attractant of several tick species have been identified, but only three examples of attempts to apply the attractant have come to our attention. SONENSHINE has demonstrated *(620)* mating disruption by dusting a tick-infested dog with a microencapsulated formulation of the pheromone, 2,6-dichlorophenol; the ticks used were the dog tick, *Dermacentor variabilis,* and the Rocky Mountain wood tick, *Dermacentor andersoni.* GLADNEY *(621)* applied a mixture of pesticide and male extract to a shaved area on the shoulders of cattle and found that the Gulf Coast tick *(Amblyoma maculatum)* could be attracted to the spots and killed by the insecticide. Rechav (622) described an aggregation pheromone in feeding males of *Amblyoma hebraeum* that attracts males, females, and nymphs. A pheromone-acaricide mixture painted on cattle was effective for four days.

G. Pheromone Formulations

Satisfactory formulations of pheromones that afford a constant release rate over a long period of time are not likely to be achieved without the cooperation of the chemical industry. Controlled release of pheromones is discussed in the Proceedings of a recent symposium (623). Fortunately, several organizations in the U. S. have been interested in the possibilities of the use of pheromone formulations on a commercial basis. One company (605) has supplied formulations for several investigators, and in addition has carried out extensive field tests with their own personnel. It has developed the use of hollow fibers, which are simply "microcapillary reservoirs that serve to contain a vaporizable material and mediate evaporation of the material into the atmosphere". The fiber wall is essentially impermeable. One end is sealed, and the release rate essentially depends on diffusion from the surface of the liquid-vapor interface to the open end of the hollow fiber. Fibers are supplied in two forms. A "tape form" consists of a parallel array of fibers on an adhesive tape; this form is used to establish point-source evaporators. The "chopped fiber form" is used for dissemination with a ground rig or from the air.

Another company (624, 625) supplies a controlled release dispenser, which is a three-layer plastic laminated sheet consisting of a bottom protective layer, a middle pheromone reservoir layer, and a top permeable layer through which the pheromone diffuses. The sheet can be cut into ribbons and applied over an area. Formulations have been field tested on a number of insect pests: gypsy moth, pink bollworm, peachtree borer, lesser peach tree borer, soybean looper, cabbage looper, European elm bark beetle, tobacco budworm, fall armyworm, tussock moth, eastern spruce budworm, and Mediterranean fruit fly.

Two companies (626, 627) provide microencapsulated formulations of pheromones. This is a very flexible system in that encapsulating materials and particle size can be varied over a wide range. The microcapsules can be applied as solid particles or sprayed as a slurry. A study with microencapsulated disparlure (536) on the gypsy moth showed that the field life of disparlure could be extended to approximately six weeks (628).

VIII. Conclusion

As we move from passive dependence on the environment to technological advances that permit both abuse and control of the environment, the consequences of our decisions become awesome. Given a reasonable level of cooperation and commitment by government agencies, the research community, farmers, and industry, integrated pest management will become the modus operandi of the future. Surely, pheromones will play an important role in such programs, if the basic research, on which such roles are based, can be strengthened and the momentum for development, mainly by governmental agencies, can be increased.

Addendum *

II. Structure Elucidation

A. Isolation

1. Collection

A method for trapping disparlure (**536**) from air with type 4 A molecular sieves and subsequent quantitative analysis of a brominated derivative by electron-capture GC has been developed by CARO et al. (*629*). It has been successfully applied in the field to disparlure air concentrations as low as 0.2 ng/m^3.

Tenax GC has been used to collect volatile compounds from fungi that are attractive to cheese mites (*630*).

B. Identification

2. Examples

Female ambrosia bark beetles *(Trypodendron lineatum)* boring in Douglas fir produce a substance, named lineatin, that attracts males. On the basis of spectral information and the results from hydrogenolysis, MACCONNELL et al. (*631*) proposed structures (**625**) or (**626**) for lineatin. The cyclobutane portion was assigned after it was observed that hydrogenolysis of grandisol (**33**) gave essentially the same products. Synthetic studies to distinguish between these structures are underway.

---

* Material in the Addendum was compiled as of October 1, 1978 and is arranged under the headings used in the main section.

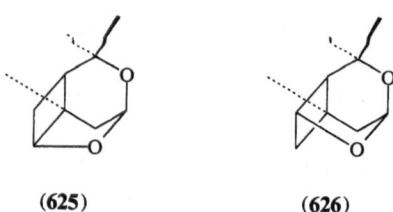

(625)   (626)

Francke *et al.* (*632*) identified spiroacetal (**627**) as an aggregation pheromone in the bark beetle *Pityogenes chalcographus*.

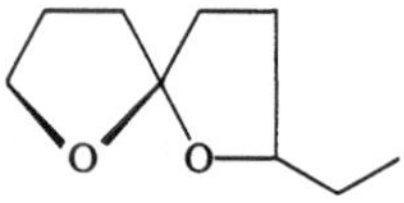

(627)

III. Synthesis

The synthesis of chiral pheromones has been recently reviewed (*633*).

A. Coleoptera

1. Bruchidae

By synthesizing both enantiomeric forms of the dried bean beetle (*Acanthoscelides obtectus*) sex pheromone, Pirkle and Boeder (*634*) determined that the natural pheromone had the R-(−) configuration (**11a**).

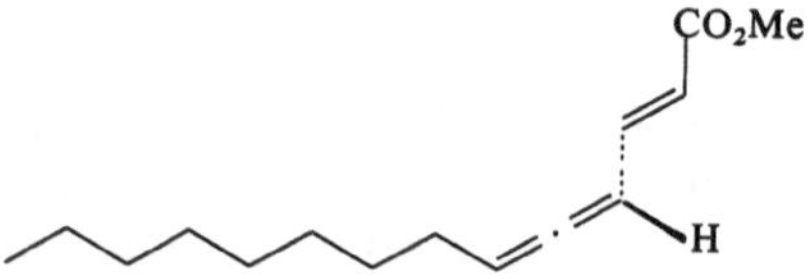

CO_2Me

(11a)

2. Curculionidae

Wenkert *et al.* (*97, 98*) have recently published (*635*) details of their synthesis of racemic grandisol (**33**), one of the sex pheromone components of the boll weevil *(Anthonomus grandis)*. Both enantiomers of **(33)** were synthesized by MORI *(636)*, DE SOUZA and GONÇALVES *(637)* reported conversion of **(106)** into the three cyclohexane sex pheromone components **(110b, 111a,** and **111b)**.

B. Diptera

An acetylenic intermediate was used by KOVALEV *et al. (638)* in their synthesis of muscalure **(318)**, the sex pheromone of the house fly *(Musca domestica)*.

C. Homoptera

A synthesis of the diene component **(333)** of the California red scale *(Aonidiella aurantii)* sex pheromone has been achieved by SNIDER and RODINI *(639)*.

D. Hymenoptera

1. Apidae

Queen substance **(338)**, the multi-purpose pheromone of the honey bee *(Apis mellifera)*, was obtained from substituted 5-bromothiophenes by TAMARU *et al. (640)*.

All four possible diastereomers of 2-methyl-5-hydroxyhexanoic acid lactone **(628)** have been synthesized by PIRKLE and ADAMS *(641)* and await comparison with natural material to determine which is the pheromone of the carpenter bee.

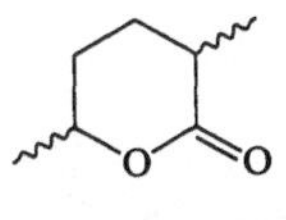

(628)

3. Diprionidae

A number of new syntheses of the sex pheromone **(373)** of the pine sawfly *(Neodiprion lecontei)* have been reported by MORI *et al. (642, 643)*, PLACE *et al. (644)*, and TAI *et al. (645)*.

F. Lepidoptera

1. Monoenes

Stereoselective reduction of β- or ω-alkynols to corresponding (E)-alkenols has been achieved by Rossi and Carpita (646) with lithium aluminum hydride. Fyles et al. (647, 648) have reported further applications of polymer supports in the synthesis of sex attractants.

5. Epoxide

Racemic disparlure (536), the sex pheromone of the gypsy moth (Lymantria dispar), has been prepared by Klünenberg and Schäfer (649) and Tolstikov et al. (650), and its geometric isomer (545) by Okada et al. (651).

6. Ketone

(Z)-1,6-Heneicosadien-11-one (629), an attractive analog of the Douglas fir tussock moth (Orgyia pseudotsugata) sex pheromone (580) has been identified and synthesized by Smith et al. (652).

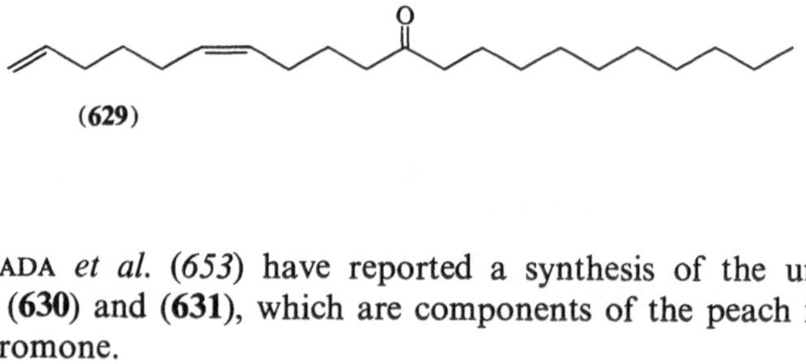

(629)

Tamada et al. (653) have reported a synthesis of the unsaturated ketones (630) and (631), which are components of the peach fruit moth sex pheromone.

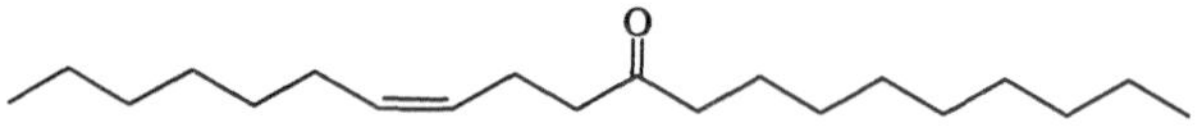

(630)

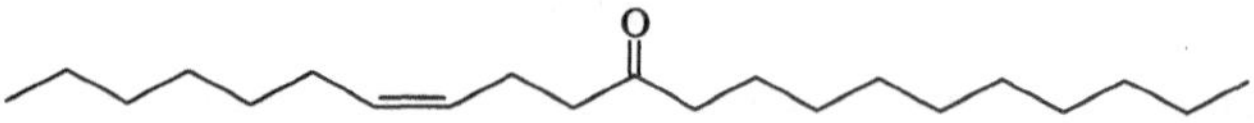

(631)

IV. Stereobiology

A. Geometric Isomers

Confirmation of the chemical structure and the natural occurrence of bombykal in *Bombyx mori* has recently been described (*654*).

Data on the attractiveness of pheromone blends and of isomer blends in field trials have been presented for the male peachtree borer (*655*) and the carpenterworm (*656*) respectively.

(*Z*)-11-Hexadecenal has been isolated from the female moth, *Heliothis armigera,* and is a potent olfactory stimulant for males in laboratory and field tests (*657*). The olfactometer response of laboratory-reared males of *H. virescens* to its pheromones, (*Z*)-11-hexadecenal and (*Z*)-9-tetradecenal and the inhibitor (*Z*)-9-tetradecen-1-ol formate has been studied (*658*).

B. Enantiomers

Trapping and behavioral tests on the attractiveness of disparlure racemates and the antipodes indicated that the (+) enantiomer is more attractive than racemic disparlure, and the (−) enantiomer exhibits an antagonistic effect (*659*). The effect of (−)-disparlure was much more apparent on in-flight than on pre-flight behavior (*660*).

The response of the bark beetle, *Ips pini,* to the attractant blend produced by conspecific males boring in ponderosa pine is inhibited by *S*-(−)-ipsenol (*661*).

C. Chemorecognition

A review relating the mechanism of pheromone perception to perception and behavior of the pheromone-stimulated insects has appeared recently (*338a*).

Many lepidopteran species produce a very precise blend of pheromone components. In addition, many male moths are captured with a considerably broader range of ratios, and, in some tests, with ratios quite different from that produced by their corresponding females. It is basically these phenomena that have led ROELOFS (*662*) to propose a hypothesis that employs threshold diagrams of binary mixtures of geometrical or positional isomers and their concentrations to illustrate activity relationships between these factors.

WRIGHT (*663*), employing the results of HAYWARD mentioned in the main review, has proposed how optical isomers may be perceived differently according to his vibrational theory.

The major trail pheromone of the ant, *Atta texana*, is methyl 4-methyl-pyrrole-2-carboxylate (*664*) and trail-following activity studies of synthetic analogs of this compound have been conducted by Sonnet and Moser (*665, 666*). It was concluded from the bioassay studies that, for activity, a molecule must have a particular shape and substitution pattern. It seems likely that the pyrrolic nitrogen atom is important in the process of chemorecognition by the receptor site and calculations of the charge of the N-atom of many of the substituted pyrroles bioassayed showed that the most active compounds all have an identical charge on the pyrrolic nitrogen (*667*). It is suggested that if behavioral activity could be related to electronic parameters then quantum calculations on as yet unsynthesized and untested structural analogs of pheromones may have predictive value for new compounds with activity.

The European corn borer (ECB) and the red banded leafroller (RBLR) both employ (Z)-11-tetradecenyl acetate as a sex pheromone. The racemate and pure enantiomers of 9-(cyclopent-2-en-1-yl)nonyl acetate were synthesized and have been found to mimic certain biological properties of the natural pheromone (*668*). The ECB responds to the S-(−)-enantiomer while the RBLR responds equally to both the S-(−)- and the R-(+)-enantiomer. Because of the behavioral response of the RBLR it is suggested that this species has two stereospecific chemoreceptors that may have different conformational requirements for the achiral phero-mone. The use of this type of stereochemical probe for the study of neuro-chemical receptor systems of achiral molecules offers an exciting metho-dology for the investigation of chemical sensing.

An interesting review on taste receptors and their specificity by V. G. Dethier (*669*) is well worth reading.

Renwick (*363*) reported on the activity of a number of frontalin analogs in field tests on *D. frontalis*, the Southern pine beetle. The only analog which showed any additive activity was 5,7-dimethyl-6,8-dioxabicyclo[3.2.1]octane when presented together with frontalin and α-pinene. The 5,7-dimethyl substitution pattern is the same as the 5-methyl-7-ethyl substitution pattern of *exo*- and *endo*-brevicomin. Renwick (*363*) noted this and also pointed out that while brevicomin is an active component of the attractant of *D. brevicomis*, it is inactive or suppresses response in *D. frontalis*. It has since been established that (−)-frontalin is the active enantiomer in this species (*46*).

The Western pine beetle, *D. brevicomis*, responds maximally to a ternary mixture of (1R,5S,7R)-(+)-*exo*-brevicomin, (1S,5R)-(−)-frontalin, and the host-produced alicyclic terpene, myrcene (*320*). Mixtures containing the antipodes of *exo*-brevicomin and frontalin were much less attractive.

Could (+)-*exo*-brevicomin and (−)-frontalin both interact with the same receptor? Let us assume that the receptor site(s) for the bicyclic ketals recognizes that face of the molecule which contains the oxygen bridges. As mentioned earlier in this review, the charge on the oxygen atoms of frontalin and *exo*- and *endo*-brevicomin are essentially the same within any one molecule, but differ slightly between molecules. Therefore (−)-frontalin may be rotated so that the oxygen atoms at positions 6 and 8 correspond to those at positions 8 and 6 of (+)-*exo*-brevicomin with the equivalence of the oxygen atoms remaining unchanged.

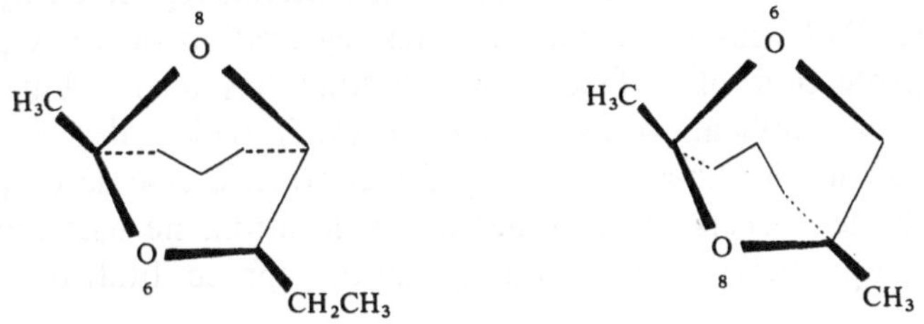

(+)-*exo*-brevicomin   (−)-frontalin

On a stereochemical basis it seems likely that a single receptor site could recognize both of these molecules. It will be interesting to establish whether the bicyclic ketals interact with a single receptor site. If they do, then it is possible that when (−)-frontalin displays activity, activity (either synergistic or inhibitory) might also be displayed by (+)-*exo*- or (+)-*endo*-brevicomin.

V. Biosynthesis

HUGHES and RENWICK (*670*) have extended their studies on hormonal and host factors that stimulate pheromone synthesis to include female Western pine beetles, *Dendroctonus brevicomis*.

The biosynthetic significance of enzymes in the defensive secretion of a bug, *Leptoglossus phylopus*, has been demonstrated (*671*), and the biosynthesis of formic acid in a formicine ant poison gland has been described (*672*).

The biosynthesis of the aggregation pheromones of the European fir engraver beetles, *Pityokteines curvidens*, *P. spinidens* and *P. vorontzovi* has recently been described in some detail by HARRING (*673*).

VII. Practical Applications

Progress in practical applications of behavior-modifying chemicals during the 1978 season was summarized at the Advanced Research Institute (ARI) on Chemical Ecology: Odour Communication in Animals, at Leeuwenhorst, Holland, September 25—29, 1978; the Proceedings will be published (674). Brooks et al. reviewed the successful commercial operation against the pink bollworm in cotton fields in 1977 and gave a preliminary report on the 1978 studies. In addition to crop protection, permeation with gossyplure with concomitant reduction by 50—80% in pesticide use resulted in lint yield improvements of as much as 20—50%. The 1978 experiments were carried out with material registered by EPA in February, 1978; this represents the first registration of a sex phero-mone for protection of a field crop. Brooks has also submitted a manuscript for publication as a chapter (675) which describes the application of hollow fiber technology for controlled release of phero-mones. As target insects for commercial exploration, he lists the pink bollworm, grape berry moth, codling moth, spruce budworm, and tussock moth.

Siddall's chapter in the ARI Proceedings is entitled "Commercial Production of Insect Pheromones: Problems and Prospects". This paper argues that industry in the USA is deterred from developing pheromones for insect control mainly because of the unrealistic and uncertain re-quirements of EPA — the very agency charged with the development of alternatives to hard pesticides. The participants in the ARI meeting resolved to request EPA to establish firm, realistic guidelines within one year. Siddall's chapter contains a number of recommendations that could form the basis for reasonable policy decisions.

Despite the usual number of inaccurate statements inherent in "popular" writings, a recent article in Harper's magazine (676) points up the ironies involved in the attempts to introduce new procedures for insect control and the frustrations of scientists who have tried to deal with EPA. At the same time, the "sympathetic" observation is made that a bureaucratic agency, faced with an option to assume a risk and possibly incur the ire of the "environmentalists" or to do nothing, will do nothing. There surely are risks involved in all of the options to hard pesticides, but the article concludes that scientific evidence and common sense rank these risks as more acceptable than continued dependence on hard pesticides by default.

References

1. LAW, J. H., and F. E. REGNIER: Pheromones. Annu. Rev. Biochem. **40**, 533—548 (1971).
2. WHITTAKER, R. H., and P. P. FEENY: Allelochemics: Chemical interactions between species. Science **171**, 757—770 (1970).
3. BROWN, W. L., JR., T. EISNER, and R. H. WHITTAKER: Allomones and kairomones, transspecific chemical messengers. Bioscience **20**, 21—22 (1970).
4. KARLSON, P., and A. BUTENANDT: Pheromones (ectohormones) in insects. Annu. Rev. Entomol. **4**, 39—58 (1959).
5. WILSON, E. O.: Chemical communication in the social insects. Science **149**, 1064—1071 (1965).
6. BEDARD, W. D., P. E. TILDEN, D. L. WOOD, R. M. SILVERSTEIN, R. G. BROWNLEE, and J. O. RODIN: Western pine beetle: field response to its sex pheromone and a synergistic host terpene, myrcene. Science **164**, 1284—1285 (1969).
7. WOOD, D. L., L. E. BROWNE, W. D. BEDARD, P. E. TILDEN, R. M. SILVERSTEIN, and J. O. RODIN: Response of *Ips confusus* [now *I. paraconfusus*] to synthetic sex pheromones in nature. Science **159**, 1373—1374 (1968).
8. EITER, K.: Insektensexuallockstoffe. Fortschr. Chem. Org. Natur. **28**, 204—255 (1970).
9. WOOD, D. L., R. M. SILVERSTEIN, and M. NAKAJIMA, Eds.: Control of Insect Behavior by Natural Products. New York: Academic Press. 1970.
10. BEROZA, M., Ed.: Chemicals Controlling Insect Behavior. New York: Academic Press. 1970.
11. JACOBSON, M.: Insect Sex Pheromones. New York: Academic Press. 1972.
12. BIRCH, M. C., Ed.: Pheromones. New York: American Elsevier. 1974.
13. JACOBSON, M., Ed.: Insecticides of the Future. New York: Marcel Dekker, Inc. 1975.
14. BEROZA, M., Ed.: Pest Management with Insect Sex Attractants. Washington D. C.: ACS Symposium Series No. 23, American Chemical Society. 1976.
15. INSCOE, M. N., and M. BEROZA: Analysis of pheromones and other compounds controlling insect behavior. In: G. ZWEIG and J. SHERMA, Eds., Analytical Methods for Pesticides and Plant Growth Regulators, Vol. VIII, Government Regulations, Pheromone Analysis, Additional Pesticides, p. 31—114. New York: Academic Press. 1976.
16. SHOREY, H. H., and J. J. McKELVEY, Eds.: Chemical Control of Insect Behavior: Theory and Application. New York: J. Wiley and Sons. 1977.
17. MacCONNELL, J. G., and R. M. SILVERSTEIN: Recent results in insect pheromone chemistry. Angew. Chem. Int. Ed. Engl. **12**, 644—654 (1973).
18. EVANS, D. A., and C. L. GREEN: Insect attractants of natural origin. Chem. Soc. Rev. **2**, 75—97 (1973).
19. SHOREY, H. H.: Behavioral responses to insect pheromones. Annu. Rev. Entomol. **18**, 349—380 (1973).
20. BAKER, R., and D. A. EVANS: Biological chemistry. Part (i). Insect chemistry. Annu. Rept. Prog. Chem. Sect. B. Org. Chem. **72**, 347—365 (1975).
21. ROELOFS, W. L., and R. T. CARDÉ: Responses of Lepidoptera to synthetic sex pheromone chemicals and their analogues. Annu. Rev. Entomol. **22**, 377—405 (1977).
22. SILVERSTEIN, R. M.: Collaborative studies of bark and ambrosia beetle pheromones. In: T. L. PAYNE, R. N. CARLSON, and R. L. THATCHER, Eds., Southern Pine Beetle Symposium. Texas Agricultural Experiment Station; Dept. of Entomology, Texas A & M University, College Station, Texas. 1974.
23. SHOREY, H. H., and L. K. GASTON: Sex pheromones of Noctuid moths. VII. Quantitative aspects of the production and release of pheromone by females of *Trichoplusia ni* (Lepidoptera: Noctuidae). Ann. Entomol. Soc. Amer. **58**, 604—608 (1965).

158   J. M. Brand, J. Chr. Young, and R. M. Silverstein:

24. Stumper, R.: Données quantitatives sur la sécrétion d'acide formique par les fourmis. C. R. hebd. séances Acad. Sci. **234**, 149—152 (1952).
25. Browne, L. E., M. C. Birch, and D. L. Wood: Novel trapping and delivery systems for airborne insect pheromones. J. Insect Physiol. **20**, 183—193 (1974).
26. Byrne, K. J., W. E. Gore, G. T. Pearce, and R. M. Silverstein: Porapak-Q collection of airborne organic compounds serving as models for insect pheromones. J. Chem. Ecol. **1**, 1—7 (1975).
27. Young, J. C., and R. M. Silverstein: Biological and chemical methodology in the study of insect communication. In: D. G. Moulton, A. Turk, and J. W. Johnson Jr., Eds., Methods in Olfactory Research, p. 75—161. London: Academic Press. 1975.
28. Roelofs, W. L.: Chemical control of insects by pheromones. In: M. Rockstein, Ed., Insect Biochemistry. New York: Academic Press. In press.
29. Schneider, D.: Insect olfaction. Deciphering system for chemical messages. Science **163**, 1031—1037 (1969).
30. Kaissling, K. E.: Insect olfaction. In: L. M. Beidler, Ed., Handbook of Sensory Physiology, Vol. IV, p. 351—431. New York: Springer. 1971.
31. Roelofs, W. L.: The scope and limitations of the electroantennogram technique in identifying pheromone components. In: N. R. McFarlane, Ed., The Evolution of Biological Activity, p. 143—161. New York: Academic Press. 1977.
32. Houx, N. W. H., S. Voerman, and W. M. F. Jongen: Purification and analysis of synthetic sex attractant by liquid chromatography on a silver-loaded resin. J. Chromatogr. **96**, 25—32 (1974).
33. Heath, R. R., J. H. Tumlinson, R. E. Doolittle, and A. T. Proveaux: Silver nitrate high pressure liquid-chromatography of geometrical isomers. J. Chromatogr. Sci. **13**, 380—382 (1975).
34. Tumlinson, J. H., and R. R. Heath: Structure elucidation of insect pheromones by microanalytical methods. J. Chem. Ecol. **2**, 87—99 (1976).
35. Houx, N. W. H., and S. Voerman: High-performance liquid-chromatography of potential insect sex attractants and other geometrical isomers on a silver-loaded ion-exchanger. J. Chromatogr. **129**, 456—459 (1976).
36. Heath, R. R., J. H. Tumlinson, and R. E. Doolittle: Analytical and preparative separation of geometrical isomers by high efficiency silver nitrate liquid chromatography. J. Chromatogr. Sci. **15**, 10—13 (1977).
37. Brownlee, R. G., and R. M. Silverstein: A micro-preparative gas chromatograph and a modified carbon skeleton determinator. Anal. Chem. **40**, 2077—2079 (1968).
38. Tumlinson, J. H., D. D. Hardee, R. C. Gueldner, A. C. Thompson, P. A. Hedin, and J. P. Minyard: Sex pheromones produced by male boll weevil: Isolation, identification, and synthesis. Science **166**, 1010—1012 (1969).
39. Beroza, M.: Microanalytical methodology relating to the identification of insect sex pheromones and related behavior control chemicals. J. Chromatogr. Sci. **13**, 314—321 (1975).
40. Inscoe, M., and M. Beroza: Insect-behavior chemicals active in field trials. In: M. Beroza, Ed., Pest Management With Insect Sex Attractants, p. 145—181. Washington, D. C.: ACS Symposium Series No. 23, American Chemical Society. 1976.
41. Mayer, M. S., and J. R. McLaughlin: An annotated compendium of insect sex pheromones. Florida Agric. Expt. Sta. Monograph Series, No. 6 (1975).
42. Tamaki, Y.: Complexity, diversity, and specificity of behavior-modifying chemicals in Lepidoptera and Diptera. In: H. H. Shorey and J. J. McKelvey, Eds., Chemical Control on Insect Behavior, p. 253—285. New York: J. Wiley and Sons. 1977.
43. Roelofs, W. L.: An overview—the evolving philosophies and methodologies of pheromone chemistry. In: H. H. Shorey and J. J. McKelvey, Eds., Chemical Control on Insect Behavior, p. 287—297. New York: J. Wiley and Sons. 1977.

44. COURNOYER, R., J. C. SHEARER, and D. H. ANDERSON: Fourier transform infrared analysis below the one-nanogram level. Anal. Chem. **49**, 2275—2277 (1977).

45. PLUMMER, E. L., T. E. STEWART, K. BYRNE, G. T. PEARCE, and R. M. SILVERSTEIN: Determination of the enantiomorphic composition of several insect pheromone alcohols. J. Chem. Ecol. **2**, 307—311 (1976).

46. STEWART, T. E., E. L. PLUMMER, L. L. McCANDLESS, J. R. WEST, and R. M. SILVERSTEIN: Determination of enantiomer composition of several bicyclic ketal insect pheromone components. J. Chem. Ecol. **3**, 27—43 (1977).

47. MORI, K.: Synthesis of optically pure (+)-*trans*-verbenol and its antipode, the pheromone of *Dendroctonus* bark beetles. Agric. Biol. Chem. **40**, 415—418 (1976).

48. MORI, K., N. MIZUMACHI, and M. MATSUI: Pheromone synthesis. 12. Synthesis of optically pure (1S,4S,5S)-2-pinen-4-ol (*cis*-verbenol), the pheromone of *Ips* bark beetles. Agric. Biol. Chem. **40**, 1611—1615 (1975).

49. PEARCE, G. T., W. E. GORE, and R. M. SILVERSTEIN: Synthesis and absolute configuration of multistriatin. J. Org. Chem. **41**, 2797—2803 (1976).

50. — — — Carbon-13 spectra of some insect pheromones and related compounds of the 6,8-dioxabicyclo-[3.2.1]octane system. J. Magnetic Reson. **27**, 497—507 (1977).

51. REECE, C. A., J. O. RODIN, R. G. BROWNLEE, W. G. DUNCAN, and R. M. SILVERSTEIN: Synthesis of the principal components of the sex attractant from male *Ips confusus* [now *I. paraconfusus*] frass. 2-Methyl-6-methylene-7-octen-4-ol, 2-methyl-6-methylene-2,7-octadien-4-ol, and (+)-*cis*-verbenol. Tetrahedron **24**, 4249—4256 (1968).

52. COOPER, M. A., J. R. SALMON, D. WHITTAKER, and U. SCHEIDEGGER: Stereochemistry of the verbenols. J. Chem. Soc. (B) 1259—1261 (1967).

53. KOVATS, E.: Gas chromatographic characterization of organic substances in the retention index system. In: J. C. GIDDINGS and R. A. KELLER, Eds., Advances in Chromatography, p. 229—247. New York: Marcel Dekker. 1966.

54. ETTRE, L. S.: The interpretation of analytical results: Qualitative and quantitative analysis. In: L. S. ETTRE and A. ZLATKIS, Eds., The Practice of Gas Chromatography, p. 381—387. New York: Wiley-Interscience. 1967.

55. MACDONALD, L. M., and J. WEATHERSTON: Gas chromatography and structural elucidation of lepidopteran pheromones. J. Chromatogr. **118**, 195—200 (1976).

56. YOUNG, J. C., R. G. BROWNLEE, J. O. RODIN, D. N. HILDEBRAND, R. M. SILVERSTEIN, D. L. WOOD, M. C. BIRCH, and L. E. BROWNE: Identification of linalool produced by two species of bark beetles of the genus *Ips*. J. Insect Physiol. **19**, 1615—1622 (1973).

57. BIERL, B. A., M. BEROZA, and C. W. COLLIER: Potent sex attractant of the gypsy moth: Its isolation, identification, and synthesis. Science **170**, 87—89 (1970).

58. BEROZA, M., and R. A. COAD: Reaction gas chromatography. J. Gas Chromatogr. **4**, 199—216 (1966).

59. — — Reaction gas chromatography. In: L. S. ETTRE and A. ZLATKIS, Eds., The Practice of Gas Chromatography, p. 461—510. New York: Wiley-Interscience. 1967.

60. BEROZA, M., and M. N. INSCOE: Precolumn reactions for structure determination. In: L. S. ETTRE and W. H. McFADDEN, Eds., Ancillary Techniques of Gas Chromatography, p. 89—144. New York: Wiley-Interscience. 1969.

61. BEROZA, M.: Determination of the chemical structure of organic compounds at the microgram level by gas chromatography. Accts. Chem. Res. **3**, 33—40 (1970).

62. SILVERSTEIN, R. M., R. G. BROWNLEE, T. E. BELLAS, D. L. WOOD, and L. E. BROWNE: Brevicomin. Principal sex attractant in the frass of the female western pine beetle. Science **159**, 889—891 (1968).

63. BIERL, B. A., M. BEROZA, and W. T. ASHTON: Reaction loops for reaction gas chromatography. Subtraction of alcohols, aldehydes, ketones, epoxides, and acids and carbon-skeleton chromatography of polar compounds. Mikrochim. Acta **1969**, 637—653.

64. Regnier, F. E., and J. C. Huang: Identification of some oxygen-containing functional groups by reaction gas chromatography. J. Chromatogr. Sci. **8**, 267—271 (1970).

65. Beroza, M., and B. A. Bierl: Apparatus for ozonolysis of microgram amounts of compound. Anal. Chem. **38**, 1976—1977 (1966).

66. — — Rapid determination of olefin position in organic compounds in microgram range by ozonolysis and gas chromatography. Anal. Chem. **39**, 1131—1135 (1967).

67. — — Ozone generator for microanalysis. Mikrochim. Acta **1969**, 720—723.

68. Moore, B. P., and W. V. Brown: Gas-liquid chromatographic identification of ozonolysis fragments as a basis for micro-scale structure determination. J. Chromatogr. **60**, 157—166 (1971).

69. Tumlinson, J. H., R. R. Heath, and R. E. Doolittle: Application of chemical ionization mass spectrometry of epoxides to the determination of olefin position in aliphatic chains. Anal. Chem. **46**, 1309—1312 (1974).

70. Morgan, E. D., and R. C. Tyler: Microchemical methods for identification of volatile pheromones. J. Chromatogr. **134**, 174—177 (1977).

71. Roelofs, W., J. Kochansky, R. Cardé, H. Arn, and S. Rauscher: Sex attractant of the grape vine moth, *Lobesia botrana*, Mitt. Schweiz. Entomol. Gesellsch. **43**, 71—73 (1973).

72. Kochansky, J., J. Tette, E. F. Taschenberg, R. T. Cardé, K. E. Kaissling, and W. L. Roelofs: Sex pheromone of the moth, *Antheraea polyphemus*. J. Insect Physiol. **21**, 1977—1983 (1975).

73. Tumlinson, J. H., M. G. Klein, R. E. Doolittle, T. L. Ladd, and A. T. Proveaux: Identification of the female Japanese beetle sex pheromone: Inhibition of male response by an enantiomer. Science **197**, 789—792 (1977).

74. Roelofs, W. L., M. J. Gieselmann, A. M. Cardé, H. Tashiro, D. S. Moreno, C. A. Henrick, and R. J. Anderson: Sex pheromone of the California red scale, *Aonidiella aurantii*. Nature **267**, 698—699 (1977).

75. — — — — — — Identification of the California red scale sex pheromone. J. Chem. Ecol. **4**, 211—224 (1978).

76. Jacobson, M., M. Beroza, and W. A. Jones: Isolation, identification and synthesis of the sex attractant of the gypsy moth. Science **132**, 1011—1012 (1960).

77. Jacobson, M.: Insect sex attractants. III. The optical resolution of d,1-10-acetoxy-*cis*-7-hexadecen-1-ol. J. Org. Chem. **27**, 2670—2671 (1962).

78. Jones, W. A., M. Jacobson, and D. F. Martin: Sex attractant of the pink bollworm moth: isolation, identification, and synthesis. Science **152**, 1516—1517 (1966).

79. Hummel, H. E., L. K. Gaston, H. H. Shorey, R. S. Kaae, K. J. Byrne, and R. M. Silverstein: Clarification of the chemical status of the pink bollworm sex pheromone. Science **181**, 873—875 (1973).

80. Jacobson, M., M. Beroza, and R. T. Yamamoto: Isolation and identification of the sex attractant of the American cockroach. Science **139**, 48—49 (1963).

81. McDonough, L. M., D. A. George, B. A. Butt, J. M. Ruth, and K. R. Hill: Sex pheromone of the codling moth: Structure and synthesis. Science **177**, 177—178 (1972).

82. Hendry, L. B., M. E. Anderson, J. Jugovich, R. O. Mumma, D. Robacker, and Z. Kosarych: Sex pheromone of the oak leaf roller: A complex chemical messenger system identified by mass fragmentography. Science **187**, 355—357 (1975).

83. Eiter, K.: Insect sex attractants. Pure Appl. Chem. **41**, 201—217 (1975).

83a. Rossi, R.: Insect pheromones. I. Synthesis of achiral components of insect pheromones. Synthesis **1977**, 817—836.

84. Landor, P. D., S. R. Landor, and S. Mukasa: Synthesis of ($\pm$)-methyl tetradeca-*trans*-2,4,5-trienoate, the allenic sex pheromone produced by the male dried bean beetle. Chem. Commun. **1971**, 1638—1639.

85. Descoins, C., C. A. Henrick, and J. B. Siddall: Synthesis of a presumed sex attractant of dried bean beetle. Tetrahedron Lett. **1972**, 3777—3780.

86. BAUDOUY, R., and J. GORE: New synthesis of a pheromone with a vinyl allenic linkage. Synthesis **1974**, 573—574.

87. MICHELOT, D., and G. LINSTRUMELLE: Cuprates alleniques I: Préparations et réactions. Synthèse stéréosélective de la phèromone de la bruche parasite du haricot. Tetrahedron Lett. **1976**, 275—276.

88. KOCIENSKI, P. J., G. CERNIGLARIO, and G. FELDSTEIN: A synthesis of (±)-methyl n-tetradeca-trans-2,4,5-trienoate, an allenic ester produced by the male dried bean beetle Acanthoscelides obtectus (Say). J. Org. Chem. **42**, 353—355 (1977).

89. KATZENELLENBOGEN, J. A.: Insect pheromone synthesis: New methodology. Science **194**, 139—148 (1976).

90. HENRICK, C. A.: The synthesis of insect sex pheromones. Tetrahedron **33**, 1845—1889 (1977).

91. TUMLINSON, J. H., R. C. GUELDNER, D. D. HARDEE, A. C. THOMPSON, P. A. HEDIN, and J. P. MINYARD: Identification and synthesis of the four compounds comprising the boll weevil sex attractant. J. Org. Chem. **36**, 2616—2621 (1971).

92. ZURFLÜH, R., L. L. DUNHAM, V. L. SPAIN, and J. B. SIDDALL: Synthetic studies on insect hormones. IX. Stereoselective total synthesis of a racemic boll weevil pheromone. J. Amer. Chem. Soc. **92**, 425—427 (1970).

93. CARGILL, R. L., and B. W. WRIGHT: A new fragmentation reaction and its application to the synthesis of (±)-grandisol. J. Org. Chem. **40**, 120—122 (1975).

94. GUELDNER, R. C., A. C. THOMPSON, and P. A. HEDIN: Stereoselective synthesis of racemic grandisol. J. Org. Chem. **37**, 1854—1856 (1972).

95. KOSUGI, H., S. SEKIGUCHI, R. SEKITA, and H. UDA: Photochemical cycloaddition reactions of α,β-unsaturated lactones with olefins, and application to synthesis of natural products. Bull. Chem. Soc. Japan **49**, 520—528 (1976).

96. AYER, W. A., and L. M. BROWNE: Transformation of carvone into racemic grandisol. Can. J. Chem. **52**, 1352-1360 (1974).

97. GOLOB, N. F.: α-Oxycyclopropylcarbinyl rearrangement and its application to the total synthesis of grandisol. Thesis. Indiana Univ., Bloomington (1974). Diss. Abstr. Int. B. **35**, 4835 (1975).

98. WENKERT, E.: Cited in reference 89.

99. BILLUPS, W. E., J. H. CROSS, and C. V. SMITH: A synthesis of (±)-grandisol. J. Amer. Chem. Soc. **95**, 3438—3439 (1973).

100. STORK, G., and J. F. COHEN: Ring size in epoxynitrile cyclization. A general synthesis of functionally substituted cyclobutanes. Application to (±) grandisol. J. Amer. Chem. Soc. **96**, 5270—5272 (1974).

101. BABLER, J. H.: Base-promoted cyclization of a δ-chloroester: Application to the total synthesis of (±) grandisol. Tetrahedron Lett. **1975**, 2045—2048.

102. TROST, B. M., and D. E. KEELEY: New synthetic methods. Secoalkylative approach to grandisol. J. Org. Chem. **40**, 2013 (1975).

103. HOBBS, P. D., and P. D. MAGNUS: Synthesis of optically active grandisol. J. Chem. Soc. Chem. Commun. **1974**, 856—857.

104. — — Studies on terpenes. 4. Synthesis of optically active grandisol, the boll weevil pheromone. J. Amer. Chem. Soc. **98**, 4594—4600 (1976).

105. BABLER, J. H., and T. R. MORTELL: Facile route to 3 of 4 terpenoid components of the boll weevil sex attractant. Tetrahedron Lett. **1972**, 669—672.

106. PELLETIER, S. W., and N. V. MODY: Facile synthesis of cyclohexyl constituents of the boll weevil sex pheromone. J. Org. Chem. **41**, 1069—1071 (1976).

107. BABLER, J. H., and M. J. COGHLAN: Facile method for bishomologation of ketones to α,β-unsaturated aldehydes. Application to synthesis of cyclohexanoid components of the boll wevil sex pheromone. Synth. Commun. **6**, 469—474 (1976).

108. Traas, P. C., H. Boelens, and H. J. Takken: Two step synthesis of a sex attractant of male boll weevil from isophorone. Rec. Trav. Chim. **95**, 308—311 (1976).

109. — — — A convenient synthesis of 3,3-dimethylcyclohexylideneacetaldehyde, a sex attractant of the male boll weevil. Synth. Commun. **6**, 489—493 (1976).

110. Bedoukian, R. H., and J. Wolinsky: Biogenetic type synthesis of cyclohexyl constituents of the boll weevil pheromone. J. Org. Chem. **40**, 2154—2156 (1975).

111. Silverstein, R. M., J. O. Rodin, W. E. Burkholder, and J. E. Gorman: Sex attractant of the black carpet beetle. Science **157**, 85—87 (1967).

112. Rodin, J. O., M. A. Laeffer, and R. M. Silverstein: Synthesis of *trans*-3,*cis*-5-tetradecadienoic acid (megatomoic acid), the sex attractant of the black carpet beetle, and its geometric isomers. J. Org. Chem. **35**, 3152—3154 (1970).

113. Rodin, J. O., R. M. Silverstein, W. E. Burkholder, and J. E. Gorman: Sex attractant of female Dermestid beetle *Trogoderma inclusum* Le Conte. Science **165**, 904—906 (1969).

114. Degraw, J. I., and J. O. Rodin: Synthesis of methyl 14-methyl-*cis*-8-hexadecenoate and 14-methyl-*cis*-8-hexadecen-1-ol. Sex attractant of *Trogoderma inclusum* Le Conte. J. Org. Chem. **36**, 2902—2903 (1971).

115. Mori, K.: Absolute configurations of (−)-14-methylhexadec-8-*cis*-en-1-ol and methyl (−)-14-methylhexadec-8-*cis*-enoate, the sex pheromone of female Dermestid beetle. Tetrahedron **30**, 3817—3820 1974).

116. Rossi, R., and A. Carpita: Insect pheromones — synthesis of chiral sex pheromone components of several species of *Trogoderma* (Coleoptera: Dermestidae). Tetrahedron **33**, 2447—2450 (1977).

117. Bellas, T. E., R. G. Brownlee, and R. M. Silverstein: Synthesis of brevicomin, principal sex attractant in the frass of the female western pine beetle. Tetrahedron **25**, 5149—5153 (1969).

118. Kocienski, P. J., and R. W. Ostrow: Stereoselective total synthesis of *exo*- and *endo*-brevicomin. J. Org. Chem. **41**, 398—400 (1976).

119. Mori, K.: Stereoselective synthesis of (±)-*endo*-brevicomin, a pheromone inhibitor produced by *Dendroctonus* bark beetles. Agric. Biol. Chem. **40**, 2499—2500 (1976).

120. Knolle, J., and H. J. Schaefer: Anodic oxidation of organic compounds. 15. Synthesis of brevicomin by Kolbe electrolysis. Angew. Chem. Int. Ed. Engl. **14**, 758 (1975).

121. Byrom, N. T., R. Grigg, and B. Kongkathip: Catalytic synthesis of *endo*-brevicomin and related di- and tri-oxybicyclo [x.2.1] systems. J. Chem. Soc. Chem. Commun. **1976**, 216—217.

122. Mori, K.: Synthesis of *exo*-brevicomin, pheromone of western bine beetle, to obtain optically active forms of known absolute configuration. Tetrahedron **30**, 4223—4227 (1974).

123. Wasserman, H. H., and E. H. Barber: Carbonyl epoxide rearrangements. Synthesis of brevicomin and related [3.2.1] bicyclic systems. J. Amer. Chem. Soc. **91**, 3674—3675 (1969).

124. Look, M.: Improved synthesis of *endo*-brevicomin for the control of bark beetles (Coleoptera: Scolytidae). J. Chem. Ecol. **2**, 83—86 (1976).

125. Coke, J. L., H. J. Williams, and S. Natarajan: A new preparation of acetylenic ketones and application to the synthesis of *exo*-brevicomin, the pheromone from *Dendroctonus brevicomis*. J. Org. Chem. **42**, 2380—2382 (1977).

126. Rodin, J. O., C. A. Reece, R. M. Silverstein, V. H. Brown, and J. I. Degraw: Synthesis of brevicomin, principal sex attractant of western pine beetle. J. Chem. Eng. Data **16**, 381—382 (1971).

127. Mundy, B. P., R. D. Otzenberger, and A. R. DeBernardis: A synthesis of frontalin and brevicomin. J. Org. Chem. **36**, 2390 (1971).

128. LIPKOWITZ, K. B., B. P. MUNDY, and D. GEESEMAN: Studies directed towards a practical synthesis of brevicomins (II). A novel synthesis of 1,5-dimethyl-8-oxo-bicyclo[3.2.1.]octane-6-one. Synth. Commun. **3**, 453—458 (1973).

129. CHAQUIN, P., J. P. MORIZUR, and J. KOSSANYI: An easy access to exo-brevicomin. J. Amer. Chem. Soc. **99**, 903—905 (1977).

130. KINZER, G. W., A. F. FENTIMAN JR., T. F. PAGE JR., R. L. FOLTZ, J. P. VITÉ, and G. B. PITMAN: Bark beetle attractants: Identification, synthesis, and field bioassay of a new compound isolated from Dendroctonus. Nature **221**, 477—478 (1969).

131. D'SILVA, T. D. J., and D. W. PECK: Convenient synthesis of frontalin — 1,5-dimethyl-6,8-dioxabicyclo[3.2.1.]octane. J. Org. Chem. **37**, 1828—1829 (1972).

132. MORI, K., S. KOBAYASHI, and M. MATSUI: Pheromone synthesis. 7. Synthesis of (±) frontalin, pheromone of Dendroctonus bark beetles. Agric. Biol. Chem. **39**, 1889—1890 (1975).

133. MORI, K.: Synthesis of optically active forms of frontalin, the pheromone of Dendroctonus bark beetles. Tetrahedron **31**, 1381—1384 (1975).

134. OHRUI, H., and S. EMOTO: A synthesis of (S)-(−)-frontalin from D-glucose. Agric. Biol. Chem. **40**, 2267—2270 (1976).

135. MORI, K.: Synthesis of optically active forms of sulcatol, the aggregation pheromone in the Scolytid beetle Gnathotrichus sulcatus. Tetrahedron **31**, 3011—3012 (1975).

136. SILVERSTEIN, R. M., J. O. RODIN, and D. L. WOOD: Sex attractants in frass produced by male Ips confusus in ponderosa pine. Science **154**, 509—510 (1966).

137. KATZENELLENBOGEN, J. A., and R. S. LENOX: The generation of allyllithium reagents by lithium-tetrahydrofuran reduction of allylic mesitoates. A new procedure for selective allylic cross coupling and allylcarbinol synthesis. J. Org. Chem. **38**, 326—335 (1973).

138. KARLSEN, S., P. FROYEN, and L. SKATTEBOL: New syntheses of the bark beetle pheromones 2-methyl-6-methylene-7-octen-4-ol (ipsenol) and 2-methyl-6-methylene-2,7-octadien-4-ol (ipsdienol). Acta Chem. Scand. B **30**, 664—668 (1976).

139. CLINET, J. C., and G. LINSTRUMELLE: An efficient method for preparation of conjugated allenic carbonyl compounds. The synthesis of two bark beetle pheromones. Nouveau J. Chim. **1**, 373—374 (1977).

140. MORI, K.: Synthesis and absolute configuration of (−)-ipsenol (2-methyl-6-methylene-7-octen-4-ol), the pheromone of Ips paraconfusus Lanier. Tetrahedron Lett. **1975**, 2187—2190.

141. — Synthesis of optically active forms of ipsenol, the pheromone of Ips bark beetles. Tetrahedron **32**, 1101—1106 (1976).

142. RILEY, R. G., R. M. SILVERSTEIN, J. A. KATZENELLENBOGEN, and R. S. LENOX: Improved synthesis of 2-methyl-6-methylene-2,7-octadien-4-ol, a pheromone of Ips paraconfusus, and an alternative synthesis of the intermediate, 2-bromomethyl-1,3-butadiene. J. Org. Chem. **39**, 1957—1958 (1974).

143. GARBERS, C. F., and F. SCOTT: Terpenoid synthesis. V. Electrophilic addition reactions in the synthesis of the ocimenones, the rose oxides, and a pheromone of Ips paraconfusus. Tetrahedron Lett. **1976**, 1625—1628.

144. MORI, K.: Absolute configuration of (+)-ipsdienol, the pheromone of Ips paraconfusus Lanier, as determined by the synthesis of its (R)-(−)-isomer. Tetrahedron Lett. **1976**, 1609—1612.

145. OHLOFF, G., and W. GIERSCH: Access to optically active ipsdienol from verbenone. Helv. Chim. Acta **60**, 1496—1500 (1977).

146. PEARCE, G. T., W. E. GORE, R. M. SILVERSTEIN, J. W. PEACOCK, R. CUTHBERT, G. N. LANIER, and J. B. SIMEONE: Chemical attractants for the smaller European elm bark beetle, Scolytus multistriatus (Coleoptera: Scolytidae). J. Chem. Ecol. **1**, 115—124 (1975).

147. Mori, K.: Absolute configuration of (−)-4-methylheptan-3-ol, a pheromone of the smaller European elm bark beetle, as determined by the synthesis of its (3R,4R)-(+)- and (3S,4R)-(+)-isomers. Tetrahedron 33, 289—294 (1977).

148. — Synthesis of (1S,2R,4S,5R)-(−)-α-multistriatin, the pheromone in the smaller European elm bark beetle, Scolytus multistriatus. Tetrahedron 32, 1979—1981 (1976).

149. Elliott, W. J., and J. Fried: Stereocontrolled synthesis of α-multistriatin, an essential component of the aggregation pheromone for the European elm bark beetle. J. Org. Chem. 41, 2475—2476 (1976).

150. Jones, G., M. A. Acquadro, and M. A. Carmody: Long chain enals via carbonyl-olefin metathesis. An application in pheromone synthesis. J. Chem. Soc. Chem. Commun. 1975, 206—207.

151. Jacobson, M., K. Ohinata, D. L. Chambers, W. A. Jones, and M. S. Fujimoto: Insect sex attractants. 13. Isolation, identification, and synthesis of sex pheromones of the male Mediterranean fruit fly. J. Med. Chem. 16, 248—251 (1973).

152. Rossi, R.: Insect pheromones. A convenient method for synthesis of (E)-6-nonen-1-ol, a component of the sex pheromone of Ceratitis capitata. Chim. Ind. Milan 59, 591 (1977).

153. Carlson, D. A., M. S. Mayer, D. L. Silhacek, J. D. James, M. Beroza, and B. A. Bierl: Sex attractant pheromone of the house fly: Isolation, identification and synthesis. Science 174, 76—78 (1971).

154. Bestmann, H. J., O. Vostrowsky, and H. Platz: Pheromone. V. Eine stereoselektive Synthese des Sexuallockstoffes der Stubenfliege (Musca domestica). Chem. Zeit. 98, 161—162 (1974).

155. Eiter, K.: Neue Synthese des Tricosen-(9-cis) ("Muscalure") eines Attraktivstoffes der Hausfliege. Naturwissenschaften 59, 468—469 (1972).

156. Carlson, D. A., R. E. Doolittle, M. Beroza, W. M. Rogoff, and G. H. Gretz: Muscalure and related compounds. I. Response of houseflies in olfactometer and pseudofly tests. J. Agric. Food Chem. 22, 194—196 (1974).

157. Cargill, R. L., and M. G. Rosenblum: Synthesis of the housefly sex attractant. J. Org. Chem. 37, 3971 (1972).

158. Richter, I., and H. K. Mangold: Synthesis of muscalure and homologous hydrocarbons. Chem. Phys. Lipids 11, 210—214 (1973).

159. Ho, T. L., and C. M. Wong: A synthesis of muscalure, the housefly sex attractant. Can. J. Chem. 52, 1923—1924 (1974).

160. Gribble, G. W., J. K. Sanstead, and J. W. Sullivan: One-step synthesis of the housefly sex attractant (Z)-tricos-9-ene (muscalure). J. Chem. Soc. Chem. Commun. 1973, 735—736.

161. Rossi, R.: Simple synthesis of sex pheromones of the housefly and tiger moths by transition metal-catalyzed olefin cross-metathesis reactions. Chim. Ind. Milan 57, 242—243 (1975).

162. Butler, C. G., R. K. Callow, and N. C. Johnston: The isolation and synthesis of queen substance, 9-oxodec-trans-2-enoic acid, a honeybee pheromone. Proc. Roy. Soc. B 155, 417—432 (1961).

163. Barbier, M., E. Lederer, and T. Nomura: Synthèse de l'acide céto-9-décène-2-trans-oique (substance royale), et de l'acide céto-8-nonène-2-trans-oique. C. R. hebd. séances Acad. Sci. 251, 1133—1135 (1960).

164. Jaeger, R. H., and R. Robinson: A simple synthesis of "queen substance". Tetrahedron 14, 320—321 (1961).

165. Kennedy, J., N. J. McCorkindale, and R. A. Raphael: A new synthesis of queen substance. J. Chem. Soc. 1961, 3813—3815.

166. Eiter, K.: Neue Synthesen der „Königinnensubstanz" und der 9-Hydroxy-2-trans-decensäure. Justus Liebig's Ann. Chem. 658, 91—99 (1962).

167. SISIDO, K., M. KAWANISI, K. KONDÔ, T. MORIMOTO, A. SAITÔ, and N. HUKUE: Syntheses of 9-keto- and 10-hydroxy-*trans*-2-decenoic acids and related compounds. J. Org. Chem. **27**, 4073—4076 (1962).

168. TROST, B. M., and T. N. SALZMANN: Applications of sulfenylations of ester enolates. Synthesis of pheromones of the honey bee. J. Org. Chem. **40**, 148—150 (1975).

169. TSUJI, J., K. MASAOKA, and T. TAKAHASHI: Simple synthesis of queen substance from the butadiene telomer. Tetrahedron Lett. **1977**, 2267—2268.

170. VINSON, S. B., R. L. JONES, P. E. SONNET, B. A. BIERL, and M. BEROZA: Isolation, identification and synthesis of host-seeking stimulants for *Cardiochiles nigriceps*, a parasitoid of tobacco budworm. Entomol. Exp. Appl. **18**, 443—450 (1975).

171. KOCIENSKI, P. J., and J. M. ANSELL: A synthesis of 3,7-dimethylpentadec-2-yl acetate. The sex pheromone of the pine sawfly *Neodiprion lecontei*. J. Org. Chem. **42**, 1102—1103 (1977).

172. MAGNUSSON, G.: Pheromone synthesis. Preparation of *erythro*-3,7-dimethylpentadecan-2-ol, the alcohol from pine sawfly sex attractant (Hymenoptera: Diprionidae). Tetrahedron Lett. **1977**, 2713—2716.

173. RILEY, R. G., R. M. SILVERSTEIN, and J. C. MOSER: Biological responses of *Atta texana* to its alarm pheromone and the enantiomer of the pheromone. Science **183**, 760—762 (1974).

174. RILEY, R. G., and R. M. SILVERSTEIN: Synthesis of *S*-(+)-4-methyl-3-heptanone, the principal alarm pheromone of *Atta texana*, and its enantiomer. Tetrahedron **30**, 1171—1174 (1974).

175. SONNET, P. E.: Synthesis of the trail marker of the Texas leaf-cutting ant, *Atta texana* (Buckley). J. Med. Chem. **15**, 97—98 (1972).

176. BRAND, J. M., R. M. DUFFIELD, J. G. MACCONNELL, and M. S. BLUM: Caste-specific compounds in male carpenter ants. Science **179**, 388—389 (1973).

177. KOCIENSKI, P. J., J. M. ANSELL, and R. W. OSTROW: A synthesis of (*E*)-4,6-dimethyl-4-octen-3-one (manicone). J. Org. Chem. **41**, 3625—3627 (1976).

178. KATZENELLENBOGEN, J. A., and T. UTAWANIT: A highly stereoselective and completely regiospecific method for the dehydration of β-hydroxy esters via β-alanoxy enolates. Application to the synthesis of trisubstituted olefins and two ant mandibular gland secretions. J. Amer. Chem. Soc. **96**, 6153—6158 (1974).

179. FALES, H. M., M. S. BLUM, R. M. CREWE, and J. M. BRAND: Alarm pheromones in the genus *Manica* derived from the mandibular gland. J. Insect Physiol. **18**, 1077—1088 (1972).

180. BANNO, K., and T. MUKAIYAMA: Stereoselective synthesis of optically-active manicones: alarm pheromone of ant. Chem. Lett. **1976**, 279—282.

181. RITTER, F., I. E. ROTGANS, E. TALMAN, P. E. VERWIEL, and F. STEIN: 5-Methyl-3-butyl-octahydroindolizine, a novel type of pheromone attractive to Pharaohs ants *(Monomorium pharaonis* (L)). Experientia **29**, 530—531 (1973).

182. OLIVER, J. E., and P. E. SONNET: Synthesis of the isomers of 3-butyl-5-methyloctahydroindolizine, a trail pheromone of Pharaoh ant. J. Org. Chem. **39**, 2662—2663 (1974).

183. IKAN, R., R. GOTTLIEB, E. D. BERGMANN, and J. ISHAY: The pheromone of the queen of the Oriental hornet, *Vespa orientalis*. J. Insect Physiol. **15**, 1709—1712 (1969).

184. COKE, J. L., and A. B. RICHON: Synthesis of optically-active δ-*n*-hexadecalactone, the proposed pheromone from *Vespa orientalis*. J. Org. Chem. **41**, 3516—3517 (1976).

185. KODAMA, M., Y. MATSUKI, and S. ITO: Syntheses of macrocyclic, terpenoids by intramolecular cyclization. I. (±)-Cembrene-A, a termite trail pheromone, and (±)-nephthenol. Tetrahedron Lett. **1975**, 3065—3068.

186. KITAHARA, Y., T. KATO, T. KOBAYASHI, and B. P. MOORE: Cyclization of polyenes. XVII. Synthesis and pheromone activity of d,l-neocembrene. Chem. Lett. **1976**, 219—222.

187. Schwarz, M., and R. M. Waters: Insect sex attractants. XII. Efficient procedure for preparation of unsaturated alcohols and acetates. Synthesis **1972**, 567—568.

188. Holan, G., and D. F. O'keefe: An improved synthesis of insect sex attractant: cis-8-dodecen-1-ol acetate. Tetrahedron Lett. **1973**, 673—674.

189. Leznoff, C. C., and T. M. Fyles: The use of polymer supports in organic synthesis. VIII. Solid-phase syntheses of insect sex attractants. Can. J. Chem. **55**, 1143—1153 (1977).

190. Bestmann, H. J., P. Range, and R. Kunstmann: Pheromone. II. Synthese des Essig-säure [cis-tetradecen-(9)-yl-esters]. Chem. Ber. **104**, 65—70 (1971).

191. Hayashi, T., and H. Midorika: Synthesis with dithiocarbamate derivatives. IV. Novel stereoselective synthesis of alkenol sex pheromones *via* 3,3'-sigmatropic rearrangement of allylic dithiocarbamates. Synthesis **1975**, 100—102.

192. Kondo, K., A. Negishi, and D. Tunemoto: A novel method for the stereoselective synthesis of *trans* olefins. Angew. Chem. Int. Ed. Engl. **13**, 407—408 (1974).

193. Fyles, T. M., C. C. Leznoff, and J. Weatherston: The use of polymer supports in organic synthesis. XII. The total stereoselective synthesis of *cis* insect sex attractants on solid phases. Can. J. Chem. **55**, 4135—4143 (1977).

194. Mori, K., M. Uchida, and M. Matsui: Synthesis of aliphatic insect pheromones from alicyclic starting materials: (Z)-6-heneicosen-11-one and (Z)-8-dodecenyl acetate. Tetrahedron **33**, 385—387 (1977).

195. Sekul, A. A., A. N. Sparks, M. Beroza, and B. A. Bierl: Natural inhibitor of corn-earworm moth sex attractant. J. Econ. Entomol. **68**, 603—604 (1975).

196. Nesbitt, B. F., P. S. Beevor, D. R. Hall, R. Lester, and V. A. Dyck: Identification of female sex pheromones of the moth, *Chilo suppressalis*. J. Insect Physiol. **21**, 1883—1886 (1975).

197. Baer, T. A., S. M. Chang, and J. W. Baum: Unpublished results cited in reference *90*.

198. Carney, R. L., S. M. Chang, R. J. Scheible, and J. W. Baum: Unpublished results cited in reference *90*.

199. Weatherston, J., W. Roelofs, and A. Comeau: Studies of physiologically active arthropod secretions. X. Sex pheromone of the eastern spruce budworm, *Choristoneura fumiferana* (Lepidoptera: Tortricidae). Can. Entomol. **103**, 1741—1747 (1971).

200. Tumlinson, J. H., D. E. Hendricks, E. R. Mitchell, R. E. Doolittle, and M. M. Brennan: Isolation, identification, and synthesis of the sex pheromone of the tobacco hornworm. J. Chem. Ecol. **1**, 203—214 (1975).

201. Leadbetter, G., and J. R. Plimmer: Synthesis and separation of geometrical isomers of insect sex pheromones. Amer. Chem. Soc. Meeting Abstr. **173**, PEST 59 (1977).

202. Truscheit, E., and K. Eiter: Synthese der vier isomeren Hexadecadien-(10.12)-ole-(1). Justus Liebig's Ann. Chem. **658**, 65—90 (1962).

203. Nesbitt, B. F., P. S. Beevor, R. A. Cole, R. Lester, and R. G. Poppi: Sex phero-mones of two Noctuid moths. Nature, New Biology **244**, 208—209 (1973).

204. Goto, G., T. Shima, H. Masuya, Y. Masuoka, and K. Hiraga: A stereoselective synthesis of (Z,E)-9,11-tetradecadienyl-1-acetate, a major component of the sex pheromone of *Spodoptera litura*. Chem. Lett. **1975**, 103—106.

205. Hall, D. R., P. S. Beevor, R. Lester, R. G. Poppi, and B. F. Nesbitt: Synthesis of the major sex pheromone of the Egyptian cotton leafworm *Spodoptera littoralis* (Boisd.). Chem. Ind. London **1975**, 216—217.

206. Doolittle, R. E., W. L. Roelofs, J. D. Solomon, R. T. Cardé, and M. Beroza: (Z,E)-3,5-Tetradecadien-1-ol acetate sex attractant for the carpenterworm moth, *Prionoxystus robiniae* (Peck) (Lepidoptera: Cossidae). J. Chem. Ecol. **2**, 399—410 (1976).

207. Babler, J. H., and M. J. Martin: A facile synthesis of the sex pheromone of the red bollworm moth from 10-undecen-1-ol. J. Org. Chem. **42**, 1799—1800 (1977).

208. BESTMANN, H. J., O. VOSTROWSKY, H. PAULUS, W. BILLMANN, and W. STRANSKY: Pheromone. XI. Eine Aufbaumethode für konjugierte $(E),(Z)$-Diene. Synthese des Bombykols, seiner Derivate und Homologen. Tetrahedron Lett. **1977**, 121—124.

209. HENRICK, C. A., M. A. GEIGEL, and W. E. WILLY: Unpublished results cited in reference 90.

210. ROELOFS, W., A. COMEAU, A. HILL, and G. MILICEVIC: Sex attractant of the codling moth: characterization with electroantennogram technique. Science **174**, 297—299 (1971).

211. NESBITT, B. F., P. S. BEEVOR, R. A. COLE, R. LESTER, and R. G. POPPI: Synthesis of both geometric isomers of the major sex pheromone of the red bollworm moth. Tetrahedron Lett. **1973**, 4669—4670.

212. HENRICK, C. A., and J. TETTE: Unpublished results cited in reference 90.

213. BUTENANDT, A., E. HECKER, M. HOPP, and W. KOCH: Über den Sexuallockstoff des Seidenspinners, IV. Die Synthese des Bombykols und der cis-trans-Isomeren Hexadecadien-(10,12)-ole-(1). Justus Liebig's Ann. Chem. **658**, 39—64 (1962).

214. DESCOINS, C., and C. A. HENRICK: Stereoselective synthesis of a sex attractant of codling moth. Tetrahedron Lett. **1972**, 2999—3002.

215. BAUM, J. W., and R. J. SCHEIBLE: Unpublished results cited in reference 90.

216. HENRICK, C. A., W. E. WILLY, J. W. BAUM, T. A. BAER, B. A. GARCIA, T. A. MASTRE, and S. M. CHANG: Stereoselective synthesis of alkyl $(2E,4E)$- and $(2Z,4E)$-3,7,11-trimethyl-2,4-dodecadienoates. Insect growth regulators with juvenile hormone activity. J. Org. Chem. **40**, 1—7 (1975).

217. TAMAKI, Y., H. NOGUCHI, and T. YUSHIMA: Sex pheromone of *Spodoptera litura* (F.) [Lepidoptera: Noctuidae]: Isolation, identification and synthesis. Appl. Entomol. Zool. **8**, 200—203 (1973).

218. VIG, O. P., A. K. VIG, A. L. GAUBA, and K. C. GUPTA: New synthesis of *trans*-8-*trans*-10-dodecadien-1-ol. J. Indian Chem. Soc. **52**, 541—542 (1975).

219. HENRICK, C. A., R. J. ANDERSON, and L. D. ROSENBLUM: Unpublished results cited in reference 90.

220. BUTT, B. A., T. P. McGOVERN, M. BEROZA, and D. O. HATHAWAY: Codling moth: Cage and field evaluations of traps baited with a synthetic sex attractant. J. Econ. Entomol. **67**, 37—40 (1974).

221. MORI, K.: Simple synthesis of sex pheromones of codling moth and red bollworm moth by the coupling of Grignard reagents with allylic halides. Tetrahedron **30**, 3807—3810 (1974).

222. GEORGE, D. A., L. M. McDONOUGH, D. O. HATHAWAY, and H. R. MOFFITT: Inhibitors of sexual attraction of male codling moths (Lepidoptera: Olethreutidae). Environ. Entomol. **4**, 606—608 (1975).

223. LABOVITZ, J. N., C. A. HENRICK, and V. L. CORBIN: Synthesis of $(7E,9Z)$-7,9-dodecadien-1-yl acetate — sex pheromone of *Lobesia botrana*. Tetrahedron Lett. **1975**, 4209—4212.

224. CARNEY, R. L., and J. W. BAUM: Unpublished results cited in reference 90.

225. NORMANT, J. F., A. COMMERCON, and J. VILLIERAS: Synthèse d'énynes et de diènes conjugués à l'aide d'organocuivreux vinyliques. Application à la synthèse du bombykol. Tetrahedron Lett. **1975**, 1465—1468.

226. DESCOINS, C., and D. SAMAIN: Ouverture acide d'alcools α cyclopropaniques α' acétyléniques complexes par le dicobaltoctacarbonyle. Obtention stéréosélective d'énynes conjugués E, précurseurs de diènes conjugués E,Z. Tetrahedron Lett. **1976**, 745—748.

227. NEGISHI, E., G. LEW, and T. YOSHIDA: Stereoselective synthesis of conjugated *trans*-enynes readily convertible into conjugated *cis,trans*-dienes and its application to the synthesis of the pheromone bombykol. J. Chem. Soc. Chem. Commun. **1973**, 874—875.

228. Labovitz, J. N., V. L. Graves, and C. A. Henrick: Unpublished results cited in reference 90.

229. Negishi, E. I., and A. Abramovitch: A highly efficient chemoselective, regioselective, and stereoselective synthesis of (7E,9Z)-dodecadien-1-yl acetate, a sex pheromone of Lobesia botrana, via a functionalized organoborate. Tetrahedron Lett. 1977, 411—414.

230. Labovitz, J. N., and C. A. Henrick: Unpublished results cited in reference 90.

231. Tanaka, S., A. Yasuda, H. Yamamoto, and H. Nozaki: A general method for the synthesis of 1,3-dienes. Simple syntheses of β- and trans-α-farnesene from farnesol. J. Amer. Chem. Soc. 97, 3252—3254 (1975).

232. Bierl, B. A., M. Beroza, R. T. Staten, P. E. Sonnet, and V. E. Adler: The pink bollworm sex attractant. J. Econ. Entomol. 67, 211—216 (1974).

233. Su, H. C., and P. G. Mahany: Synthesis of the sex pheromone of the female Angoumois grain moth and its geometric isomers. J. Econ. Entomol. 67, 319—321 (1974).

234. Sonnet, P. E.: A practical synthesis of the sex pheromone of the pink bollworm. J. Org. Chem. 39, 3793—3794 (1974).

235. Mori, K., M. Tominaga, and M. Matsui: Stereoselective synthesis of the pink bollworm sex pheromone, (Z,Z)-7,11-hexadecadienyl acetate and its (Z,E)-isomer. Tetrahedron 31, 1846—1848 (1975).

236. Disselnkotter, H., K. Eiter, W. Karl, and D. Wendisch: Trennung und analytische Bestimmung synthetischer Pheromone am Beispiel der Isomeren 7,11-Hexadecadien-1-yl acetate (Gossyplure). Tetrahedron 32, 1591—1595 (1976).

237. Sonnet, P. E., B. A. Bierl, and M. Beroza: Effects of hexamethylphosphoric triamide (Hempa) upon allylic Grignard reagents: Synthesis of long chain alkenol acetates. J. Amer. Oil Chem. Soc. 51, 371—372 (1974).

238. Bestmann, H. J., K. H. Koschatzky, W. Stransky, and O. Vostrowsky: Pheromone. IX. Stereoselektive Synthesen von (Z)-7, (Z)-11- und (Z)-7,(E)-11-Hexadecadienylacetat, dem Sexualpheromon von Pectinophora gossypiella (Gelechiidae, Lepid.). Tetrahedron Lett. 1976, 353—356.

239. Sonnet, P. E.: Geometrical isomerization of 1,5-dienes: Isomers of gossyplure, the pink bollworm sex attractant. J. Amer. Oil Chem. Soc. 53, 36—38 (1976).

240. Anderson, R. J., and C. A. Henrick: Stereochemical control in Wittig olefin synthesis. Preparation of the pink bollworm sex pheromone mixture, gossyplure. J. Amer. Chem. Soc. 97, 4327—4334 (1975).

241. Trave, R., L. Garanti, A. Marchesini, and M. Pavan: Chemical nature of components of the odorous secretion of Lepidopterus Cossus cossus. Chim. Ind. Milan 48, 1167—1176 (1966).

242. Garanti, L., A. Marchesini, U. M. Pagnoni, and R. Trave: Synthesis of (5Z)-5,13-tetradecadien-1-yl acetate and (3E,5Z)-3,5,13-tetradecatrien-1-yl acetate (secretion of larvae of Lepidopterus Cossus cossus) and of their geometrical isomers. Gazz. Chim. Ital. 106, 187—196 (1976).

243. Bestmann, H. J., and O. Vostrowsky: Pheromone. III. Eine stereoselektive Synthese von 7,8-Z-Epoxy-2-methyloctadecan, dem Sexuallockstoff des Schwammspinners (Lymantria dispar, Porthetria dispar, Lepidoptera). Tetrahedron Lett. 1974, 207—208.

244. Bestmann, H. J., O. Vostrowsky, and W. Stransky: Pheromone. X. Eine stereoselektive Synthese des (Z)-7,8-Epoxy-2-methyloctadecans (Disparlure). Chem. Ber. 109, 3375—3378 (1976).

245. Eiter, K.: Novel total synthesis of sex attractant of Porthetria dispar (7,8-cis-epoxy-2-methyloctadecane). Angew. Chem. Int. Ed. Engl. 11, 60—61 (1972).

246. Sheads, R. E., and M. Beroza: Preparation of tritium-labelled disparlure, the sex attractant of the gypsy moth. J. Agric. Food Chem. 21, 751—753 (1973).

247. Shamshur, A. A., M. A. Rekhter, and L. A. Vlad: (Chemistry of pheromones. 2.

New synthesis of disparlure, sexual attractant of unpaired *Bombyx mori.*) Khim. Prir. Soed. Akad. Nauk Uzb. **1973**, 545—548.

248. KOVALEV, B. G., R. I. ISHCHENKO, and V. A. MARCHENKO: (Synthetic studies on sexual attractants of insects. I. Synthesis of 2-methyl-7-octadecene oxide (disparlure) as a sexual attractant of an unpaired silkworm *Porthetria dispar.*) Zh. Org. Khim. **9**, 6—8 (1973).

249. MYCHAJLOWSKIJ, M., and T. H. CHAN: The synthesis of alkenes from carbonyl compounds and carbanions α to silicon. V. Stereoselective synthesis of *E*- and *Z*-disubstituted alkenes. Tetrahedron Lett. **1976**, 4439—4442.

250. CHAN, T. H., and E. CHANG: The synthesis of alkenes from carbonyl compounds and carbanions α to silicon. III. A full report and a synthesis of the sex pheromone of gypsy moth. J. Org. Chem. **39**, 3264—3268 (1974).

251. KUPPER, F. W., and R. STRECK: Synthese von Insektenlockstoffen an Metathese-katalysatoren. Z. Naturforsch. **31b**, 1256—1264 (1976).

252. IWAKI, S., S. MARUMO, T. SAITO, M. YAMADA, and K. KATAGIRI: Synthesis and activity of optically active disparlure. J. Amer. Chem. Soc. **96**, 7842—7844 (1974).

253. MORI, K., T. TAKIGAWA, and M. MATSUI: Stereoselective synthesis of optically active disparlure, the pheromone of the gypsy moth *(Porthetria dispar L.).* Tetrahedron Lett. **1976**, 3953—3956.

254. FARNUM, D. G., T. VEYSOGLU, A. M. CARDÉ, B. DUHLEMSWILER, and T. A. PANCOAST: A stereospecific synthesis of (+)-disparlure, sex attractant of the gypsy moth. Tetrahedron Lett. **1977**, 4009—4012.

255. SMITH, R. G., G. E. DATERMAN, and G. D. DAVES: Douglas-fir tussock moth: Sex pheromone identification and synthesis. Science **188**, 63—64 (1975).

256. SMITH, R. G., G. D. DAVES JR., and G. E. DATERMAN: Synthesis of (*Z*)-6-heneicosen-11-one. Douglas-fir tussock moth sex attractant. J. Org. Chem. **40**, 1593—1595 (1975).

257. KOCIENSKI. P. J., and G. J. CERNIGLIARIO: A synthesis of (*Z*)-6-heneicosen-11-one. The sex pheromone of the Douglas fir tussock moth. J. Org. Chem. **41**, 2927—2928 (1976).

258. LABOVITZ, J. N., V. L. GRAVES, and C. A. HENRICK: Unpublished results cited in reference *90.*

259. TAMAKI, Y., K. HONMA, and K. KAWASAKI: Sex pheromone of the peach fruit moth, *Carposina niponensis* Walsingham (Lepidoptera: Carposinidae): Isolation, identification and synthesis. Appl. Entomol. Zool. **12**, 60—68 (1977).

260. SATO, T., R. NISHIDA, Y. KUWAHARA, H. FUKAMI, and S. ISHII: Syntheses of female sex pheromone analogs of the German cockroach and their biological activity. Agric. Biol. Chem. **40**, 391—394 (1976).

261. BURGSTAHLER, A. W., M. E. SANDERS, C. G. SHAEFFER, and L. O. WEIGEL: Phase-transfer methylation of benzyl 3-oxobutanoate as a route to 3-methyl-2-alkanones. Improved syntheses of two female sex pheromones of the German cockroach. Synthesis **1977**, 405—407.

262. SCHWARZ, M., J. E. OLIVER, and P. E. SONNET: Synthesis of 3,11-dimethyl-2-nonacosanone, a sex pheromone of the German cockroach. J. Org. Chem. **40**, 2410—2411 (1975).

263. ROSENBLUM, L. D., R. J. ANDERSON, and C. A. HENRICK: Synthesis of 3,11-dimethyl-2-nonacosanone, a sex pheromone of the German cockroach, *Blatella germanica.* Tetrahedron Lett. **1976**, 419—422.

264. NISHIDA, R., T. SATO, Y. KUWAHARA, H. FUKAMI, and S. ISHII: Synthesis of 29-hydroxy-3,11-dimethyl-2-nonacosanone, sex pheromone of the German cockroach. Agric. Biol. Chem. **40**, 1407—1410 (1976).

265. SILVERSTEIN, R. M., and J. C. YOUNG: Insects generally use multicomponent pheromones. In: M. BEROZA, Ed., Management with insect sex attractants, p. 1—29. ACS Symposium Series, No. 23. 1976.

266. SILVERSTEIN, R. M.: Complexity diversity, and specificity of behavior-modifying chemicals: examples mainly from Coleoptera and Hymenoptera. In: H. H. SHOREY and J. J. MCKELVEY, Eds., Chemical Control of Insect Behavior, p. 231—251. New York: J. Wiley and Sons. 1977.

267. KLUN, J. A., O. L. CHAPMAN, K. C. MATTES, P. W. WOJTKOWSKI, M. BEROZA, and P. E. SONNET: Insect sex pheromones: minor amount of opposite geometrical isomer critical to attraction. Science 181, 661—663 (1973).

268. KLUN, J. A., and G. A. JUNK: Iowa European corn borer sex pheromone: isolation and identification of four C_{14} esters. J. Chem. Ecol. 3, 447—459 (1977).

269. CARDÉ, R. T., and W. L. ROELOFS: Attraction of redbanded leafroller moths, Argyrotaenia velutinana, to blends of (Z)- and (E)-11-tridecenyl acetates. J. Chem. Ecol. 3, 143—149 (1977).

270. ROELOFS, W. L., and R. T. CARDÉ: Sex pheromones in the reproductive isolation of lepidopterous species. In: M. C. BIRCH, Ed., Pheromones, p. 96—114. Amsterdam: North-Holland Publishing Co. 1974.

271. SUGIE, H., K. YAGINUMA, and Y. TAMAKI: Sex pheromone of the Asiatic leafroller, Archippus breviplicanus Walsingham (Lepidoptera: Tortricidae): isolation, and identification. Appl. Ent. Zool. 12, 69—74 (1977).

272. HONMA, K.: Isolating factors between the smaller tea tortrix and the summer-fruit tortrix (Lepidoptera: Tortricidae) I. Distribution of the two species in Japan. Appl. Ent. Zool. 9, 143—146 (1974).

273. SATO, R., and Y. TAMAKI: Isolating factors between the smaller tea tortrix and the summer-fruit tortrix (Lepidoptera: Tortricidae) IV. Role of the pheromonal components, (Z)-9- and (Z)-11-tetradecen-1-ol acetates. Appl. Ent. Zool. 12, 50—59 (1977).

274. TAMAKI, Y., H. NOGUCHI, T. YUSHIMA, and C. HIRANO: Two sex pheromones of the smaller tea tortrix: isolation, identification and synthesis. Appl. Ent. Zool. 6, 139—141 (1971).

275. TAMAKI, Y., H. NOGUCHI, T. YUSHIMA, C. HIRANO, K. HONMA, and H. SUGAWARA: Sex pheromone of the summer-fruit tortrix: isolation and identification. Kontyu. 39, 338—340 (1971).

276. MEIJER, G. M., F. J. RITTER, C. J. PERSOONS, A. K. MINKS, and S. VOERMAN: Sex pheromones of summer-fruit tortrix moth, Adoxophyes orana; two synergistic isomers. Science 175, 1469—1460 (1972).

277. VOERMAN, S., A. K. MINKS, and E. A. GOEWIE: Specificity of the pheromone system of Adoxophyes orana and Clepsis spectrana. J. Chem. Ecol. 1, 423—429 (1975).

278. HONMA, K., and Y. TAMAKI: Isolating factors between the smaller tea tortrix and the summer-fruit tortrix (Lepidoptera: Tortricidae) II. Sexual isolation. Appl. Ent. Zool. 11, 202—208 (1976).

279. BEROZA, M., G. M. MUSCHIK, and C. R. GENTRY: Small proportion of opposite isomer increases potency of synthetic pheromone of Oriental fruit moth. Nature, New Biology 244, 149—150 (1973).

280. ROELOFS, W. L., A. COMEAU, and R. SELLE: Sex pheromone of the Oriental fruit moth. Nature 224, 723 (1969).

281. ROELOFS, W. L., and R. T. CARDÉ: Oriental fruit moth and lesser appleworm attractant mixtures refined. Environ. Entomol. 3, 586—588 (1974).

282. BARROWS, E. M., W. J. BELL, and C. D. MICHENER: Individual odor differences and their social functions in insects. Proc. Natl. Acad. Sci. USA 72, 2824—2828 (1975).

283. AVERHOFF, W. W., and R. H. RICHARDSON: Pheromonal control of mating patterns in Drosophila melanogaster. Behav. Gen. 4, 207—225 (1974).

284. — — Multiple pheromone system controlling mating in Drosophila melanogaster. Proc. Natl. Acad. Sci. USA 73, 591—593 (1976).

285. CARDÉ, R. T., T. C. BAKER, and W. L. ROELOFS: Sex attractant responses of male

Oriental fruit moths to a range of component ratios: pheromone polymorphism? Experientia **32**, 1406—1407 (1976).

286. ROTHSCHILD, G. H. L.: Attractants for monitoring *Pectinophora scutigera* and related species of Australia. Environ. Entomol. **4**, 983—985 (1975).

287. FLINT, H. M., R. L. SMITH, D. E. FOREY, and B. R. HORN: Pink bollworm response of males to (*Z,Z-*) and (*Z,E-*) isomers of Gossyplure. Environ. Entomol. **6**, 274—275, (1977).

288. UNDERHILL, E. W., W. F. STECK, and M. D. CHISHOLM: Sex pheromone of the clover cutworm moth, *Scotogramma trifolii;* isolation, identification and field studies. Environ. Entomol. **5**, 307—310 (1976).

289. STRUBLE, D. L., and G. E. SWAILES: A sex attractant for the clover cutworm, *Scotogramma trifolii* (Rottenberg), a mixture of Z-11-hexadecen-1-ol acetate and Z-11-hexadecen-1-ol. Environ. Entomol. **4**, 632—636 (1975).

290. — — Sex attractant for clover cutworm, *Scotogramma trifolii;* field tests with various ratios of Z-11-hexadecen-1-yl acetate and Z-11-hexadecen-1-ol, and with various quantities of attractant on two types of carriers. Can. Entomol. **109**, 369—373 (1977).

291. STECK, W., E. W. UNDERHILL, B. K. BAILEY, and M. D. CHISHOLM: A sex attractant for male moths of the glassy cutworm *Crymodes devastator* (Brace): a mixture of Z-11-hexadecen-1-yl acetate, Z-11-hexadecenal and Z-7-dodecen-1-yl acetate. Environ. Entomol. **6**, 270—273 (1977).

292. UNDERHILL, E. W., M. D. CHISHOLM, and W. STECK: Olefinic aldehydes as constituents of sex attractants for noctuid moths. Environ. Entomol. **6**, 333—337 (1977).

293. TAMAKI, Y., K. KAWASAKI, H. YAMADA, T. KOSHIHARA, N. OSAKI, T. ANDO, S. YOSHIDA, and H. KAKINOHANA: (*Z*)-11-Hexadecenal and (*Z*)-11-hexadecenyl acetate: sex-pheromone components of the diamond back moth (Lepidoptera: Plutellidae). Appl. Ent. Zool. **12**, 208—210 (1977).

294. TAMAKI, Y.: Division of Entomology, National Institute of Agriculture Sciences, Nishigahara, Tokyo 114, private communication.

295. ROELOFS, W. L., A. CARDÉ, A. HILL, and R. CARDÉ: Sex pheromone of the three lined leafroller, *Pandemis limitata.* Environ. Entomol. **5**, 649—652 (1976).

296. ROELOFS, W. L., R. F. LAGIER, and S. C. HOYT: Sex pheromone of the moth, *Pandemis pyrusana.* Environ. Entomol. **6**, 353—354 (1977).

297. HILL, A. S., R. T. CARDÉ, W. M. BODE, and W. L. ROELOFS: Sex pheromone components of the variegated leafroller moth, *Platynota flavedana.* J. Chem. Ecol. **3**, 369—376 (1977).

298. BUTLER, L. I., J. E. HALFHILL, L. M. McDONOUGH, and B. A. BUTT: Sex attractant of the alfalfa looper, *Autographa californica,* and the celery looper, *Anagrapha falcifera* (Lepidoptera: Noctuidae). J. Chem. Ecol. **3**, 65—70 (1977).

299. KAAE, R. S., H. H. SHOREY, and K. K. GASTON: Pheromone concentration as a mechanism for reproductive isolation between two lepidopterous species. Science **179**, 487—488 (1973).

300. CROSS, J. H., R. C. BYLER, R. F. CASSIDY, R. M. SILVERSTEIN, R. E. GREENBLATT, W. E. BURKHOLDER, A. R. LEVINSON, and H. Z. LEVINSON: Porapak Q collection of pheromone components and isolation of (*Z*)- and (*E*)-14-methyl-8-hexadecenal, sex pheromone components, from the females of a few species of *Trogoderma* (Coleoptera: Dermestidae). J. Chem. Ecol. **2**, 457—468 (1976).

301. SILVERSTEIN, R. M.: Spectrometric identification of insect sex attractants. J. Chem. Educ. **45**, 794—797 (1968).

302. BARAK, A. V., and W. E. BURKHOLDER: Behavior and pheromone studies with *Attagenus elongatulus* Casey (Coleoptera: Dermestidae). J. Chem. Ecol. **3**, 219—237 (1977).

303. FUKUI, H., F. MATSUMURA, A. V. BARAK, and W. E. BURKHOLDER: Isolation and identification of a major sex-attracting component of *Attagenus elongatulus* (Casey) (Coleoptera: Dermestidae). J. Chem. Ecol. **3**, 539—548 (1977).

304. BOWERS, W. S., L. R. NAULT, R. E. WEBB, and S. R. DUTKY: Aphid alarm pheromone: isolation, identification, synthesis. Science **177**, 1121—1122 (1972).

305. EDWARDS, J. S., J. B. SIDDALL, L. L. DUNHAM, P. UDEN, and C. J. KISLOW: Trans-β-farnesene, alarm pheromone of the green peach aphid, *Myzus persicae* (Sulzer). Nature **241**, 126—127 (1973).

306. WEINTJENS, W. H. J. M., A. C. LAKWIJK, and T. VAN DERMARREL: Alarm pheromones of grain aphids. Experientia **29**, 658—660 (1973).

307. NAULT, L. R., and W. S. BOWERS: Multiple alarm pheromones in aphids. Ent. exp. & Appl. **17**, 455—457 (1974).

308. BOWERS, W. S., C. NISHINO, M. E. MONTGOMERY, and L. R. NAULT: Structure-activity relationships of analogs of the aphid alarm pheromone, (*E*)-β-farnesene. J. Insect Physiol. **23**, 697—701 (1977).

308a. KAISSLING, K. E.: Max-Planck Institut für Verhaltensphysiologie, 8131 Seewiesen, Deutsche Bundesrepublik, private communication.

309. KAFKA, W. A., G. OHLOFF, D. SCHNEIDER, and E. VARESHI: Olfactory discrimination of two enantiomers of 4-methylhexanoic acid by the migratory locust and the honey-bee. J. Comp. Physiol. **87**, 277—284 (1973).

310. LENSKY, Y., and M. S. BLUM: Chirality in insect receptors. Life Sci. **14**, 2045—2049 (1974).

311. STÄDLER, E.: Host plant stimuli affecting oviposition behavior of the eastern spruce budworm. Ent. exp. & Appl. **17**, 176—188 (1974).

312. RILEY, R. G., R. M. SILVERSTEIN, and J. C. MOSER: Isolation, identification, synthesis, and biological activity of volatile compounds from the heads of *Atta* ants. J. Insect Physiol. **20**, 1629—1637 (1974).

313. McGURK, D. J., J. FROST, E. J. EISENBRAUN, K. VICK, W. A. DREW, and J. YOUNG: Volatile compounds in ants: identification of 4-methyl-3-heptanone from *Pogonomyrmex* ants. J. Insect Physiol. **12**, 1435—1441 (1966).

314. BENTHUYSEN, J. L., and M. S. BLUM: Quantitative sensitivity of the ant *Pogonomyrmex barbatus* to the enantiomers of its alarm pheromone. J. Ga. Entomol. Soc. **9**, 235—238 (1974).

315. BYRNE, K. W., A. SWIGAR, R. M. SILVERSTEIN, J. H. BORDEN, and E. STOKKINK: Sulcatol: population aggregation pheromone in *Gnathotrichus sulcatus* (Coleoptera: Scolytidae). J. Insect Physiol. **20**, 1895—1900 (1974).

316. BORDEN, J. H., L. CHONG, J. A. McLEAN, K. N. SLESSOR, and K. MORI: *Gnathotrichus sulcatus;* synergistic response to enantiomers of the aggregation pheromone sulcatol. Science **192**, 894—896 (1976).

317. BORDEN, J. H.: Biological Sciences Department, Simon Fraser University, Burnaby, B. C. Canada V5A 1S6, private communication.

318. LANIER, G. N., W. E. GORE, G. T. PEARCE, J. W. PEACOCK, and R. M. SILVERSTEIN: Response of the European elm bark beetle, *Scolytus multistriatus* (Coleoptera: Scolytidae), to isomers and components of its pheromone. J. Chem. Ecol. **3**, 1—8 (1977).

319. ACKLIN, W., V. PRELOG, F. SCHENKER, B. SERDAREVIĆ, and P. WALTER: Reaktionen mit Mikroorganismen. Reduktion von stereoisomeren Dekalonen-(1) und einigen verwandten bicyclischen und tricyclischen Ketonen durch *Curvularia falcata*. Helv. Chim. Acta **48**, 1725—1746 (1965).

320. WOOD, D. L., L. E. BROWNE, B. EWING, K. KINDAHL, W. D. BEDARD, P. E. TILDEN, K. MORI, G. B. PITMAN, and P. R. HUGHES: Western pine beetle: specificity among enantiomers of male and female components of an attractant pheromone. Science **192**, 896—898 (1976).

321. DICKENS, J. C., and T. L. PAYNE: Bark beetle olfaction: pheromone receptor system in *Dendroctonus frontalis.* J. Insect Physiol. **23,** 481—489 (1977).

322. VITÉ, J. P., and J. A. A. RENWICK: Anwendbarkeit von Borkenkäferpheromonen: Konfiguration und Konsequenzen. Z. ang. Ent. **82,** 112—116 (1976).

323. RENWICK, J. A. A., and J. P. VITÉ: Pheromones and host volatiles that govern aggregation of the six-spined engraver beetle, *Ips calligraphus.* J. Insect Physiol. **18,** 1215—1219 (1972).

324. VITÉ, J. P., D. KLIMETZEK, G. LOSKANT, R. HEDDEN, and K. MORI: Chirality of insect pheromones: response interruption by inactive antipodes. Naturwiss. **63,** 582—583 (1976).

325. KRAWIELITZKI, S., D. KLIMETZEK, A. BAKKE, J. P. VITÉ, and K. MORI: Field and laboratory response of *Ips typographus* to optically pure pheromonal components. Z. ang. Ent. **83,** 300—302 (1977).

326. PLIMMER, J. R., C. P. SCHWALBE, E. C. PASZEK, B. A. BIERL, R. E. WEBB, S. MARUMO, and S. IWAKI: Contrasting effectiveness of (+) and (−) enantiomers of disparlure for trapping native populations of gypsy moth in Massachusetts. Environ. Entomol. **6,** 518—522 (1977).

327. BIRCH, M. C.: Department of Entomology, University of California, Davis, CA 95616, private communication.

328. LANIER, G. N.: Department of Entomology, SUNY College of Environmental Science and Forestry, Syracuse, N. Y. 13210, private communication.

329. LANIER, G. N., M. C. BIRCH, R. F. SCHMITZ, and M. M. FURNISS: Pheromones of *Ips pini* (Coleoptera: Scolytidae): variation in response among three populations. Can. Entomol. **104,** 1917—1923 (1972).

330. HARRING, C. M., and K. MORI: *Pityokteines curvidens* Germ. (Coleoptera: Scolytidae): aggregation in response to optically pure ipsenol. Z. ang. Ent. **82,** 327—329 (1977).

331. TASHIRO, H., and D. L. CHAMBERS: Reproduction in the California red scale, *Aonidiella aurantii* (Homoptera: Diaspididae). I. Discovery and extraction of a female sex pheromone. Ann. Entomol. Soc. Am. **60,** 1166—1170 (1967).

332. NISHINO, C., W. S. BOWERS, M. E. MONTGOMERY, L. R. NAULT, and M. W. NIELSON: Alarm pheromone of the spotted alfalfa aphid, *Therioaphis maculata* Buckton (Homoptera: Aphididae). J. Chem. Ecol. **3,** 349—357 (1977).

333. SCHNEIDER, D.: Insect antennae. Annu. Rev. Entomol. **9,** 103—122 (1964).

334. — Olfactory receptors for the sexual attractant (bombykol) of the silk moth. In: F. O. SCHMITT, Ed., The Neurosciences: Second Study Program, p. 511—518. New York: Rockefeller Univ. Press. 1970.

335. KAFKA, W. A.: Molekulare Wechselwirkungen bei der Erregung einzelner Reichzellen. Z. Vergl. Physiol. **70,** 105—143 (1970).

336. KAISSLING, K. E.: Kinetics of olfactory receptor potentials. In: C. PFAFFMANN, Ed., Olfaction and Taste, p. 52—70. New York: Rockefeller Univ. Press. 1969.

337. SEABROOK, W. D.: Insect chemosensory responses to other insects. In: H. H. SHOREY and J. J. MCKELVEY, Eds., Chemical Control of Insect Behavior, p. 15—43. New York: J. Wiley and Sons. 1977.

338. KAISSLING, K. E.: Control of insect behavior via chemoreceptor organs. In: H. H. SHOREY and J. J. MCKELVEY, Eds., Chemical Control of Insect Behavior, p. 45—65. New York: J. Wiley and Sons. 1977.

338a. SEABROOK, W. D.: Neurobiological contributions to understanding insect pheromone systems. Annu. Rev. Entomol. **23,** 471—485 (1978).

339. EHRLICH, P.: Über den jetzigen Stand der Chemotherapie. Ber. **42,** 17—47 (1909).

340. RIDDIFORD, L. M.: Antennal proteins of Saturniid moths—their possible role in olfaction. J. Insect Physiol. **16,** 653—660 (1970).

341. Ferkovich, S. M., M. S. Mayer, and R. R. Rutter: Conversion of the sex pheromone of the cabbage looper. Nature 242, 53—55 (1973).
342. Mayer, M. S., S. M. Ferkovich, and R. R. Rutter: Localization and reactions of
a pheromone degradative enzyme isolated from an insect antenna. Chemical Senses
and Flavor 2, 51—61 (1976).
343. Aksamit, R., and D. E. Koshland: Identification of the ribose binding protein as
the receptor for ribose chemotaxis in Salmonella typhimurium. Biochemistry 13,
4473—4478 (1974).
344. Hazelbaer, G. L., and J. Adler: Role of the galactose binding protein in chemotaxis
of Escherichia coli toward galactose. Nature, New Biol. 230, 101—104 (1971).
345. Zukin, R., P. Strange, L. Heavey, and D. E. Koshland: Properties of the galactose
binding protein of Salmonella typhimurium and Escherichia coli. Biochemistry 16,
381—386 (1977).
346. Wright, R. H.: 6822 Blenheim Street, Vancouver, B. C., Canada V6N 1R7,
private communication.
347. Alworth, W. L.: Stereochemistry and its Application in Biochemistry, 311 p. New
York: Wiley-Interscience. 1972.
347a. Aihara, Y., and T. Shibuya: Responses of single olfactory receptor cells to sex
pheromones in the tobacco cutworm moth, Spodoptera litura. J. Insect Physiol. 23,
779—783 (1977).
348. Hayward, L. D., and R. N. Totty: Optical activity of symmetric compounds in
chiral media. I. Induced circular dichroism of unbound substrates. Can. J. Chem.
49, 624—631 (1971).
349. Hayward, L. D.: A symmetry rule and mechanism for optical activity induced in
achiral ketones by chiral solvents. Chem. Phys. Letters 33, 53—56 (1975).
350. Hayward, L. D., and S. Claesson: The effect of temperature and high pressure on
intermolecularly induced optical activity. Chemica Scripta 9, 21—23 (1976).
351. Hayward, L. D.: A new theory of olfaction based on dispersion-induced optical
activity. Nature 267, 554—555 (1977).
352. Schipper, P. E.: Vibronic contributions to DICD in achiral molecules. Chemical
Physics 12, 15—23 (1976).
353. — Associate-induced circular dichroism (AICD): a theory of the circular dichroism
of associated or bonded chromophores in the dipole approximation. Chemical
Physics 23, 159—166 (1977).
354. Blum, M. S., S. L. Warter, and J. G. Traynham: Chemical releasers of social
behavior. VI. The relation of structure to activity of ketones as releasers of alarm for
Irdomyrmex pruinosus (Roger). J. Insect Physiol. 12, 419—427 (1966).
355. Hayward, L. D.: Department of Chemistry, University of British Columbia, Vancouver, B. C., Canada V6T 1W5, private communication.
356. Wright, R. H.: Odor and molecular vibration: neural coding of olfactory information. J. theor. Biol. 64, 473—502 (1977).
357. Payne, T. L., and J. C. Dickens: Adaptation to determine receptor system specificity
in insect olfactory communication. J. Insect Physiol. 22, 1569—1572 (1976).
358. Payne, T. L.: Bark beetle olfaction — III. Antennal olfactory responsiveness of
Dendroctonus frontalis Zimmerman and D. brevicomis LeConte (Coleoptera:
Scolytidae) to aggregation pheromones and host tree terpene hydrocarbons. J.
Chem. Ecol. 1, 233—242 (1975).
359. Seibt, W., D. Schneider, and T. Eisner: Duftpinsel, Flügeltaschen und Balz des
Tagfalters Danaus chrysippus (Lepidoptera: Danaidae). Z. Tierpsychol. 31, 513—530
(1972).
360. Grant, G. G.: Electroantennogram responses to the scent brush secretions of
several male moths. Ann. Entomol. Soc. Am. 64, 1428—1431 (1971).

361. GRANT, G. G., U. E. BRADY, and J. M. BRAND: Male armyworm scent brush secretions: identification and electroantennogram study of major components. Ann. Entomol. Soc. Am. **65**, 1224—1227 (1972).

362. NAGAI, T., A. N. STARRATT, D. G. R. McLEOD, and G. R. DRISCOLL: Electroantennogram responses of the European corn borer, *Ostrinia nubilalis,* to (Z)- and (E)-tetradecenyl acetates. J. Insect Physiol. **23**, 591—597 (1977).

363. RENWICK, J. A. A.: Chemical aspects of bark beetle aggregation. Contrib. Boyce Thompson Inst. **24**, 337—341 (1970).

364. PAYNE, T. L., and W. E. FINN: Pheromone receptor system in the females of the greater wax moth *Galleria mellonella.* J. Insect Physiol. **23**, 879—881 (1977).

365. RENCE, B., and W. LOHER: Contact chemoreceptive sex recognition in the male cricket, *Teleogryllus commodus.* Physiol. Entomol. **2**, 225—236 (1977).

366. KOSHLAND, D. E.: A response regulator model in a simple sensory system. Science **196**, 1055—1063 (1977).

367. BORDEN, J. H.: Aggregation pheromones in the Scolytidae. In: M. C. BIRCH, Ed., Pheromones, p. 135—160. Amsterdam: North-Holland Publishing Co. 1974.

368. MOORE, B. P.: Volatile terpenes from *Nasutitermes* soldiers (Isoptera: Termitidae). J. Insect Physiol. **10**, 371—375 (1964).

369. — Studies on the chemical composition and function of the cephalic gland secretion in Australian termites. J. Insect Physiol. **14**, 33—39 (1968).

370. GRÜNANGER. P., A. QUILICO, and M. PAVAN: Sul secreto del Formicide *Myrmicaria natalensis* Fred. Accad. Nazion. Lincei **28**, 293—300 (1960).

371. HAYASHI, N., H. KOMAE, and H. HIYAMA: Monterpene hydrocarbons from ants. Z. Naturforsch. **28c**, 626 (1973).

372. BRAND, J. M., M. S. BLUM, H. A. LLOYD, and D. J. C. FLETCHER: Monoterpene hydrocarbons in the poison gland secretion of the ant *Myrmicaria natalensis* (Hymenoptera: Formicidae). Ann. Entomol. Soc. Am. **67**, 525—526 (1974).

373. RENWICK, J. A.: Identification of two oxygenated terpenes from the bark beetles *Dendroctonus frontalis* and *Dendroctonus brevicomis.* Contrib. Boyce Thompson Inst. **23**, 355—360 (1967).

374. VITÉ, J. P., A. BAKKE, and J. A. A. RENWICK: Pheromones in *Ips* (Coleoptera: Scolytidae): occurrence and production. Can. Entomol. **104**, 1967—1975 (1972).

375. HUGHES, P. R.: Effect of α-pinene exposure on *trans*-verbenol synthesis in *Dendroctonus ponderosae* Hopk. Naturwiss. **60**, 261—262 (1973).

376. — *Dendroctonus;* Production of pheromones and related compounds in response to host monoterpenes. Z. ang. Ent. **73**, 294—312 (1973).

377. BORDEN, J. H., K. K. NAIR, and C. E. SLATER: Synthetic juvenile hormone: induction of sex pheromone production in *Ips confusus.* Science **166**, 1626—1627 (1969).

378. RENWICK, J. A. A., P. R. HUGHES, and T. DEJ. TANLETIN: Oxidation products of pinene in the bark beetle *Dendroctonus frontalis.* J. Insect Physiol. **19**, 1735—1740 (1973).

379. RENWICK, J. A. A., P. R. HUGHES, and I. S. KRULL: Selective production of *cis*- and *trans*-verbenol from (−)- and (+)-α-pinene by a bark beetle. Science **191**, 199—201 (1976).

380. FRANKLIN, E. C.: Within-tree variation of monoterpene composition and yield in slash pine clones and families. Forest Sci. **22**, 185—191 (1976).

381. MIROV, N. T.: Composition of gum turpentines of pines. U. S. Forest Serv. Techn. Bul. No. 1239, p. 159 (1961).

382. RENWICK, J. A. A., and P. R. HUGHES: Oxidation of unsaturated cyclic hydrocarbons by *Dendroctonus frontalis.* Insect Biochem. **5**, 459—463 (1975).

383. VITÉ, J. P., G. B. PITMAN, A. F. FENTIMAN, JR., and G. W. KINZER: 3-Methyl-2-cyclohexen-1-ol isolated from *Dendroctonus.* Naturwiss. **59**, 469 (1972).

384. Kinzer, G. W., A. F. Fentiman, Jr., R. L. Foltz, and J. A. Rudinsky: Bark beetle attractants: 3-methyl-2-cyclohexene-1-one isolated from *Dendroctonus pseudotsugae*. J. Econ. Entomol. **64**, 970—971 (1971).

385. Rudinsky, J. A.: Masking of the aggregating pheromone in *Dendroctonus pseudotsugae* Hopk. Science **166**, 884—885 (1969).

386. Rudinsky, J. A., M. E. Morgan, L. M. Libbey, and R. R. Michael: Sound production in Scolytidae: 3-methyl-2-cyclohexene-1-one released by the female Douglas-fir beetle in response to male sonic signal. Environ. Entomol. **2**, 505—509 (1973).

387. Libbey, L. M., M. E. Morgan, T. B. Putnam, and J. A. Rudinsky: Pheromones released during inter- and intra-sex response of the scolytid beetle, *Dendroctonus brevicomis*. J. Insect Physiol. **20**, 1667—1671 (1974).

388. Rudinsky, J. A., M. E. Morgan, L. M. Libbey, and T. B. Putnam: Release of frontalin by male Douglas-fir beetle. Z. ang. Ent. **81**, 267—269 (1976).

389. Rudinsky, J. A., L. C. Ryker, R. R. Michael, L. M. Libbey, and M. E. Morgan: Sound production in Scolytidae: female sonic stimulus of male pheromone release in two *Dendroctonus* beetles. J. Insect Physiol. **22**, 1675—1681 (1976).

390. Rudinsky, J. A., M. E. Morgan, L. M. Libbey, and T. B. Putnam: Limonene released by the Scolytid beetle *Dendroctonus pseudotsugae*. Z. ang. Ent. **82**, 376—379 (1977).

391. Pitman, G. B., and J. P. Vité: Biosynthesis of methylcyclohexenone by male Douglas-fir beetle. Environ. Entomol. **3**, 886—887 (1974).

392. Hughes, P. R.: Myrcene: a precursor of pheromones in *Ips* beetles. J. Insect Physiol. **20**, 1271—1275 (1974).

393. Bakke, A.: Spruce bark beetle, *Ips typographus;* pheromone production and field response to synthetic pheromones. Naturwiss. **63**, 92 (1976).

394. Chararas, C.: Physiologie des insectes. — L'action synergique des constituants glucidiques et des constituants terpéniques dans le processus d'attraction secondaire et le mécanisme de l'élaboration des phéromones chez le Scolytidae parasites des conifères. C. R. Acad. Sci. hebd. Séances **284**, 1545—1548 (1977).

395. Hughes, P. R., and J. A. A. Renwick: Neural and hormonal control of pheromone biosynthesis in the bark beetle, *Ips paraconfusus*. Physiol. Entomol. **2**, 117—123 (1977).

396. Renwick, J. A. A., P. R. Hughes, G. B. Pitman, and J. P. Vité: Oxidation products of terpenes identified from *Dendroctonus* and *Ips* bark beetles. J. Insect Physiol. **22**, 725—727 (1976).

397. Hughes, P. R.: Pheromones of *Dendroctonus;* origin of α-pinene oxidation products present in emergent adults. J. Insect Physiol. **21**, 687—691 (1975).

398. Bridges, J. R.: Southern Forest Experiment Station, Pineville, Louisiana 71360, private communication.

399. Taskinen, J.: The acid catalysed reaction of some monoterpene alcohols in aqueous ethanol. Int. Flavours Food Addit. **7**, 235—236 (1977).

400. Gerken, B., and P. Hughes: Hormonale Stimulation der Biosynthese geschlechtsspezifischer Duftstoffe bei Borkenkäfern. Z. ang. Ent. **82**, 108—110 (1976).

401. Barth, R. H., Jr.: Hormonal control of sex attractant production in the Cuban cockroach. Science **133**, 1598—1599 (1961).

402. Barth, R. H.: The endocrine control of mating behavior in the cockroach *Byrsotria fumigata* (Guérin). Gen. Comp. Endrocrinol. **2**, 53—69 (1962).

403. Barth, R. H., Jr.: Endocrine-exocrine mediated behavior in insects. Proc. Int. Congr. Zool., 16th, 1963 Vol. 3, p. 3—5 (abstr.) (1963).

404. Barth, R. H.: Insect mating behavior: endocrine control of a chemical communication system. Science **149**, 882—883 (1965).

405. BARTH, R. H., JR.: The comparative physiology of reproductive processes in cock-roaches. Part I. Mating behavior and its endocrine control. Adv. Reprod. Physiol. **3**, 167—207 (1968).

406. — Pheromone-endocrine interactions in insects. Mem. Soc. Endocrinol. **18**, 373—404 (1970).

407. BARTH, R. H., and L. J. LESTER: Neuro-hormonal control of sexual behavior in insects. Annu. Rev. Entomol. **18**, 445—472 (1973).

408. BELL, W. J., and R. H. BARTH, JR.: Quantitative effect of juvenile hormone on reproduction in the cockroach *Byrsotria fumigata*. J. Insect Physiol. **16**, 2303—2313 (1970).

409. HUGHES, P. R.: Boyce Thompson Institute for Plant Research, 1086 N. Broadway, Yonkers, N. Y. 10701, private communication.

410. TURNQUIST, R. L., and W. A. BRINDLEY: Microsomal oxidase activities in relation to age and chlorcyclizine induction in American cockroach, *Periplaneta americana*, fat body, midgut, and hindgut. Pesticide Biochem. Physiol. **5**, 211—220 (1975).

411. FORGASH, A. J., and S. AHMAD: Hydroxylation and demethylation by gut microsomes of gypsy moth larvae. Int. J. Biochem. **5**, 11—15 (1974).

412. GILBERT, M. D., and C. F. WILKINSON: An inhibitor of microsomal oxidation from gut tissues of the honey bee *(Apis mellifera)*. Comp. Biochem. Physiol. **50 B**, 613—619 (1975).

413. BRATTSTEN, L. B., C. F. WILKINSON, and T. EISNER: Herbivore-plant interactions: mixed function oxidases and secondary plant substances. Science **196**, 1349—1352 (1977).

414. BAKER, J. E.: Substrate specificity in the control of digestive enzymes in larvae of the black carpet beetle. J. Insect Physiol. **23**, 749—753 (1977).

415. HENRY, S. M., Ed.: Symbiosis, Vol. II. 443 p. New York: Academic Press. 1967.

416. BLUM, M. S., and J. M. BRAND: Social insect pheromones; their chemistry and function. Am. Zoologist **12**, 553—576 (1972).

417. KAMIŃSKI, E., L. M. LIBBEY, S. STAWICKI, and E. WASOWICZ: Identification of the predominant volatile compounds produced by *Aspergillus flavus*. Appl. Microbiol. **24**, 721—726 (1972).

418. CAVILL, G. W. K., and H. HINTERBERGER: The chemistry of ants. IV. Terpenoid constituents of some *Dolichoderus* and *Iridomyrmex* species. Aust. J. Chem. **13**, 514—519 (1960).

419. DUFFIELD, R. M., J. M. BRAND, and M. S. BLUM: 6-Methyl-5-hepten-2-one in *Formica* species: identification and function as an alarm pheromone (Hymenoptera: Formicidae). Ann. Entomol. Soc. Am. **3**, 309—310 (1977).

420. SCHILDKNECHT, H., D. BERGER, D. KRAUSS, J. CONNERT, J. GEHLHAUS, and H. ESSENBREIS: Defense chemistry of *Stenus comma* (Coleoptera: Staphylinidae). LXI. J. Chem. Ecol. **2**, 1—11 (1976).

421. MILLER, M. W.: The Pfizer handbook of microbial metabolites, 772 p. New York: McGraw-Hill Book Co., Inc. 1961.

422. SHIBATA, S., S. NATORI, and S. UDAGAWA: List of fungal products, 170 p. Tokyo: University of Tokyo Press. 1964.

423. BRAND, J. M., and S. J. BARRAS: The major volatile constituents of a Basidiomycete associated with the southern pine beetle. Lloydia **40**, 398—400 (1977).

424. KATAYAMA, T.: Volatile constituents of algae. XX. Pharmacological action of volatile constituents and biochemical significance of the existence of acrylic acid. Chem. Abstr. **62**, 10822h (1965).

425. BLUM, M. S., and G. E. BOHART: Neral and geranial: identification in a Colletid bee. Ann. Entomol. Soc. Am. **65**, 274—275 (1972).

426. COLLINS, R. P., and A. F. HALIM: Production of monoterpenes by the filamentous fungus *Ceratocystis variospora*. Lloydia **33**, 481—482 (1970).

427. Hoyt, C. P., G. O. Osborne, and A. P. Nulcock: Production of an insect sex attractant by symbiotic bacteria. Nature 230, 472—473 (1971).
428. Matsumura, F., H. C. Coppel, and A. Tai: Isolation and identification of termite trail following pheromone. Nature 219, 963—964 (1968).
429. Howard, R., F. Matsumura, and H. C. Coppel: Trail-following pheromones of the Rhinotermitidae: approaches to their authentication and specificity. J. Chem. Ecol. 2, 147—166 (1976).
430. Matsumura, F., K. Nishimoto, T. Ikeda, and H. C. Coppel: Influence of carbon sources on the production of the termite trail-following substance by Gloephyllum trabeum. J. Chem. Ecol. 2, 299—305.
431. Eckenrode, C. J., G. E. Harman, and D. R. Webb: Seed-borne microorganisms stimulate seedcorn maggot egg laying. Nature 256, 487—488 (1975).
432. Moeck, H. A.: Ethanol as the primary attractant for the ambrosia beetle Trypodendron lineatum (Coleoptera: Scolytidae). Can. Entomol. 102, 985—995 (1970).
433. Pitman, G. B., R. L. Hedden, and R. I. Gara: Synergistic effects of ethyl alcohol on the aggregation of Dendroctonus pseudotsugae (Col., Scolytidae) in response to pheromones. Z. ang. Eng. 78, 203—208 (1975).
434. Renwick, J. A. A., J. P. Vité, and R. F. Billings: Aggregation pheromones in the ambrosia beetle Platypus flavicornis. Naturwiss. 64, 226 (1977).
435. Rudinsky, J. A.: Multiple functions of the southern pine beetle pheromone, verbenone. Environ. Entomol. 2, 511—514 (1973).
436. Fonken, G. S., and R. A. Johnson: Chemical oxidations with microorganisms, 292 p. New York: Marcel Dekker, Inc. 1972.
437. Bhattacharyya, P. K., B. R. Prema, B. D. Kulkarni, and S. K. Pradhan: Microbiological transformation of terpenes: hydroxylation of α-pinene. Nature 187, 689—690 (1960).
438. Prema, B. R., and P. K. Bhattacharyya: Microbiological transformation of terpenes. II. Transformation of α-pinene. Appl. Microbiol. 10, 524—528 (1962).
439. Brand, J. M., J. W. Bracke, A. J. Markovetz, D. L. Wood, and L. E. Browne: Production of verbenol pheromone by a bacterium isolated from bark beetles. Nature 254, 136—137 (1975).
440. Barras, S. J., and T. Perry: Fungal symbionts in the prothoracic mycangium of Dendroctonus frontalis (Coleoptera: Scolytidae). Z. ang. Ent. 71. 95—104 (1972).
441. Barras, S. J., and J. J. Taylor: Varietal Ceratocystis minor identification from mycangium of Dendroctonus frontalis. Mycopath. Mycol. Applic. 50, 293—305 (1973).
442. Brand, J. M., J. W. Bracke, L. N. Britton, A. J. Markovetz, and S. J. Barras: Bark beetle pheromones: production of verbenone by a mycangial fungus of Dendroctonus frontalis. J. Chem. Ecol. 2, 195—199 (1976).
443. Renwick, J. A. A., and J. P. Vité: Systems of chemical communication in Dendroctonus. Contrib. Boyce Thompson Inst. 24, 283—292 (1970).
444. Vité, J. P., and W. Francke: The aggregation pheromones of bark beetles: progress and problems. Naturwiss. 63, 550—555 (1976).
445. Suomalainen, H.: Trends in physiology and biochemistry of yeasts. Antonie van Leeuwenhoek, Vol. 35, Supplement: Yeast Symposium, p. 83—111 (1969).
446. Brand, J. M., J. Schultz, S. J. Barras, L. J. Edson, T. L. Payne, and R. L. Hedden: Bark beetle pheromones: enhancement of Dendroctonus frontalis (Coleoptera: Scolytidae) aggregation pheromone by yeast metabolites in laboratory bioassays. J. Chem. Ecol. 3, 657—666 (1977).
447. Renwick, J. A. A., G. B. Pitman, and J. P. Vité: 2-Phenylethanol isolated from bark beetles. Naturwiss. 63, 198 (1976).
448. Payne, T. L., E. R. Hart, L. J. Edson, F. A. McCarthy, P. M. Billings, and

J. E. Coster: Olfactometer for assay of behavioral chemicals for the southern pine beetle, *Dendroctonus frontalis* (Coleoptera: Scolytidae). J. Chem. Ecol. **2,** 411—419 (1976).

449. Goeden, R. D., and D. M. Norris: Some biological and ecological aspects of ovipositional attack in *Carya* spp. by *Scolytus quadrispinosus* (Coleoptera: Scolytidae). Ann. Entomol. Soc. Am. **58,** 771—777 (1965).

450. Greany, P. D., J. H. Tumlinson, D. L. Chambers, and G. M. Boush: Chemically mediated host finding by *Biosteres (Opius) longicaudatus*, a parasitoid of tephritid fruit fly larvae. J. Chem. Ecol. **3,** 189—195 (1977).

451. Fales, H. M., M. S. Blum, R. M. Crewe, and J. M. Brand: Alarm pheromones in the genus *Manica* derived from the mandibular gland. J. Insect Physiol. **18,** 1077—1088 (1972).

452. Meinwald, J., A. F. Kluge, J. F. Carrel, and T. Eisner: Acyclic ketones in the defensive secretion of a "daddy long legs" *(Leiobunum vittatum)*. Proc. Natl. Acad. Sci. USA **68,** 1467—1468 (1971).

453. Gore, W. E., G. T. Pearce, G. N. Lanier, J. B. Simeone, R. M. Silverstein, J. W. Peacock, and R. A. Cuthbert: Aggregation attractant of the European elm bark beetle, *Scolytus multistriatus;* production of individual components and related aggregation behavior. J. Chem. Ecol. **3,** 429—446 (1977).

454. Tumlinson, J. H., D. D. Hardee, J. P. Minyard, A. C. Thompson, R. T. Gast, and P. A. Hedin: Boll weevil sex attractant: isolation studies. J. Econ. Entomol. **61,** 470—474 (1968).

455. Tumlinson, J. H., R. C. Gueldner, D. D. Hardee, A. C. Thompson, P. A. Hedin, and J. P. Minyard: The boll weevil sex attractant. In: M. Beroza, Ed., Chemicals controlling insect behavior, p. 41—59. New York: Academic Press. 1970.

456. Hardee, D. D.: Pheromone production by male boll weevil as affected by food and host factors. Contrib. Boyce Thompson Inst. **24,** 315—322 (1970).

457. Mitlin, N., and P. A. Hedin: Biosynthesis of grandlure, the pheromone of the boll weevil, *Anthonomus grandis*, from acetate, mevalonate, and glucose. J. Insect Physiol. **20,** 1825—1831 (1974).

458. Hedin, P. A.: A study of factors that control biosynthesis of the compounds which comprise the boll weevil pheromone. J. Chem. Ecol. **3,** 279—289 (1977).

459. Gueldner, R. C., P. P. Sikorowski, and J. M. Wyatt: Bacterial load and pheromone production in the boll weevil, *Anthonomus grandis*. J. Invert. Pathol. **29,** 397—398 (1977).

460. Gordon, H. T., D. F. Waterhouse, and A. R. Gilbey: Incorporation of ^{14}C-acetate into scent constituents by the green vegetable bug. Nature **197,** 818 (1963).

461. Happ, G. M., and J. Meinwald: Biosynthesis of arthropod secretions. I. Monoterpene synthesis in an ant. J. Am. Chem. Soc. **87,** 2507—2508 (1965).

462. Leyrer, R. L., and R. E. Monroe: Isolation and identification of the scent of the moth, *Galleria mellonella*, and a revaluation of its sex pheromone. J. Insect Physiol. **19,** 2267—2271 (1973).

463. Schmidt, S. P., and R. E. Monroe: Biosynthesis of the waxmoth sex attractants. Insect Biochem. **6,** 377—380 (1976).

464. Yendol, W. G.: Fatty acid composition of *Galleria* larvae, hemolymph, diet (Lepidoptera: Galleriidae). Ann. Entomol. Soc. Am. **63,** 339—341 (1970).

465. Clearwater, J. R.: Pheromone metabolism in male *Pseudoletia separata* (Walk.) and *Mamestra configurata* (Walk.) (Lepidoptera: Noctuidae). Comp. Biochem. Physiol. **50 B,** 77—82 (1975).

466. Weatherston, J., and J. E. Percy: The biosynthesis of phenethyl alcohol in the male bertha army worm, *Mamestra configurata*. Insect Biochem. **6,** 413—417 (1976).

467. Ayrapaa, T.: Formation of phenethyl alcohol from ^{14}C-labelled phenylalanine. J. Inst. Brewing **71,** 341—347 (1965).

468. DREWS, B., H. SPECHT, and E. SCHWARTZ: Über die bei der alkoholischen Gärung der Hefe entstehenden aromatischen Alkohole. Monatsschr. Brau. **19**, 76—87 (1966).
469. LINGAPPA, B. T., M. PRASAD, and Y. LINGAPPA: Phenethyl alcohol and tryptophol: autoantibiotics produced by the fungus *Candida albicans*. Science **163**, 192—193 (1969).
470. ROSAZZA, J. P., R. JUHL, and P. DAVIS: Tryptophol formation by *Zygosaccharomyces priorianus*. Appl. Microbiol. **26**, 98—105 (1973).
471. CASNATI, G., A. RICCA, and M. PAVAN: Sulla secrezione defensiva delle glandole mandibulari di *Paltothyreus tarsatus*. Chemica Ind., Milano **49**, 57—58 (1967).
472. CREWE, R. M., and D. J. C. FLETCHER: Ponerine ant secretions: the mandibular gland secretion of *Paltothyreus tarsatus*. J. Entomol. Soc. Sth. Afr. **37**, 291—298 (1974).
473. CREWE, R. M., and F. P. ROSS: Biosynthesis of alkyl sulphides by an ant. Nature **254**, 448—449 (1975).
474. — — Pheromone biosynthesis: the formation of sulphides by the ant *Paltothyreus tarsatus*. Insect Biochem. **5**, 839—843 (1975).
475. HENDRY, L. B., J. K. WICHMANN, D. M. HINDENLANG, R. O. MUMMA, and M. E. ANDERSON: Evidence for origin of insect sex pheromone: presence in food plants. Science **188**, 59—63 (1975).
476. HENDRY, L. B.: Insect pheromones: diet related. Science **192**, 143—145 (1976).
477. MILLER, J. R., T. C. BAKER, R. T. CARDÉ, and W. L. ROELOFS: Reinvestigation of oak leaf roller sex pheromone components and the hypothesis that they vary with diet. Science **192**, 140—143 (1976).
478. HINDENLANG, D. M., and J. K. WICHMANN: Reexamination of tetradecenyl acetates in oak leaf roller sex pheromone and in plants. Science **195**, 86—89 (1977).
479. MINKS, A. K.: Decreased sex pheromone production in an in-bred stock of the summer-fruit tortrix moth, *Adoxophyes orana*. Ent. exp. & Appl. **14**, 361—364 (1971).
480. RICHERSON, J. V., and E. A. CAMERON: Differences in pheromone release and sexual behavior between laboratory-reared and wild gypsy moth adults. Environ. Entomol. **3**, 475—481 (1974).
481. SUGIE, H., S. YAMAZAKI, and Y. TAMAKI: On the origin of the sex pheromone components of the smaller tea tortrix moth, *Adoxophyes fasciata* Walsingham (Lepidoptera: Tortricidae). Appl. Ent. Zool. **11**, 371—373 (1976).
482. PLISKE, T. E., and T. EISNER: Sex pheromone of the queen butterfly: biology. Science **154**, 1170—1172 (1969).
483. EDGAR, J. A., and C. C. J. CULVENOR: Pyrrolizidine alkaloids in *Parsonsia* species (family Apocynaceae) which attract Danaid butterflies. Experientia **31**, 393—394 (1975).
484. EDGAR, J. A., C. C. J. CULVENOR, and G. S. ROBINSON: Hairpencil dihydropyrrolizines of Danainae from the New Hebrides. J. Aust. Entomol. Soc. **12**, 144—150 (1973).
485. SCHNEIDER, D., M. BOPPRÉ, H. SCHNEIDER, W. R. THOMPSON, C. J. BORIACK, R. L. PETTY, and J. MEINWALD: A pheromone precursor and its uptake in male *Danaus* butterflies. J. Comp. Physiol. **97**, 245—256 (1975).
486. EDGAR, J. A., C. C. J. CULVENOR, and T. E. PLISKE: Isolation of a lactone, structurally related to the esterifying acids of pyrrolizidine alkaloids, from the costal fringes of male Ithomiinae. J. Chem. Ecol. **2**, 263—270 (1976).
487. NOLTE, D. J.: A pheromone for melanization of locusts. Nature **200**, 660—661 (1963).
488. NOLTE, D. J., S. H. EGGERS, and I. R. MAY: A locust pheromone: locustol. J. Insect Physiol. **19**, 1547—1554 (1973).
489. NOLTE, D. J.: Locustol and its analogues. J. Insect Physiol. **22**, 833—838 (1976).
490. MAY, I. R.: Physiological aspects of gregarising locusts. Ph. D. Thesis, University of Witwatersrand, Johannesburg, South Africa (1973).

491. NOLTE, D. J.: The action of locustol. J. Insect Physiol. 23, 899—903 (1977).
492. VRKOC, J., K. UBIK, J. ZDAREK, and C. KONTEV: Ethyl acrylate and vanillin as components of the male sex pheromone complex in *Eurygaster integriceps* (Heteroptera: Scutelleridae). Acta ent. bohemoslov. 74, 205—206 (1977).
493. LANIER, G. N., and W. E. BURKHOLDER: Pheromones in speciation of Coleoptera. In: M. C. BIRCH, Ed., Pheromones, p. 161—189. Amsterdam: North-Holland Publishing Co. 1974.
494. VICK, K. W., W. E. BURKHOLDER, and J. E. GORMAN: Interspecific response of sex pheromones of *Trogoderma* species (Coleoptera: Dermestidae). Ann. Entomol. Soc. Am. 63, 379—381 (1970).
495. LEVINSON, H. Z., and A. R. BAR ILAN: Olfactory and tactile behavior of the khapra beetle, *Trogoderma granarium,* with special reference to its assembling scent. J. Insect Physiol. 16, 561—572 (1970).
496. GREENBLATT, R. E., W. E. BURKHOLDER, J. H. CROSS, R. F. CASSIDY, R. M. SILVERSTEIN, A. R. LEVINSON, and H. Z. LEVINSON: Chemical basis for interspecific responses to sex pheromones of *Trogoderma* species (Coleoptera: Dermestidae). J. Chem. Ecol. 3, 337—347 (1977).
497. YARGER, R. G., R. M. SILVERSTEIN, and W. E. BURKHOLDER: Sex pheromone of the female dermestid beetle *Trogoderma glabrum* (Herbst). J. Chem. Ecol. 1, 323—334 (1975).
498. RODIN, J. O., R. M. SILVERSTEIN, W. E. BURKHOLDER, and J. E. GORMAN: Sex attractant of female dermestid beetle *Trogoderma inclusum* Le Conte. Science 165, 904—906 (1969).
499. CROSS, J. H., R. C. BYLER, R. M. SILVERSTEIN, R. E. GREENBLATT, J. E. GORMAN, and W. E. BURKHOLDER: Sex pheromone components and calling behavior of the female dermestid beetle, *Trogoderma variabile* Ballion (Coleoptera: Dermestidae). J. Chem. Ecol. 3, 115—125 (1977).
500. CARDÉ, R. T., A. M. CARDÉ, A. S. HILL, and W. L. ROELOFS: Sex pheromone specificity as a reproductive isolating mechanism among sibling species *Archips argyrospilus* and *A. mortuanus* and other sympatric tortricine moths (Lepidoptera: Tortricidae). J. Chem. Ecol. 3, 71—84 (1977).
501. STECK, W., E. W. UNDERHILL, and M. D. CHISHOLM: Attraction and inhibition in moth species responding to sex-attractant lures containing Z-11-hexadecen-1-yl acetate. J. Chem. Ecol. 3, 603—612 (1977).
502. KLUN, J. A.: Organic Chemicals Synthesis Laboratory, Room 311, Bldg. 306, ARC-East, USDA, ARS, Beltsville, MD 20705, private communication and presentation at Entomological Society of America Meeting, Washington, D.C., November 1977.
502a. CARDÉ, R. T., W. L. ROELOFS, R. G. HARRISON, A. T. VAWTER, P. F. BRUSSARD, A. MUTUURA, and E. MUNROE: European corn borer: pheromone polymorphism or sibling species? Science 199, 555—556 (1978).
503. HÖLLDOBLER, B., and U. MASCHWITZ: Der Hochzeitsschwarm der Rossameisen *Camponotus herculeanus* L. (Hym. Formicidae). Z. vergl. Physiol. 50, 551—568 (1965).
504. BRAND, J. M., R. M. DUFFIELD, J. G. MACCONNELL, M. S. BLUM, and H. M. FALES: Caste-specific compounds in male carpenter ants. Science 179, 388—389 (1973).
505. BRAND, J. M., H. M. FALES, E. A. SOKOLOSKI, J. G. MACCONNELL, M. S. BLUM, and R. M. DUFFIELD: Identification of mellein in the mandibular gland secretions of carpenter ants. Life Sciences 13, 201—211 (1973).
506. DUFFIELD, R. M.: A comparative study of the mandibular gland chemistry of formicine and ponerine ant species. Ph. D. Thesis, University of Georgia, Áthens, GA 30602 (1976).

507. Flint, M. L., and R. van den Bosch: A Source Book on Integrated Pest Management. International Center for Integrated and Biological Control, University of California, Berkeley. 1977.

508. Dethier, V. G.: Man's Plague? Insects and Agriculture. Princeton, New Jersey: Darwin Press. 1976.

509. Carson, R.: Silent Spring. Boston: Houghton Mifflin Co. 1962.

510. Stark, R. W., and A. R. Gittins, Eds.: Pest Management for the 21st Century. Idaho Research Foundation, Inc., University Station, Box 3367, Moscow, ID 83843. 1973.

511. Ayeres, J. H., and 11 co-authors: New Innovative Pesticides: An Evaluation of Incentives and Disincentives for Commercial Development by Industry. Environmental Protection Agency, Office of Pesticide Programs, Criteria, and Evaluation Division, Washington, D. C. 20460. 1977.

512. Franz, J. M.: Toward integrated control of forests pests in Europe. In: J. F. Anderson and H. K. Kaya, Eds., Perspectives in Forest Entomology. New York: Academic Press. 1976.

513. Pest Control Strategies—Understanding and Action. Symposium Proceedings, June 22—23, 1977, Cornell University, Ithaca, N. Y.

514. Campbell, R. W., and M. W. McFadden: Design of a pest management research and development program. Bull. Entomol. Soc. 23, 216—220 (1977).

515. Wood, D. L., and 33 collaborators: Integrated pest management of the western pine beetle. Research approach and principal results. In: C. B. Huffaker, Ed., New Technology of Pest Control. New York: J. Wiley. In press.

516. Good, J. M.: Progress Report, Pest Management Pilot Projects, NAR-5-21 (4/77). Pest Management Programs, Extension Service, USDA, Washington, D. C. 20250. 1977.

517. — Establishing and Operating Grower-Owned Organizations for Integrated Pest Management, PA-1180. Extension Service, USDA, Washington, D. C. 20250. 1977.

518. Koehler, C. S., J. J. McKelvey, Jr., W. L. Roelofs, H. H. Shorey, R. M. Silverstein, and D. L. Wood: Advancing toward operational behavior-modifying chemicals. In: H. H. Shorey and J. J. McKelvey, Jr., Eds., Chemical control of Insect Behavior, Theory and Application. New York: Wiley-Interscience. 1977.

519. Daterman, G. E.: Forestry Science Laboratory, P. O. Box 887, Corvallis, Ore., 97330.

520. Phillips, W. G.: EPA's Registration Requirements for Insect Behavior Controlling Chemicals—Philosophy and Mandates. In: M. Beroza, Ed., Pest Management with Insect Sex Attractants. ACS Symposium Series, No. 23, American Chemical Society, Washington, D. C. 1976.

521. Djerassi, C., C. Shih-Coleman, and J. Diekman: Insect control of the future: Operational and policy aspects. Science 186, 596 (1974).

522. Siddall, J. B., and C. M. Olsen: Pheromones in Agriculture—from Chemical Synthesis to Commercial Use. In: M. Beroza, Ed., Pest Management with Insect Sex Attractants. ACS Symposium Series, No. 23, American Chemical Society, Washington, D. C. 1976.

523. Gary, N. E.: Chemical mating attractants in the honey bee. Science 136, 773—774 (1962).

524. Butler, C. G., and E. M. Fairey: Pheromones of the honeybee: biological studies of the mandibular gland secretion of the queen. J. Apic. Res. 3, 65—76 (1964).

525. Silverstein, R. M., D. L. Wood, and J. O. Rodin: Abstracts 16th Annual Meeting Entomological Society of Canada, Sept. 1966, Banff, Canada.

526. Wood, D. L., R. W. Stark, R. M. Silverstein, and J. O. Rodin: Unique synergistic effect produced by the principal sex attractant compunds of Ips confusus (Le Conte) [now I. paraconfusus] (Coleoptera: Scolytidae). Nature 215, 206 (1967).

527. BIRCH, M., and 10 co-authors: Programs utilizing pheromones in survey or control. In: M. BIRCH, Ed., Pheromones. New York: North-Holland, American Elsevier. 1974.

528. WOOD, D. L.: Manipulation of forest insect pests. In: H. H. SHOREY and J. J. MCKELVEY, Jr., Eds., Chemical Control of Insect Behavior. New York: J. Wiley. 1977.

529. ROELOFS, W. L.: Chemical control of insects by pheromones. In: M. ROCKSTEIN, Ed., Insect Biochemistry. New York: Academic Press. In press.

530. MITCHELL, E. R.: Recent advances in the use of sex pheromones for control of insect pests. In: Proceedings, 1977 International Controlled Release Pesticide Symposium, Aug. 22—24, 1977, Corvallis, Oregon.

531. MINKS, A. K.: Trapping with behavior-modifying chemicals: feasibility and limitations. In: H. H. SHOREY and J. J. MCKELVEY, JR., Eds., Chemical Control of Insect Behavior. Theory and Application. New York: Wiley-Interscience. 1977.

532. BONESS, M., K. EITER, and H. DISSELNKÖTTER: Untersuchung über Sexuallockstoff von Lepidopteran und ihre Verwendung im Pflanzenschutz. Pflanzenschutz-Nachrichten Bayer 30, 212—234 (1977).

533. BONESS, M.: Bayer AG, 509 Leverkusen, BRD, private communication.

534. WOOD, D. L., and W. D. BEDARD: The role of pheromones in the population dynamics of the western pine beetle. In: Proceedings of the XV International Congress of Entomology, Washington, D. C. Entomological Society of America, 4603 Calvert Road, Box AJ, College Park, MD 20742. 1976.

535. BEDARD, W. D., D. L. WOOD, and P. E. TILDEN: Use of behavior modifying chemicals to lower western pine beetle-caused tree mortality and to protect trees. In: W. E. WATERS, Ed., Forest Insect Ecology and Management. U. S. Forest Service, Washington, D. C. In press.

536. BEDARD, W. D.: U. S. Forest Service, P. O. Box 245, Berkeley, CA 94720, private communication.

537. VITÉ, J. P.: Möglichkeiten und Grenzen der Pheromonanwendung in der Borkenkäferbekämpfung. Zeit. angew. Entomol. 77, 325—329 (1974).

538. WOOD, D. L.: Application of pheromones for manipulating forest insect pests. In: Proceedings of a Symposium on Insect Pheromones and their Applications. Nagaoka and Tokyo, Japan, December 8—11, 1976. 1977.

539. MCLEAN, J. A., and J. H. BORDEN: Survey for Gnathotrichus sulcatus (Coleoptera: Scolytidae) in a commercial sawmill with the pheromone, sulcatol. Canadian J. Forest Research 5, 586—591 (1975).

540. — — Suppression of Gnathotrichus sulcatus with sulcatol-baited traps in a sawmill and notes on the occurrence of G. retusus and Trypodendron lineatum. Can. J. Forest Research 7, 348—356 (1977).

541. PEACOCK, J. W., R. A. CUTHBERT, W. E. GORE, G. N. LANIER, G. T. PEARCE, and R. M. SILVERSTEIN: Collection on Porapak Q of the aggregating pheromone of Scolytus multistriatus (Coleoptera: Scolytidae). J. Chem. Ecol. 1, 149—160 (1975).

542. LANIER, G. N., R. M. SILVERSTEIN, and J. W. PEACOCK: Attractant pheromone of the European elm bark beetle (Scolytus multistriatus); Isolation, identification, synthesis, and utilization studies. In: J. F. ANDERSON and H. K. KAYA, Eds., Perspectives in Forest Entomology. New York: Academic Press. 1976.

543. PEACOCK, J. W., and R. A. CUTHBERT: U. S. Forest Service, Box 365, Delaware, OH 43015, private communication.

544. ARCIERO, M.: Use of multilure-baited traps in the California Dutch Elm Disease Program for Survey and Detection of Scolytus multistriatus. Presented at 25 th Annual Meeting of the Entomological Society of America, Nov. 27—Dec. 1, 1977, Washington, D. C.

545. FURNISS, M. M., G. E. DATERMAN, L. N. KLINE, M. D. MCGREGOR, G. C. TROSTLE,

L. F. Pettinger, and J. A. Rudinsky: Effectiveness of the Douglas-fir beetle anti-aggregative pheromones methylcyclohexanone at three concentrations and spacings around felled host trees. Can. Entomol. **108**, 381—392 (1974).

546. Furniss, M. M., J. W. Young, M. D. McGregor, R. L. Livington, and D. R. Hamel: Effectiveness of controlled-release formulations of MCH for preventing Douglas-fir beetle infestations in felled trees. Can. Entomol. **104**, 1063—1069 (1977).

547. Furniss, M. M.: Forestry Science Laboratory, 1221 S. Main, Moscow, ID 83843, private communication.

548. Knopf, J. A. E., and G. B. Pitman: Aggregation pheromone for manipulation of the Douglas-fir beetle. J. Econ. Entomol. **65**, 723—726 (1972).

549. Payne, T. L., J. E. Coster, and P. C. Johnson: Effects of a slow-release formulation of synthetic *endo-* and *exo-*brevicomin on southern pine beetle flight and landing behavior. J. Chem. Ecol. **3**, 133—141 (1977).

550. Richerson, J. V.: Dept. of Entomology, Texas A & M University, College Station, TX 77843, private communication.

551. Vité, J. P.: Silviculture and the management of bark beetle pests. Proc. Tall Timbers Conf. on Ecological Animal Control by Habitat Management **3**, 155—168 (1971).

552. Copony, J. A., and C. L. Morris: Southern pine beetle suppression with frontalure and cacodylic acid treatment. J. Econ. Entomol. **65**, 754—757 (1972).

553. Dyer, E. D. A.: Spruce beetles aggregated by the synthetic pheromone frontalin. Can. J. Forest Res. **3**, 486—494 (1973).

554. Dyer, E. D. A., and L. Safranyik: Assessment of the impact of pheromone-baited trees on a spruce beetle population (Coleoptera: Scolytidae). Can. Entomol. **109**, 77—80 (1977).

555. Daterman, G. E., G. D. Daves, Jr., and R. G. Smith: Comparison of sex pheromone versus an inhibitor for disruption of pheromone communication in *Rhyacionia buolania*. Envir. Entomol. **4**, 944—946 (1975).

556. Leonard, D. E.: Recent developments in ecology and control of the gypsy moth. Annu. Rev. Entomol. **19**, 197—229 (1974).

557. Cameron, E. A.: Disparlure: a potential tool for gypsy moth population manipulation. Bull. Entomol. Soc. Amer. **19**, 15—19 (1973).

558. Cameron, E. A., and V. C. Mastro: Effectiveness of disparlure in combination with Sevin 4 Oil for gypsy moth suppression in Pennsylvania 1974. Forest Insect and Disease Management/evaluation report. Northeastern Area State and Private forestry, USDA Forest Service, Upper Darby, PA 19082. 1976.

559. Cameron, E. A.: Integrated control and the role of chemical ecology: the gypsy moth (Lepidoptera). Paper presented to the 46th annual Meeting of the Eastern Branch, Entomol. Soc. Am., Hershey, Pa., Sept. 25—27 (1974).

560. Cameron, E. A., C. P. Schwalbe, L. J. Stevens, and M. Beroza: Field tests of the olefin precursor of disparlure for suppression of mating in the gypsy moth. J. Econ. Entomol. **68**, 158—160 (1975).

561. Cardé, R. T., W. L. Roelofs, and C. C. Doane: Natural inhibitor of the gypsy moth sex attractant. Nature **241**, 474—475 (1973).

562. Beroza, M.: Control of the Gypsy moth and other insects with behavior-modifying chemicals. In: M. Beroza, Ed., Pest Management with insect Sex Attractants and other behavior-controlling chemicals. ACS Symposium Series No. 23. American Chemical Society, Washington, D. C. 1976.

563. Richerson, J. V.: Pheromone-mediated behavior of the gypsy moth. J. Chem. Ecol. **3**, 291—308 (1977).

564. Cameron, E. A.: Use of disparlure for gypsy moth mating disruption. Presented at the Eastern Branch Meeting, Entomological Society of America, September 14—16, 1977, Boston, Mass.

565. PLIMMER, J. R., B. A. BIERL, R. E. WEBB, and C. P. SCHWALBE: Controlled release of pheromone in gypsy moth programs. In: H. B. SCHER, Ed., Controlled release pesticides. ACS Symposium Series No. 53. American Chemical Society, Washington, D. C. 1977.

566. PLIMMER, J. R., J. H. CARO, and H. P. FREEMAN: Distribution and dissipation of aerially applied disparlure under a woodland canopy. Envir. Entomol., in press.

567. PLIMMER, J. R.: Agricultural Research Service, Beltsville, MD 20705, private communication.

568. BONESS, M.: Disparlure: Comparison of effectiveness in *Lymantria dispar* and *Lymantria monacha* as shown in field tests. VIII International Plant Protection Congress, Section V, Biological and Genetic Control, 41—47 (1975).

569. SANDERS, C. J.: Forest lepidoptera—the spruce budworm. In: M. BIRCH, Ed., Pheromones. New York: North-Holland/American Elsevier. 1974.

570. — Canadian Forestry Service, Great Lakes Forest Research Centre, Sault Ste. Marie, Ontario, Canada P6A 5M7, private communication.

571. SOWER, L. L., and G. E. DATERMAN: Evaluation of synthetic sex pheromone as a control agent for Douglas-fir tussock moth. Envir. Entomol. **6,** 889—892 (1977).

572. DATERMAN, G. E., and L. L. SOWER: Douglas-fir tussock moth pheromone research using controlled-release systems. In: Proceedings 1977 Controlled Release Pesticide Symposium. Oregon State Univ. Corvallis, Oreg.

573. ROELOFS, W. L., R. T. CARDÉ, E. F. TASCHENBERG, and R. W. WEIRES, JR.: Pheromone research for the control of lepidopterous pests in New York. In: M. BEROZA, Ed., Pest Management with Insect Sex Attractants and other behavior-controlling chemicals. ACS Symposium Series No. 23. American Chemical Society, Washington, D. C. 1976.

574. CARDÉ, R. T.: Utilization of pheromone in the population management of moth pests. Environmental Health Perspectives **14,** 133—144 (1976).

575. TETTE, J. P., E. H. GLASS, J. L. BRANN, JR., and P. A. ARNESON: A brief summary of the New York State Apple Pest Management—Application Project 1973—1975. New York State Agricultural Experiment Station, Geneva, N. Y. 1976.

576. TETTE, J. P., and E. H. GLASS: A progress survey of New York State Fruit Pest Management Project. New York State Agricultural Experiment Station, Geneva, N. Y. 1977.

577. TETTE, J. P.: N. Y. State Agricultural Experiment Station, Geneva, N. Y. 14456, private communication.

578. TASCHENBERG, E. F., and W. L. ROELOFS: Male redbanded leafroller moth: orientation disruption in vineyards. Envir. Entomol. **7,** 103—106 (1978).

579. TUMLINSON, J. H., E. R. MITCHELL, and D. L. CHAMBERS: Manipulating complexes of insect pests with various combinations of behavior-modifying chemicals. In: M. BEROZA, Ed., Pest Management with Insect Sex Attractants and other behavior-controlling chemicals. ACS Symposium Series No. 23. American Chemical Society, Washington, D. C. 1976.

580. ROELOFS, W. L.: Communication disruption by pheromone components. In: Insect Pheromones and their Applications. Symposium Proceedings, Nagaoka and Tokyo, Japan, Dec. 8—11, 1976. National Institute of Agricultural Sciences, Nishigahara, Kita-ku, Tokyo 114, Japan. 1977.

581. MADSEN, H. F., and J. M. VAKENTI: Codling moth: use of Codlemone-baited traps and visual detection of entries to determine need of sprays. Envir. Entomol. **2,** 677—679 (1973).

582. MADSEN, G. F., J. M. VAKENTI, and F. E. PETERS: Codling moth: suppression by male removal with sex pheromone traps in an isolated apple orchard. J. Econ. Entomol. **69,** 597—599 (1976).

583. Audemard, H., and H. G. Milaire: Le piégage du carpocapse *(Laspeyresia pomonelle*
L.) avec une pheromone sexuelle de synthèse: premiers resultats utilisables pour
l'estimation des populations et la conduite de la lutte. Ann. Zool.-Ecol. anim. **7,**
61—80 (1975).

584. Audemard, H., F. Beauvais, and C. Descoins: La lutte contre le carpocapse
(Laspeyresia pomonella L.) avec une pheromone sexuelle de synthèse par la méthode
de confusion des males: premier essay en verger commercial de pommiers. Rev. Zool.
Agric. et Pathol. Vég. **76,** 15—24 (1977).

585. Madsen, H. F., and F. E. Peters: Pest management: monitoring populations of
Archips argyrospilus and *Archips rosanus* (Lepidoptera: Tortricidae) with sex pheromone traps. Can. Entomol. **108,** 1281—1284 (1976).

586. Moffit, H. R.: USDA, ARS, 3706 Nob Hill Blvd., Yakima, WA 98902, private
communication.

587. Minks, A. K.: Experience with pheromone trapping and disruption of moths.
Pestic. Sci. **7,** 642—646 (1976).

588. Minks, A. K., S. Voerman, and J. A. Klun: Disruption of pheromone communication
with micro-encapsulated antipheromones against *Adoxophyes orana.* Ent. exp. & appl.
20, 163—169 (1976).

589. Arn, H., B. Delley, M. Bagiolini, and P. J. Charmillot: Communication disruption
with sex attractant for control of the plum fruit moth, *Grapholitha funebrans;* A two-
year field study. Entomol. Exp. Appl. **19,** 139—147 (1976).

590. Neilson, W. T., A. I. Rivard, R. Trottier, and R. J. Whitman: Pherocon AM
standard traps and their use to determine spray dates for control of the apple maggot.
J. Econ. Entomol. **69,** 527—532 (1976).

591. Chambers, D. L.: Attractants for fruit fly survey and control. In: H. H. Shorey
and J. J. McKelvey, Jr., Eds., Chemical Control of Insect Behavior. Theory and
Application. New York: J. Wiley. 1977.

592. Hedin, P. A., R. C. Gueldner, and A. C. Thompson: Utilization of the boll weevil
pheromone for insect control. In: M. Beroza, Ed., Pest Management with Insect
Sex Attractants and other behavior-controlling chemicals. ACS Symposium Series
N. 23. American Chemical Society, Washington, D. C. 1976.

593. Knipling, E. F.: Biomathematical basis for suppression and elimination of boll
weevil populations. Conference on Boll Weevil Suppression, Management, and
Elimination Technology, Memphis, Tennessee, Feb. 13—15, 1974.

594. Mitchell, E. B., E. P. Lloyd, D. D. Hardee, W. H. Cross, and T. B. Davich:
In-field traps and insecticides for suppression and elimination of populations of boll
weevils. J. Econ. Entomol. **69,** 83—88 (1976).

595. Huddleston, P. M., E. B. Mitchell, and N. M. Wilson: Disruption of boll weevil
communication. J. Econ. Entomol. **70,** 83—85 (1977).

596. Lloyd, G. W.: USDA, ARS, Pesticides Registration Division, Washington, DC 20250,
private communication.

597. Mitchell, E. R.: Disruption of pheromonal communication among coexistent pest
insects with multichemical formulations. Bioscience **25,** 493—499 (1975).

598. Mitchell, E. R., M. Jacobson, and A. H. Baumhover: *Heliothis* spp. Disruption
of pheromonal communication with (Z)-9-tetradecen-1-ol formate. Envir. Entomol. **4,**
577—579 (1975).

599. Mitchell, E. R., A. H. Baumhover, and M. Jacobson: Reduction of mating potential
of male *Heliothis* spp. and *Spodoptera frugiperda* in field plots treated with disruptants.
Envir. Entomol. **5,** 484—486 (1976).

600. Hendricks, D. E., A. W. Hartstack, and J. R. Raulston: Compatibility of
virelure and looplure dispensed from traps for cabbage looper and tobacco budworm
survey. Envir. Entomol. **6,** 556—558 (1977).

601. TOSCANO, N. C., R. K. SHARMA, R. A. VAN STEENWYCK, V. SEVACHERIAN, C. HER-
    NANDEZ, H. T. REYNOLDS, K. KIDO, and A. J. MUELLER: Sexlure traps reduce
    insecticide treatments for pink bollworm. Calif. Agric. **30**, 12—13 (1976).
602. MCLAUGHLIN, J. R., H. H. SHOREY, L. K. GASTON, R. S. KAAE, and F. D. STEWART:
    Sex pheromones of lepidoptera. XXXI. Disruption of sex pheromone communication
    in *Pectinophora gossypiella* with hexalure. Envir. Entom. **1**, 645—650 (1972).
603. SHOREY, H. H., R. S. KAAE, and L. K. GASTON: Sex pheromone of lepidoptera. Develop-
    ment of a pheromonal control of *Pectinophora gossypiella* in cotton. J. Econ. Entomol.
    67, 347—350 (1974).
604. GASTON, L. K., R. S. KAAE, H. H. SHOREY, and D. SELLERS: Controlling the pink
    bollworm by disrupting sex pheromone communication between adult moths.
    Science **196**, 904—905 (1977).
605. BROOKS, T. W., and R. KITTERMAN: Controlled release of insect pheromone form-
    ulations based on hollow fibers, and methods of applications. Paper presented at
    the June 26—29, 1977 meeting of the American Society of Agricultural Engineers.
    Available from Conrel Co., 110 A. St., Needham Hts., MA 02194. 1977.
606. SWENSON, D. W.: Conrel Co., 110 A. St., Needham, Mass. 02194, private communic-
    ation.
607. MARKS, R. J.: The influence of behavior modifying chemicals on mating success of
    the red bollworm, *Diparopsis castanea* Hmps. (Lepidoptera: Noctuidae) in Malawi.
    Bull. Entomol. Res. **66**, 279—300 (1976).
608. NEUMARK, S., R. M. WATERS, and M. JACOBSON: Improvement of the attractiveness
    of *Spodoptera littoralis* sex pheromone, and its possible use in safety belts around
    cultivated areas to control the pest in Israel. Envir. Letters **10**, 97—120 (1975).
609. CHERRETT, J. M.: Dept. of Applied Zoology, University College of North Wales,
    Bangor, Gwynedd, North Wales LL57 2UW, private communication.
610. BURKHOLDER, W. E.: Application of pheromones for manipulating insect pests of
    stored products. In: Proceedings of a Symposium on Insect Pheromones and their
    Applications, Nagaoka and Tokyo, Japan, Dec. 8—11, 1976. Agriculture, Forestry,
    and Fisheries Research Council, Ministry of Agriculture and Forestry, Kasumigaseki,
    Chiyoda-ku, Tokyo, Japan. Available from National Institute of Agricultural
    Sciences, Nishiga-hara, Kita-ku, Tokyo, 114, Japan. 1976.
611. LEVINSON, H. Z.: Possibilities of using insectistatics and pheromones in pest control.
    Naturwiss. **62**, 272—282 (1975).
612. SHAPAS, T. J., W. E. BURKHOLDER, and G. M. BOUSH: Population suppression of
    T. glabrum by using pheromone luring for protozoan pathogen dissemination.
    J. Econ. Entomol. **70**, 469—474 (1977).
613. REICHMUTH, CH., R. WOHLGEMUTH, A. R. LEVINSON, and H. Z. LEVINSON: Unter-
    suchung über den Einsatz von Pheromon-geköderten Klebefallen zur Bekämpfung
    von Motten im Vorratsschutz. Z. angew. Entomol. **82**, 95—102 (1976).
614. SOWER, L. L., W. K. TURNER, and J. C. FISH: Population-density-dependent mating
    frequency among *Plodia interpunctella* (Lepidoptera: Phycitidae) in the presence of syn-
    thetic sex pheromones with behavioral observations. J. Chem. Ecol. **1**, 335—342 (1975).
615. SOWER, L. L., and G. P. WHITMER: Population growth and mating success of Indian
    meal moths and almond moths in the presence of synthetic sex pheromones. Envir.
    Entomol. **6**, 17—20 (1977).
616. WHEATLEY, P. E.: Research being undertaken by the Tropical Stored Products Center
    in the United Kingdom and Overseas. EPPO Bull. **4**, 495—500 (1975).
617. HAINES, C. P.: The use of sex pheromones for pest management in stored products.
    Pestic. Sci. **7**, 647—649 (1976).
618. CARLSON, D. A., and M. BEROZA: Field evaluation of (*Z*)-9-tricosene, a sex attractant
    pheromone of the house fly. Envir. Entomol. **2**, 555—559 (1973).

619. A better attractant for house flies. Agric. Res. **21** (9), 8—9. USDA, Agricultural Research Service, Washington, DC 20250. 1973.

620. SONENSHINE, D. E.: Dept. of Biological Sciences, Old Dominion University, Norfolk, VA 23508, private communication.

621. GLADNEY, W. J., S. E. ERNST, and R. R. GRABBE: The aggregation response of the Gulf Coast Tick on cattle. Ann. Entomol. Soc. Am. **67**, 750—752 (1974).

622. RECHAV, Y., and G. B. WHITEHEAD: Field trials with pheromone-acaricide mixtures for control of *Amblyomma hebraeum.* In preparation.

623. YOUNG, J., T. M. GRANES, R. CURTIS, and M. M. FURNISS: Controlled release of pheromones and insect growth regulators. In: H. B. SCHER, Ed., Controlled Release Pesticides. ACS Symposium Series No. 53. American Chemical Society, Washington, D. C. 1977.

624. KYDONIEUS, A. F., J. K. SMITH, and M. BEROZA: Controlled release of pheromones through multi-layered polymeric dispensers. In: D. R. PAUL and F. W. HARRIS, Eds., Controlled Release Formulations. ACS Symposium Series No. 33. American Chemical Society, Washington, D. C. 1976.

625. Herculite Protective Fabrics Corp., 1107 Broadway, New York, N. Y. 10010.

626. NCR Corp., Capsular Products Marketing and Sales, Dayton, Ohio 45479.

627. Pennwalt Corp., Three Parkway, Philadelphia, Pa. 19102.

628. CAMERON, E. A., C. P. SCHWALBE, M. BEROZA, and E. F. KNIPPLING: Disruption of gypsy moth mating with microencapsulated disparlure. Science **183**, 972—973 (1974).

629. CARO, J. H., B. A. BIERL, H. P. FREEMAN, and P. E. SONNET: A method for trapping disparlure from air and its determination by electron-capture gas chromatography. J. Agric. Food Chem. **26**, 461—463 (1978).

630. VANHAELEN, M., R. VANHAELEN-FASTRÉ, and J. GEERAERTS: Isolation and characterization of trace amounts of volatile compounds affecting insect chemosensory behaviour by combined pre-concentration on Tenax GC and gas chromatography. J. Chromatogr. **144**, 108—112 (1977).

631. MACCONNELL, J. G., J. H. BORDEN, R. M. SILVERSTEIN, and E. STOKKINK: Isolation and tentative identification of lineatin, a pheromone from the frass of *Trypodendron lineatum* (Coleoptera: Scolytidae). J. Chem. Ecol. **3**, 549—561 (1977).

632. FRANCKE, W., V. HEEMANN, B. GERKEN, J. A. A. RENWICK, and J. P. VITÉ: 2-Ethyl-1,6-dioxaspiro[4.4]nonane, principal aggregation pheromone of *Pityogenes chalcographus* (L.). Naturwissenschaften **64**, 590—591 (1977).

633. ROSSI, R.: Insect pheromones. II. Synthesis of chiral components of insect pheromones. Synthesis (Stuttgart) **1978**, 413—434.

634. PIRKLE, W. H., and C. W. BOEDER: Synthesis and absolute configuration of (−)-methyl (*E*)-2,4,5-tetradecatrienoate, the sex attractant of the male dried bean weevil. J. Org. Chem. **43**, 2091—2093 (1978).

635. WENKERT, E., D. A. BERGES, and N. F. GOLOB: Oxycyclopropanes in organochemical synthesis. Total synthesis of (−)-valeranone and (±)-grandisol. J. Amer. Chem. Soc. **100**, 1263—1267 (1978).

636. MORI, K.: Synthesis of the both enantiomers of grandisol, the boll weevil pheromone. Tetrahedron **34**, 915—920 (1978).

637. DE SOUZA, J. P., and A. M. R. GONÇALVES: Alternative route to three of the four terpenoid components of the boll weevil sex pheromone. J. Org. Chem. **43**, 2068—2069 (1978).

638. KOVALEV, B. H., V. V. STAN, T. K. ANTOCH, V. P. KONYUKHOV, and S. F. NEDOPEKINA: Synthesis in field of attractants (sexual attractants) of insects. III. Easy method for preparation of higher dialkylacetylenes. Synthesis of muscalure, an indoor fly attractant for *Musca domestica.* Zh. Org. Khim. **13**, 2049—2052 (1977).

639. SNIDER, B. B., and D. RODINI: Synthesis of the Al component of the female sex pheromone of the California red scale. Tetrahedron Letters **1978**, 1399—1400.

640. TAMARU, Y., Y. YAMADA, and Z. YOSHIDA: Palladium catalyzed thienylation of α-methallyl alcohol with 5-substituted 2-bromothiophenes and its application to synthesis of queen substance. Tetrahedron Letters **1978**, 919—922.

641. PIRKLE, W. H., and P. E. ADAMS: Synthesis of the carpenter bee pheromone. Chiral 2-methyl-5-hydroxyhexanoic acid lactone. J. Org. Chem. **43**, 378—379 (1978).

642. MORI, K., S. MASUDA, and M. MATSUI: Pheromone synthesis 22. New synthesis of a stereoisomeric mixture of 3,7-dimethylpentadec-2-yl acetate, sex pheromone of pine sawflies. Agric. Biol. Chem. **42**, 1015—1018 (1978).

643. MORI, K., S. TAMADA, and M. MATSUI: Stereocontrolled synthesis of the four possible stereoisomers of *erythro*-3,7-dimethylpentadec-2-yl acetate and propionate, the sex pheromone of the pine sawflies. Tetrahedron Letters **1978**, 901—904.

644. PLACE, P., M.-L. ROUMESTANT, and J. GORE: New synthesis of 3,7-dimethylpentadec-2-yl acetate sex pheromone of the pine sawfly *Neodiprion lecontei*. J. Org. Chem. **43**, 1001—1002 (1978).

645. TAI, A., M. IMAIDA, T. ODA, and H. WATANABE: Synthesis of the optically active common precursor of the sex pheromone of pine sawflies. Application of enantioface differentiating hydrogenation with modified nickel. Chemistry Lett. **1978**, 61—64.

646. ROSSI, R., and A. CARPITA: Insect pheromones: Stereoselective reduction of β- or ω-alkynols to corresponding (*E*)-alkenols by lithium tetrahydroaluminate. Synthesis **1977**, 561—562.

647. FYLES, T. M., C. C. LEZNOFF, and J. WEATHERSTON: Use of polymer supports in organic synthesis. XIII. Some solid phase syntheses of the sex attractant of the spruce budworm: *trans*-11-tetradecenal. J. Chem. Ecol. **4**, 109—116 (1978).

648. — — — Use of polymer supports in organic synthesis. XV. Bifunctionalized resins. Applications to synthesis of insect sex attractants. Can. J. Chem. **56**, 1031—1041 (1978).

649. KLÜNENBERG, H., and H. J. SCHÄFER: Synthesis of disparlure by Kolbe electrolysis. Angew. Chem. Int. Ed. Engl. **17**, 47—48 (1978).

650. TOLSTIKOV, G. A., V. N. ODINOKOV, R. I. GALEEVA, and R. S. BAKEEVA: New stereoselective synthesis of racemic disparlure, sex pheromone of the gypsy moth (*Porthetria dispar* L.). Tetrahedron Letters **1978**, 1857—1858.

651. OKADA, K., K. MORI, and M. MATSUI: Pheromone synthesis. 18. Stereoselective synthesis of racemic (*E*)-7,8-epoxy-2-methyloctadecane, geometrical isomer of racemic disparlure. Agric. Biol. Chem. **41**, 2485—2486 (1977).

652. SMITH, L. M., R. G. SMITH, T. M. LOEHR, G. D. DAVES, JR., G. E. DATERMAN, and R. H. WOHLEB: Douglas fir tussock moth pheromone: Identification of a diene analogue of the principal attractant and synthesis of stereochemically defined 1,6-, 2,6-, and 3,6-heneicosadien-11-ones. J. Org. Chem. **43**, 2361—2366 (1978).

653. TAMADA, S., K. MORI, and M. MATSUI: Pheromone synthesis. 19. Simple synthesis of (*Z*)-7-eicosen-11-one and (*Z*)-7-nonadecen-11-one, pheromone of the peach fruit moth. Agric. Biol. Chem. **42**, 191—192 (1978).

654. KASANG, G., K.-E. KAISSLING, O. VOSTROWSKY, and H. J. BESTMANN: Bombykal, eine zweite Pheromonkomponente des Seidenspinners *Bombyx mori* L. Angew. Chem. **90**, 74—75 (1978).

655. BARRY, M. W., D. G. NIELSEN, F. F. PURRINGTON, and J. H. TUMLINSON: Attractivity of pheromone blends to male peachtree borer, *Synanthedon exitiosa*. Environ. Entomol. **7**, 1—3 (1978).

656. SOLOMON, J. D., M. E. DIX, and R. E. DOOLITTLE: Attractiveness of the synthetic carpenterworm sex attractant increased by isomeric mixtures and prolonged by preservatives. Environ. Entomol. **7**, 39—41 (1978).

657. Piccardi, P., A. Capizzi, G. Cassani, P. Spinelli, E. Arsura, and P. Massardo: A sex pheromone component of the old world bollworm *Heliothis armigera*. J. Insect Physiol. **23**, 1443—1445 (1977).

658. Tingle, F. C., and E. R. Mitchell: Response of *Heliothis virescens* to pheromonal components and an inhibitor in olfactometers. Experientia **34**, 153—154 (1978).

659. Cardé, R. T., C. C. Doane, T. C. Baker, S. Iwaki, and S. Marumo: Attractancy of optically active pheromone for male gypsy moths. Environ. Entomol. **6**, 768—772 (1977).

660. Miller, J. R., and W. L. Roelofs: Gypsy moth responses to pheromone enantiomers as evaluated in a sustained-flight tunnel. Environ. Entomol. **7**, 42—44 (1978).

661. Birch, M. C., D. M. Light, and K. Mori: Selective inhibition of response of *Ips pini* to its pheromone by the (S)-(−)-enantiomer of ipsenol. Nature **270**, 738—739 (1977).

662. Roelofs, W. L.: Threshold hypothesis for pheromone perception. J. Chem. Ecol. (in press).

663. Wright, R. H.: Odor and molecular vibration: Optical isomers. Chemical Senses and Flavour **3**, 35—37 (1978).

664. Tumlinson, J. H., R. M. Silverstein, J. C. Moser, R. G. Brownlee, and J. M. Ruth: Identification of the trail pheromone of a leaf-cutting ant, *Atta texana*. Nature **234**, 348—349 (1971).

665. Sonnet, P. E., and J. C. Moser: Synthetic analogs of the trail pheromone of the leaf-cutting ant, *Atta texana* (Buckley). J. Agric. Food Chem. **20**, 1191—1194 (1972).

666. — — Trail pheromones: responses of the Texas leafcutting ant, *Atta texana* to selected halo- and cyanopyrrole-2-aldehydes, ketones, and esters. Environ. Entomol. **2**, 851—854 (1973).

667. Caputo, J. F., R. E. Caputo, and J. M. Brand: Significance of the pyrrolic nitrogen atom in receptor recognition of *Atta texana* (Buckley) (Hymenoptera: Formicidae) trail pheromone and parapheromones. J. Chem. Ecol. (in press).

668. Chapman, O. L., K. C. Mattes, R. S. Sheridan, and J. R. Klun: Stereochemical evidence of dual chemoreceptors for an achiral sex pheromone in Lepidoptera. J. Amer. Chem. Soc. **100**, 4878—4884 (1978).

669. Dethier, V. G.: Other tastes, other worlds. Science **201**, 224—228 (1978).

670. Hughes, P. R., and J. A. A. Renwick: Hormonal and host factors stimulating pheromone synthesis in female western pine beetles, *Dendroctonus brevicomis*. Physiol. Entomol. **2**, 289—292 (1977).

671. Aldrich, J. R., M. S. Blum, A. Hefetz, H. M. Fales, H. A. Lloyd, and P. Roller: Proteins in a nonvenomous defensive secretion: biosynthetic significance. Science **201**, 452—454 (1978).

672. Hefetz, A., and M. S. Blum: Biosynthesis and accumulation of formic acid in the poison gland of the carpenter ant *Camponotus pennsylvanicus*. Science **201**, 454—455 (1978).

673. Harring, C. M.: Aggregation pheromones of the European fir engraver beetles *Pityokteines curvidens*, *P. spinidens* and *P. vorontzovi* and the role of juvenile hormone in pheromone biosynthesis. Z. ang. Ent. **85**, 281—317 (1978).

674. Ritter, F. J. (ed.): Chemical Ecology: Odour Communication in Animals. Amsterdam: Elsevier/North Holland Press. In press.

675. Kydonieus, A. F. (ed.): Controlled Release Technologies. Cleveland: CRC Press, Inc. In press.

676. Tucker, W.: Of mites and men. Harpers **257** (No. 1539), 43—58 (Aug. 1978).

(Received February 16, 1978)

The Structural Polymers of the Primary Cell Walls of Dicots

By M. McNeil, A. G. Darvill, and P. Albersheim, Department of Chemistry, University of Colorado, Boulder, Colorado, U.S.A.

With 8 Figures

Contents

Acknowledgement. Support by Department of Energy Contract EY-76-S-02-1426 is gratefully acknowledged.

I. Introduction

A. Why Study the Structure of Cell Walls?

The cell walls of plants are analogous to the skeletons of animals. The walls control the rate of growth of plant cells and thus of the plants. The walls are a structural barrier to some molecules and to invading pathogens. Cell walls are also a source of food, fiber and energy. Thus, knowledge of the mode of synthesis of cell walls, of the structure of cell walls, and of the function of cell walls is of great importance.

Plant cell walls are of two general types: primary cell walls and secondary cell walls. Primary cell walls are laid down by undifferentiated cells that are still growing and it is these primary walls that control cell growth. Secondary walls are derived from primary cell walls by cells which have stopped growing and are differentiating. This review is concerned with the structure of primary cell walls and will be limited to a consideration of the primary cell walls of dicotyledonous plants. The primary cell walls of dicots differ, at least to some degree, from the primary cell walls of monocots and from the cell walls of lower plants (3, 107).

B. Availability of Homogeneous Cell Wall Preparations

Success in structural studies of cell walls depends largely on the purity and homogeneity of the cell wall preparations examined. Some studies have been made of primary cell walls extracted from whole plant tissues. Since all tissues of intact plants contain a variety of cell types, the walls prepared from these tissues are not homogeneous. The desire for homogeneous wall preparations has been satisfied by use of suspension-cultured cells. Suspension-cultures, of at least some types of plant cells, can be maintained in a totally undifferentiated state. These types of suspension cultures provide a source of homogeneous primary cell walls.

Suspension-cultured plant cells represent a somewhat artificial situation since one is interested in studying the cells of the intact plant. The structures of the walls of cells in culture and the walls of cells in the plant may conceivably differ and it is essential when studying suspension-cultured cells to keep this possibility in mind. However, the available data indicate that the walls of suspension-cultured cells are very similar in structure to the walls of intact plant tissues. Cell wall glycosyl and glycosyl-linkage compositions of pea (58) and red kidney bean (105) hypocotyl tissues show that these tissues contain walls which are very similar to the walls isolated from suspension-cultured sycamore (123) and suspension-cultured red kidney bean cells (105). In addition, it has been demonstrated that the primary walls of cambial cells which were prepared from the branches of sycamore trees are very similar to the walls of suspension-cultured sycamore cells (104).

The hypothesis that the walls of suspension-cultured cells reflect the walls of intact plants is further substantiated by the fact the xyloglucan, a hemicellulosic component of the cell walls of bean and pea plants, is extremely similar to the xyloglucan of suspension-cultured sycamore cells (133). Indeed, amyloids, polysaccharides which can not be structurally differentiated from xyloglucans, have been obtained from a wide variety of seeds (see Section IV A).

The structural similarity of the walls isolated from a variety of suspension-cultured cells supports the hypothesis that the primary walls of all dicots are structurally similar. The similarity of the walls of different suspension-cultured dicot cells has been demonstrated by comparing the glycosyl and glycosyl-linkage compositions of the walls of suspension-cultured sycamore, red kidney bean, tomato and soybean cells (3). The results show that although quantitative differences exist, qualitatively all of the walls contain the same glycosyl residues and the same glycosyl-linkages. It is difficult to see how the walls of different suspension-cultured dicot cells could be so similar to each other and similar to the walls of intact tissues, if the walls of these cells did not reflect the structures of the cell walls of intact tissues.

C. Goals of Cell Wall Structural Research

The general structure of the primary cell walls of plants has been envisioned for many years to be composed of cellulose fibers embedded in an amorphous mixture of polysaccharides and glycoproteins. Although this picture of primary walls appears to be accurate, it obviously lacks considerable detail. A more detailed description of the primary cell wall will eventually include the following:

1) isolation and identification of each of the individual macromolecular components of the cell wall

2) determination of the primary structure of each of these macromolecules

3) determination of the three dimensional structures of these macromolecules

4) determination of how these macromolecules are attached to one another or how they are interrelated

5) determination of how the interrelated macromolecules are distributed throughout the thickness and length of the wall

Plant cell wall research is basically at the stage of identifying and elucidating the covalent structures of the macromolecular components of the walls (points 1 and 2 above). Hence, this review will primarily consider the identity and structure of the major macromolecular components of primary cell walls.

The known structural features of each of the proposed wall components and the methods used to obtain this structural information will be described. A later part of this review will discuss the available information about the chemical bonding (covalent or otherwise) which exists between the structural components (point 4). Although the three dimensional structures of polysaccharides have received attention lately (*47, 61, 108, 109*), little is known about the secondary, tertiary and quaternary structures of plant cell wall polymers (point 3). Further, although the distribution of the polysaccharides throughout the wall has been studied at the ultrastructural level, no generally accepted description of this important phenomenon is available, and it will not be considered in this review.

Most of the plant polysaccharides that have been described have not been obtained from isolated cell walls, but rather from other plant organelles and tissues. We believe that in order to describe the structure of primary cell walls, it will eventually be necessary to study polysaccharides that have been isolated directly from primary cell walls. This review will comprehensively consider all those polysaccharides which have been isolated from primary cell walls. This review will also consider a selection of those studies of plant polysaccharides which have been isolated from sources other than primary cell walls, but which are thought to resemble primary cell wall polysaccharides.

D. Problems Associated With Cell Wall Structural Research

There are significant technical problems facing those who wish to study the structure of primary cell walls. An important and sometimes overlooked problem is the purification of the cell walls. Generally, cell walls are purified by being insoluble in buffered salt solutions and in organic solvents. Such purification procedures will undoubtedly remove some of the molecules present in the walls of intact plant tissues, and some of the discarded molecules may have a structural function within the wall. Any molecule solubilized by the wall purification procedures employed is not considered, in this review, as a structural component of the wall.

The purity of the wall preparations that have been studied can also be questioned. Even though elaborate washing of the wall preparations is customary, the walls may be contaminated by some of the cytoplasmic components which attach to or sediment with the walls following tissue homogenation. Starch grains are an example of a difficult to remove cell wall contaminant.

The insolubility of the cell wall structural components is the cause of another technical problem. In order to study the structure of the individual wall components, these components must be solubilized and purified. In essence, it is impossible to do this without alteration of the structures of the wall components. All of the methods currently used for solubilization of the structural components have some disadvantages. Most chemical solubilization techniques are suspected or known to break covalent bonds and to solubilize a size-heterogeneous mixture of cell wall components. Enzymes which solubilize wall polymers do so by hydrolyzing covalent bonds and, thereby, alter the polymers which they are solubilizing. Purified enzymes do have the advantage of degrading the wall polymers in a predictable manner. However, the possibility always exists that even highly purified enzymes are contaminated with undetected degradative activites. This problem is minimized by very careful screening of the enzymes on model substrates. Enzymes have the same deficiency as most chemical extraction procedures in that enzymes generally fail to extract all of their substrate from the walls. It is usually not apparent why an extraction procedure fails to completely solubilize a particular wall component. The difficulties associated with the solubilization of cell wall polymers remain, perhaps, the major impediment to progress in cell wall analysis.

It is no easy task to purify a wall component to homogeneity even after the component has been successfully extracted from the wall. The available methods for the purification of polysaccharides are being improved dramatically, but even so, the heterogeneity and complexity of the interconnected cell wall polymers makes this a very difficult problem.

Once the wall polysaccharides and proteins are purified, there remains the problem of determining their primary structures. The evolution of powerful and sensitive methods for determining the sequence of glycosyl residues in a polysaccharide and the sequence of amino acids residues in a protein makes structural analysis one of the most tractable problems facing those who study cell wall structure. Nevertheless, the complexity of the wall and the problems associated with its study suggest that complete elucidation of the structure of primary cell walls will not come for some time in the future. But those of us working in the area of cell wall structural research are excited by the knowledge that the methods available today and the methods which are being developed make this very necessary research feasible and attractive.

E. Types of Cell Wall Polysaccharides

Early workers considered the wall to be composed of three polysaccharide fractions; namely cellulose, the hemicelluloses and the pectic polysaccharides. Grouped in the pectic fraction are all of the polysaccharides extracted from cell walls by hot water, ammonium oxalate, weak acid or chelating agents. Hemicelluloses are not extracted by weak acids but are extracted by relatively strong alkali. The alkali-extracted wall residue is mostly composed of cellulose. These extraction techniques have led to some confusion and contradictions in the literature. This is mainly due to incomplete and overlapping extraction of the wall polymers by the chemical procedures employed. Nevertheless, recent work has shown that the classification of the wall polysaccharides into cellulose, hemicelluloses and pectic polysaccharides is reasonably accurate.

Today, we classify the pectic polysaccharides as those polymers found in covalent association with galacturonosyl-containing polysaccharides, namely the rhamnogalacturonans, the arabans, the galactans and possibly the arabinogalactans. The hemicelluloses are those polysaccharides non-covalently associated with cellulose. It has been proposed and is accepted by us that the hemicelluloses are capable of hydrogen-bonding strongly to cellulose (31). In dicot primary cell walls, these polysaccharides are the xyloglucans and glucuronoarabinoxylans (31, 46, 75, 123).

The original classification of wall polysaccharides can be related to the modern terminology, as mild acid preferentially extracts the pectic polysaccharides, while subsequent extraction with alkali preferentially solubilizes the hemicelluloses. Therefore, in this review, discussion of the non-cellulosic cell wall polysaccharides will be found under the general headings of the pectic polysaccharides and the hemicelluloses. Cellulose as well as the non-polysaccharide components of primary cell walls will also be considered.

II. Sugar Nomenclature and Abbreviations Used in this Review

A. Glycosyl Residues

A sugar residue glycosidically linked through its reducing carbon (C-1) is called a glycosyl residue, e.g., 4-linked glucosyl residues are glucosyl residues glycosidically linked at C-1 and which also have another glycosyl residue attached to them at C-4. Sugars with their reducing carbons free, whether or not the sugars have other glycosyl residues attached to them, are called glycoses, e.g., 4-linked glucose indicates a glucose which is located at the reducing end of an oligo- or polysaccharide and which has another glycosyl residue attached to it at C-4.

B. Abbreviations

Glc = glucose; Gal = galactose; Man = mannose; Xyl = xylose; Api = apiose; Ara = arabinose; Rha = rhamnose; Fuc = fucose; GlcA = glucuronic acid; GalA = galacturonic acid; p = pyranose ring form; f = furanose ring form.

C. Absolute Configuration

All of the sugars of which the plant cell wall polymers are composed, except for arabinose, rhamnose, galactose and fucose, are invariably found in the D configuration. Arabinose, rhamnose and fucose have consistently been found in the L configuration. Galactose is almost always in the D configuration, but has been found in the L configuration in plant tissue. The D or L configuration is omitted from the nomenclature used in this review. Those instances where the absolute configuration has been specifically determined are generally noted in the text.

D. Anomeric Configuration

The anomeric configuration, α or β, of the glycosidic linkages is designated when known.

E. Ring Size

Except for most of the arabinosyl residues, all of the primary cell wall glycosyl residues of dicots have been found to be in the pyranose ring form. It is possible that some of the glycosyl residues that have been determined to be 4-linked pyranosyl residues are, in fact, 5-linked furanosyl residues. This is true because methylation analysis does not distinguish between the two possibilities. The ring forms for all of the glycosyl residues, except arabinose, are not designated. The ring form of arabinosyl residues is designated when known.

F. Linkage Analysis

Methylation data is expressed using a simplified "linkage" notation. The linkages of the glycosyl residues are determined from the position of the O-methyl groups introduced during methylation analysis. Methyl groups are never attached to either C-1 (protected from O-methylation by its participation in glycosidic linkage) or C-5 (protected from O-methylation by its participation in the pyranose ring). Arabinofuranosyl residues are an exception and may have methyl groups on C-5 but never on C-4. In the notation adopted, all carbons designated as "linked" do not have O-methyl groups attached (protected from methylation by glycosidic linkage with another sugar), whereas all of the remaining carbons except C-1 and C-5 (or C-4 in the case of arabinofuranosyl residues), do have O-methyl groups attached. For example, a glycosyl residue designated as "terminal" (T) is glycosidically linked to another glycosyl or sugar residue only through C-1 and contains no glycosyl residues linked to other carbons. A glycosyl residue designated as 2-linked is glycosidically linked to another glycosyl or sugar residue through C-1 and has another glycosyl residue linked to it at C-2. A glycosyl residue designated as 3,6-linked is glycosidically linked to another sugar through C-1 and has glycosyl residues linked to it at C-3 and C-6.

G. Polymer Names

Cell wall polymers are often referred to by the quantitatively dominant glycosyl residues of which they are composed (*e. g.*, xyloglucan). These names do not mean the polymers are composed solely of the glycosyl residues referred to in the polymer's name (*e. g.*, xyloglucan also contains arabinosyl, fucosyl, and galactosyl residues).

III. Methods Used in the Structural Analysis of Cell Wall Polysaccharides

A. Introduction

Many of the structural studies discussed in this review have relied upon the use of a few well-defined experimental techniques. These techniques will be described in this section. Particular emphasis is placed on those methods which have proved useful in studying the structures of primary cell wall polysaccharides and, in particular, the wall polysaccharide of suspension-cultured sycamore cells.

B. Solubilization and Fractionation of Cell Wall Polysaccharides

A successful first step in fractionating primary cell walls has been the treatment of the walls with a purified endo-α-1,4-galacturonase (45, 123). This enzyme hydrolyzes α-4-linked galacturonosyl linkages resulting in the solubilization of approximately 18% of the mass of the wall. The cell wall residue remaining after endopolygalacturonase treatment can be extracted with alkali to yield additional pectic polysaccharides and the hemicelluloses. On the other hand, the endopolygalacturonase-treated cell walls can be extracted with a second enzyme; an endo-β-1,4-glucanase has been of value in this regard. The endoglucanase specifically fragments the cell wall xyloglucan (31).

The endopolygalacturonase and alkali-solubilized cell wall polysaccharides have each been fractionated by ion exchange and gel filtration chromatography (45). Ion exchange chromatography is particularly valuable in separating the acidic polysaccharides from the neutral polysaccharides and is also useful in separating those polysaccharides which contain differing amounts of acidic residues. For example, anion exchange chromatography can separate the xylan and xyloglucan polysaccharides solubilized by alkali treatment of endopolygalacturonase-treated walls (46). Five non-cellulosic primary cell wall polysaccharides have been highly purified by a combination of anion exchange and gel filtration chromatography. These five polysaccharides are xyloglucan (31), glucuronoarabinoxylan (46), homogalacturonan (45), rhamnogalacturonan I (45), and rhamnogalacturonan II (44).

It is important to be able to determine which chromatography column fractions contain polysaccharides and, specifically, which fractions contain hexosyl, pentosyl, or uronosyl residues. It is also important to detect in the fractions the presence of proteins and the presence of the specific amino acid characteristic of wall proteins, hydroxyproline. The detection of these substances is carried out by facile and sensitive colorimetric procedures. Although these reactions are not discussed here, the most frequently used colorimetric assays in our laboratory are the anthrone assay for detection of hexosyl residues (50), the orcinol assay for detecting pentosyl residues (50), the m-hydroxy-diphenyl assay for detection of uronosyl residues (35), the Lowry assay for detection of proteins (90), and the Kivirikko and Liesmaa assay for the detection of hydroxyprolyl residues (76).

C. Quantitative Analysis of the Glycosyl Residues of Oligo- or Polysaccharides

The most frequent procedure used by those studying polysaccharide structures is the determination of the glycosyl composition of the sample being investigated. This assay is used both for determining the purity of polysaccharides and for identifying which polysaccharides are present in a particular chromatographic fraction. The most commonly used and the most accurate method for the quantitative analysis of the glycosyl residues involves the conversion of the glycosyl residues into their corresponding volatile alditol acetates. The alditol acetates are conveniently separated and quantitated by flame ionization gas chromatography (5).

An efficient procedure for converting polysaccharides into their monosaccharide constituents is by hydrolysis at 121° for 1 hr with 2N trifluoroacetic acid. Trifluoroacetic acid is volatile and can be easily removed from the hydrolyzed samples by evaporation in a stream of air. The monosaccharides are converted to the corresponding alditols by reduction with sodium borohydride and the alditols are acetylated with acetic anhydride in the presence of a weak base, such as sodium acetate. The steps involved in converting a polysaccharide into its constituent alditol acetates are summarized in Fig. 1.

D. Uronic Acid Quantitation

Polysaccharides which contain uronosyl residues, such as the acidic pectic polysaccharides and the glucuronoarabinoxylans, pose a particular problem in quantitative determinations of glycosyl compositions. Uronosidic bonds are resistant to acid hydrolysis. Conditions sufficiently harsh to hydrolyze the uronosidic bonds often result in significant degradation of the uronic acids as well as degradation of some of the neutral sugars. Uronic acids, even when converted into their monomeric form, do not form stable alditol acetates as the reduced form of an uronic acid is an aldonic acid which cannot be acetylated by the standard procedures.

The problems associated with the presence of uronosyl residues can be bypassed by converting the uronosyl residues to the corresponding hexosyl residues. This reduction is most successfully carried out by converting the uronosyl residues into lactones by reaction with water soluble carbodiimides. The lactones are reduced with sodium borodeuteride (124). It is important to use sodium borodeuteride rather than sodium

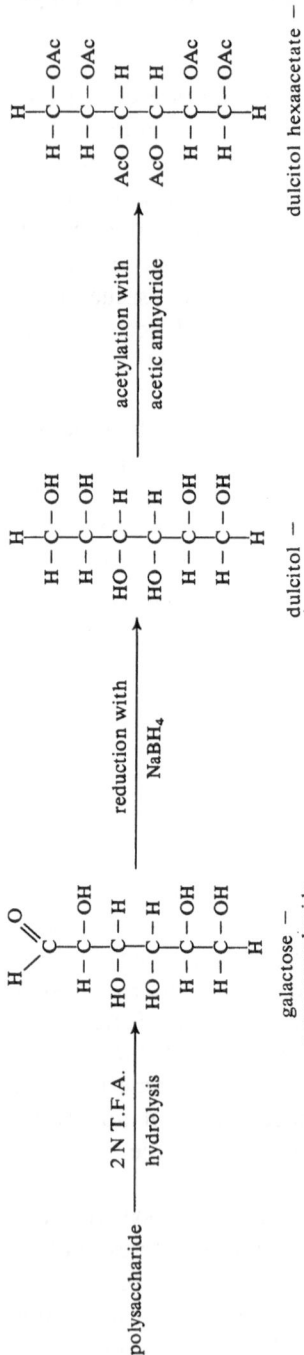

Fig. 1. Glycosyl composition analysis is achieved by conversion of the glycosyl residues of oligo- and polysaccharides into volatile alditol acetate derivatives. Experimental details are in ref. (5)

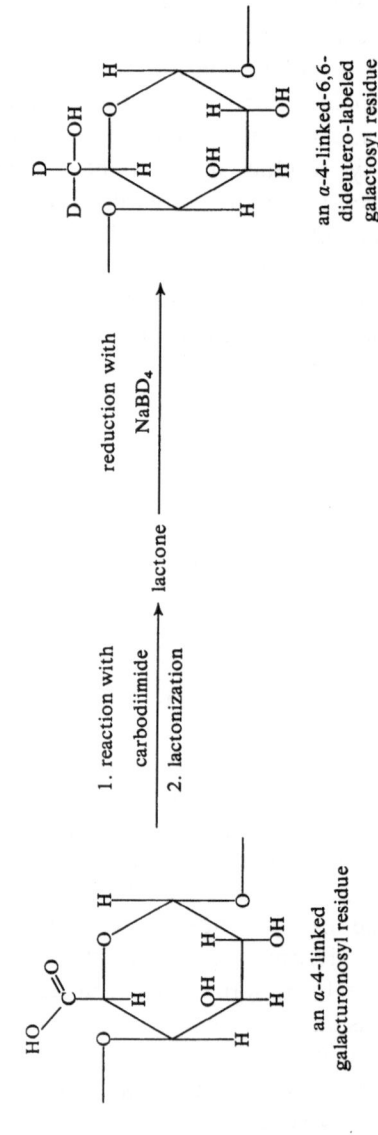

Fig. 2. Conversion of uronosyl residues to dideuterohexosyl residues. Experimental details are in ref. (124)

borohydride in order to label, with 2 deuterium atoms, the C-6 primary alcohol of the hexosyl residues which have been formed from the uronosyl residues. Acidic polysaccharides are converted by this procedure into neutral polysaccharides and the neutral polysaccharides can then be analyzed by formation of the corresponding alditol acetates. The alditol acetates derived from deuterium labeled hexosyl residues are quantitatively distinguished from the unlabeled hexosyl residues by combined gas chromatography-mass spectrometry. The steps involved in the reduction of uronosyl residues are summarized in Fig. 2.

E. Glycosyl-Linkage Composition Analysis

Once the glycosyl composition of a polysaccharide is known, the next step is to determine the glycosyl-linkage compositions. This analysis allows one to quantitatively determine the amounts of the differently linked glycosyl residues, such as the amount of a polysaccharide composed of 3-linked glucosyl residues, 4-linked glucosyl residues and 3,4-linked glucosyl residues. The method of choice for glycosyl-linkage analysis involves formation of partially methylated partially acetylated alditols. This technique substitutes methyl groups for the hydrogen atoms of all exposed hydroxyl groups in the polysaccharide (111). Wherever a glycosyl residue is connected to another glycosyl residue, the hydroxyl group is masked, and so in these positions, no methyl group is attached. After permethylation, the polysaccharide is hydrolyzed with trifluoroacetic acid to yield the partially methylated aldoses. The partially methylated aldoses are reduced to the corresponding alditols and then acetylated. The partially methylated alditol acetates are volatile and are quantitated and tentatively identified by flame ionization gas chromatography (123). The identity of the partially methylated alditol acetates is confirmed by combined gas chromatography-mass spectrometry (33). Chemical ionization mass spectrometry (93) has proven to be of value in augmenting electron ionization mass spectrometry (33) in identifying these derivatives.

Uronosyl residues of the polysaccharides offer the same problems for glycosyl linkage analysis as they do for glycosyl composition analysis. Before the linkages to the uronosyl residues can be ascertained, the uronosyl residues must be reduced by the carbodiimide method (124) to their corresponding deuterium-labeled alditols.

The partially methylated alditol acetate glycosyl-linkage analysis method is summarized in Fig. 3.

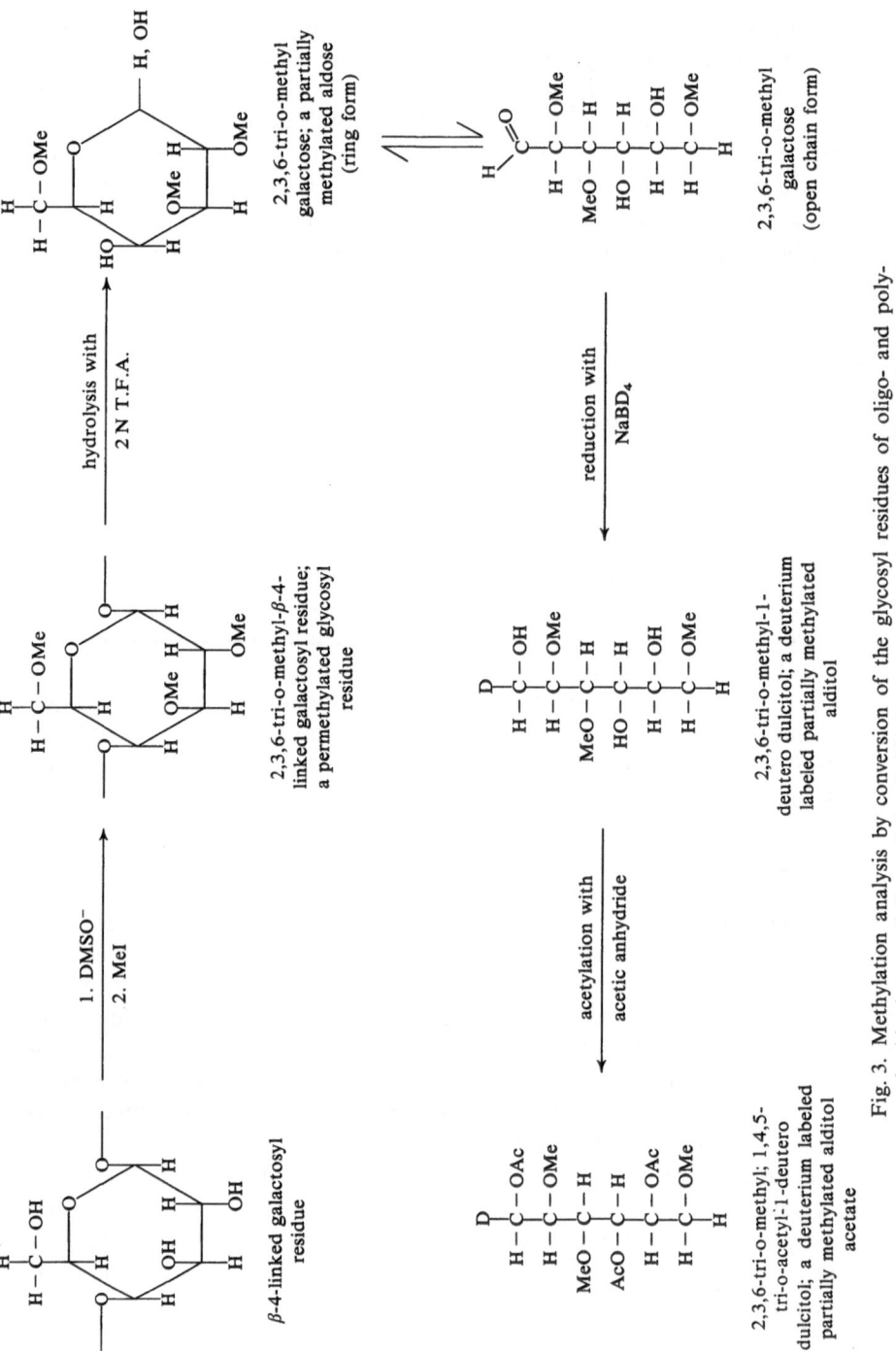

Fig. 3. Methylation analysis by conversion of the glycosyl residues of oligo- and poly-saccharides into partially methylated alditol acetates. Experimental details are in ref. (123)

F. Sequencing the Glycosyl Residues in Polysaccharides

New methods for sequencing the glycosyl residues of oligo- and polysaccharides are now being developed. The new methods are sophisticated elaborations of the rather commonplace procedure of converting polysaccharides by partial hydrolysis into structurally analyzable oligosaccharides. The conversion of polysaccharides into oligosaccharides is generaly achieved by partial acid hydrolysis or by acetolysis (43). These two methods differ in the rate at which they catalyze the hydrolysis of different glycosidic linkages and therefore, the two methods yield different sets of oligosaccharides from the same polysaccharide.

The goal of converting polysaccharides into manageable oligosaccharides can also be achieved with the assistance of highly purified endoglycanases. It is frequently rather laborious to obtain these enzymes, but their value cannot be questioned. One example of the value of such enzymes is the formation of a set of identifiable oligosaccharides from xyloglucan with the aid of an endo-β-1,4-glucanase (31). Other useful enzymes which have been purified with a goal of studying the structures of the primary cell wall polysaccharides include the widely used endo-α-1,4-galacturonase (45, 123), an endo-β-1,4-galactanase (80) and an endo-α-1,5-arabanase (69, 130).

Polysaccharides may also be specifically cleaved into sets of analyzable oligosaccharides by periodate oxidation and, in particular, by using the procedure known as the SMITH degradation (43, 113). This procedure involves periodate oxidation of those glycosyl residues of an oligo- or polysaccharide which possess vicinal hydroxyls. The glycosidic bonds of the glycosyl residues that have been modified by periodate oxidation are preferentially cleaved by acid hydrolysis. Analysis of the SMITH degradation products gives information on the sequence of glycosyl residues in the original polysaccharide.

The SMITH degradation products may contain oligosaccharides which are resistant to the periodate oxidation as well as fragments of the modified glycosyl residues. An example of the SMITH degradation of a polymer containing 3- and 6-linked galactosyl residues is shown in Fig. 4. Methylation analysis of the oxidation products would show the presence of a Gal-(1→3)-Gal disaccharide in the original polysaccharide. These results indicate that the polysaccharide possesses a repeating sequence of Gal-(1→6)-Gal-(1→3)-Gal-(1→3)-Gal-(1→6)-Gal.

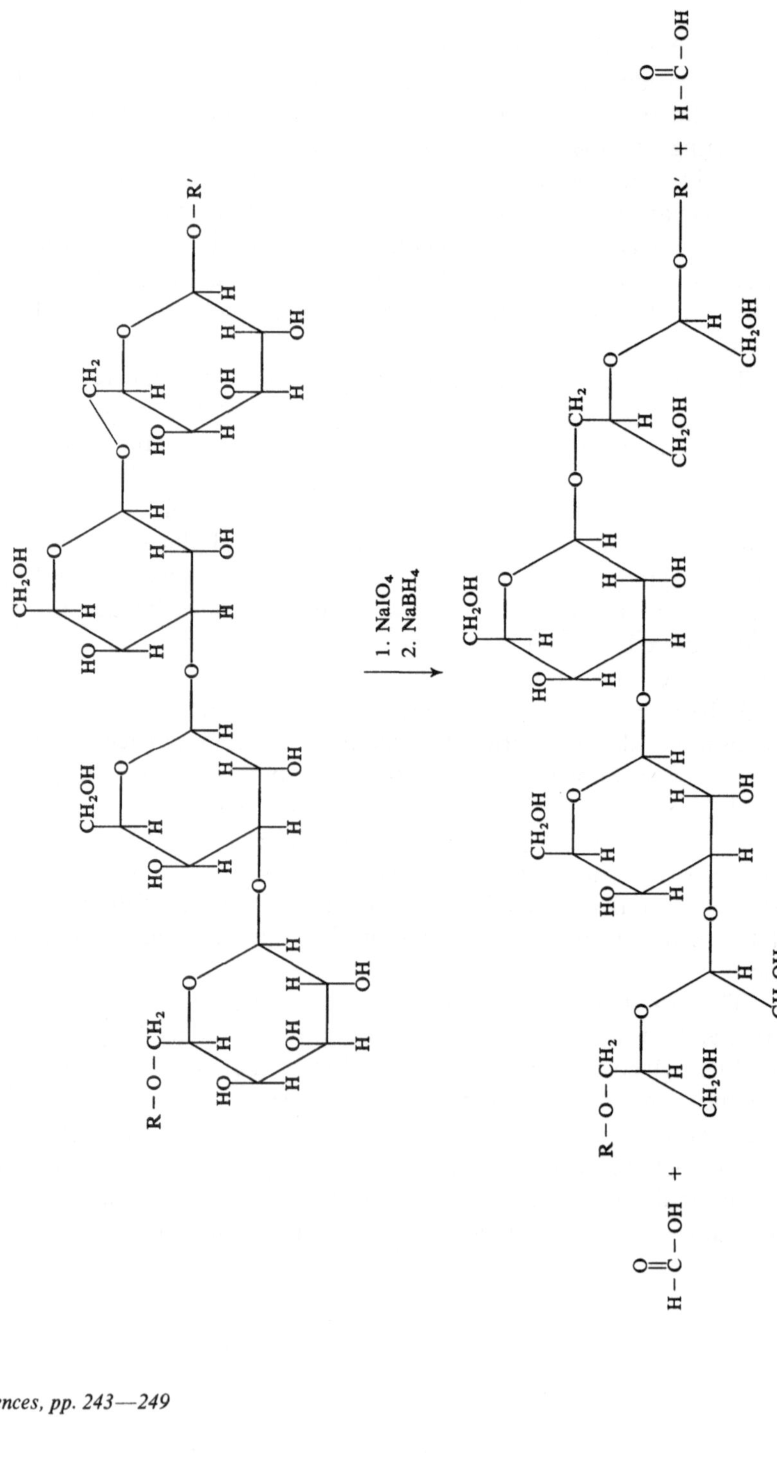

$$\xrightarrow[\text{2. NaBH}_4]{\text{1. NaIO}_4}$$

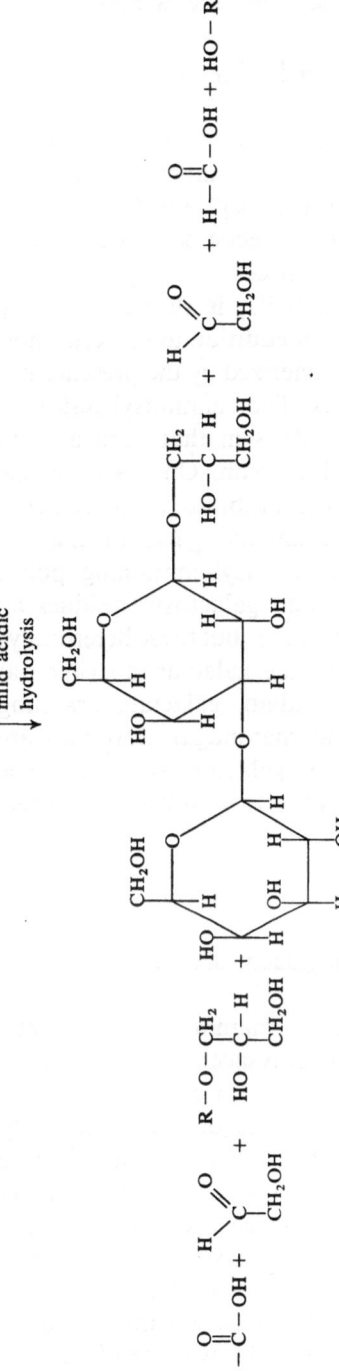

Fig. 4. Smith Degradation of a portion of a β-galactan containing two β-3-linked galactosyl residues between two β-6-linked galactosyl residues. The products of the reaction, Gal-(1 $\overset{β}{\rightarrow}$ 3)-Gal-(1 $\overset{β}{\rightarrow}$ 3)-Glycerol, formic acid and glycolaldehyde, can be isolated and quantitated. Consideration of the products obtained by Smith degradation in conjunction with the results of methylation analysis would lead to the correct sequence of the variously-linked glycosyl residues. Different Smith degradation products would be obtained from a polymer containing 3-linked and 6-linked galactosyl residues in the same ratio but in different sequence. Experimental details are in ref. (113)

IV. The Pectic Polysaccharides

A. Introduction

Primary cell walls are characterized by a relatively high content (35%) of pectic polysaccharides. The most characteristic components of the pectic polysaccharides are galacturonosyl residues (*135*). The most characteristic physical property of the pectic polysaccharides is an ability to form gels (*61, 109*). The area between primary cell walls of adjoining cells, known as the middle lamella, is thought to be particularly rich in pectic polysaccharides (*62*). In addition to galacturonosyl residues, the pectic polysaccharides are characterized by the presence of rhamnosyl, arabinosyl, and galactosyl residues. The rhamnosyl residues are closely associated with galacturonosyl residues in that both are integral components of the same polysaccharide chain. On the other hand, a considerable proportion of the cell wall arabinosyl and galactosyl residues appear to be components of araban and galactan chains, which are covalently attached to the galacturonosyl-containing polysaccharides. Some of the cell wall arabinosyl and galactosyl residues are likely to be constituents of arabinogalactan chains, but these heteropolysaccharides may not be covalently attached to the galacturonosyl-containing polymers. Discussions of the pectic araban, galactan, arabinogalactan as well as the homogalacturonan and rhamnogalacturonans are presented below. The evidence that the araban, galactan, and rhamnogalacturonan polymers are covalently linked to one another is discussed in Section IX.

B. Rhamnogalacturonan I

Polysaccharides containing only rhamnose and galacturonic acid have never been isolated. Such polysaccharides always have other sugars covalently attached to them. However, rhamnogalacturonans are thought to be the backbone chains of the pectic polymers. The rhamnogalacturonan described in this section is called rhamnogalacturonan I to distinguish it from rhamnogalacturonan II, an entirely different type of pectic polysaccharide which is discussed in Section IV G.

Rhamnose has long been known to be associated with the galacturonosyl residues of the pectic polysaccharides (*135*). More recently, rhamnosyl residues have been found to be glycosidically-linked to galacturonosyl residues. The available structural knowledge of rhamnogalacturonan I has been obtained by isolation of oligosaccharide fragments of the polymer (*12, 13, 16, 19, 23, 118, 123*) and by methylation analysis of the intact polymer (*13, 19, 23, 123*). The oligosaccharides have been obtained by either partial acid hydrolysis or by acetolysis. Partial acid hydrolysis of rhamnogalacturonan I has yielded the disaccharide GalA-(1 $\xrightarrow{2}$ 2)-Rha as well as the tetrasaccharide GalA-(1→2)-Rha-(1→4)-GalA-(1→2)-Rha. Acetolysis has yielded the trisaccharides GalA-(1→4)-GalA-(1→2)-Rha and GalA-(1→2)-Rha-(1→2)-Rha as well as the tetrasaccharide GalA-(1→4)-GalA-(1→2)-Rha-(1→2)-Rha (*12, 16*).

Methylation analysis was used to demonstrate that the rhamnosyl residues of the above oligosaccharides are 2-linked. Methylation analysis of the intact polysaccharides has demonstrated that about 50% of the rhamnosyl residues are 2,4-linked (*19, 23, 123*). No aldobiuronic acid with a galacturonosyl residue attached to C-4 of a rhamnose has been isolated. It has generally been assumed that the C-4 of rhamnose is a point of attachement of other neutral glycosyl residues (*12, 123*). The manner in which the oligosaccharides that have been characterized are arranged in the intact rhamnogalacturonan has not yet been established (*12, 123*). Clearly, there are regions of alternating galacturonosyl and rhamnosyl residues and regions of alternating galacturonosyl-galacturonosyl-rhamnosyl-rhamnosyl residues.

Further information about the structure of rhamnogalacturonan I has recently been obtained in this laboratory. The walls of suspension-cultured sycamore cells were exhaustively treated with endopolygalacturonase. This enzyme solubilizes rhamnogalacturonans as well as homogalacturonans (*45*). These acidic polysaccharides were separated from one another by ion exchange chromatography on DEAE Sephadex. The rhamnogalacturonan contains, in addition to rhamnosyl and galacturonosyl residues substantial amounts of arabinosyl and galactosyl residues (as noted above, rhamnogalacturonans have never been isolated free of other neutral glycosyl residues). The ratio of rhamnosyl to galacturonosyl to arabinosyl to galactosyl residues is 1 : 2 : 1.5 : 1.5.

Rhamnogalacturonan I is very large as it partially includes in an Agarose 5 m column. This suggests a degree of polymerization of about 2,000. The apparent size of rhamnogalacturonan I is not altered when chromatographed in 0.5 M NaCl containing 5 mM EDTA. This suggests that the apparent size of rhamnogalacturonan I is not due to non-covalent aggregation. If the backbone of rhamnogalacturonan I is a single linear chain, then this chain contains about 300 residues of rhamnose and 600 residues of galacturonic acid uninterrupted by regions of homogalacturonans. This is a major upward revision of the size and, indeed, of the structure of the rhamnogalacturonans compared with that envisioned earlier by this laboratory (*123*).

Linkage analysis of the rhamnogalacturonan I isolated from the walls of suspension-cultured sycamore cells shows the presence of 2- and 2,4-linked rhamnosyl residues and 4-linked galacturonosyl residues in a ratio of 1 : 1 : 4. The manner in which the galacturonosyl and rhamnosyl residues are arranged is assumed to be as in the oligosaccharides discussed above; indeed, the disaccharide GalA-(1→2)-Rha has been isolated from sycamore cell walls (123).

The basic structural features of rhamnogalacturonan I are illustrated in Fig. 5. Note that this is not an exact structure. Rather, the structure presented is a pictorial summation of the data presented above. No information is available on whether the rhamnosidic bonds are in the α or β configuration. The attachment of araban and galactan to this polymer is discussed in Section IX.

$$[\longrightarrow 4) - GalA - (1\xrightarrow{a} 4) - GalA - (1\xrightarrow{a} 2) - Rha - (1\longrightarrow 4) - GalA - (1\xrightarrow{a} 2) - Rha -$$
$$- (1\longrightarrow 2) - Rha - (1\longrightarrow]_{100-200}$$

Fig. 5. One possible sequence of rhamnosyl and galacturonosyl residues in rhamnogalacturonan I. Approximately half the rhamnosyl residues have an unidentified glycosyl residue attached to C-4 as well as having a galacturonosyl or another rhamnosyl residue attached to C-2. Approximately 5% of the galacturonosyl residues have an unidentified glycosyl residue attached to C-3. It is known that arabans and galactans are covalently linked to the rhamnogalacturonan backbone. It is not known how the arabans and galactans are attached. It is possible that one of these polysaccharides is attached to the C-4 of rhamnosyl residues and the other polysaccharide is attached to the C-3 of galacturonosyl residues

C. Homogalacturonan

The acidic pectic polysaccharides are characterized not only by large molecular weight rhamnogalacturonan I regions, but also by regions of unbranched α-4-linked galacturonosyl residues. Like the rhamnogalacturonan regions, the homogalacturonan regions are larger than was predicted (123). The existence of the homogalacturonan regions was established by examining the polysaccharides released by purified endopolygalacturonase from the walls of suspension-cultured sycamore cells. Approximately 5% of the wall is converted by endopolygalacturonase into mono-, di-, and trigalacturonic acid. These are the expected products of the action of the endopolygalacturonase on an α-4-linked galacturonan. The endopolygalacturonase also releases an α-4-linked galacturonan from sycamore walls which accounts for 1 to 2% of the starting wall material (45). The galacturonan is stable to further enzyme degradation due to esterification of the uronosyl carboxyl groups (54, 123). Similar α-4-linked galacturonans have been isolated from sun flower seeds (136) as well as from apple pectin (30). The sycamore homogalacturonan has an apparent degree of polymerization greater than 25 as deduced by gel filtration chromatography (45). It seems probable that the homogalacturonan regions of the pectic polysaccharides are considerably longer than 25, since these polysaccharides have been exposed to the action of the endopolygalacturonase which would hydrolyze any region of the homogalacturonans possessing sufficient de-esterification to be susceptible to the enzyme. Certainly, four consecutive unesterified galacturonosyl residues are susceptible to the action of the enzyme (54).

The fact that the galacturonosyl residues of pectic polymers are 4-linked has been established by converting the galacturonan to the corresponding galactan by reduction of the uronosyl carboxyl groups. The resulting galactan was then subjected to methylation analysis (45). In addition, ASPINALL and KOU-SHII JIANG (19) have methylated unreduced pectic polymers and have isolated from these polymers 2,3-dimethyl-galacturonic acid. These workers also have methylated carboxyl-reduced pectins and have isolated 2,3,6-tri-methyl galactose (19), the product expected from 4-linked galacturonosyl residues. Further evidence of the 4-linked nature of the galacturonosyl residues is provided by the successful hydrolysis of sycamore cell wall galacturonans with the endopolygalacturonase specific for α-4-linked galacturonosyl residues (123).

The fact that the 4-linked galacturonosyl residues of the cell wall galacturonans are in the α-anomeric configuration has been demonstrated by characterization of the galacturonosyl-containing oligosaccharides derived from wall polymers (18), and by the fact that both the galacturonan and the derived oligomers have highly positive optical rotations (19).

The carboxyl groups of the galacturonosyl residues of the cell wall pectic polysaccharides are known to be highly methyl esterified (13, 15, 16, 19, 23, 118). The degree of esterification of the carboxyl groups varies depending on the source of the pectic polymers (13, 19, 23, 118). It is not known how the methyl esters are distributed along the polygalacturonan backbone. However, it is clear that there are regions which are highly methyl esterified, and therefore, are not susceptible to the endopolygalacturonase which requires free carboxyl groups (45), as well as regions which are relatively free of methyl esters and therefore, are susceptible to the endopolygalacturonase.

D. Araban

Arabans have been isolated from the cell walls of many dicotyledonous plants. Until recently, no homo-araban has been isolated specifically from primary cell walls. However, methylation analysis of the walls of suspension-cultured sycamore (*123*) and pea (*58*) cells strongly suggested that these primary cell walls possess arabans which are structurally similar to arabans obtained from other tissues or organelles. Recently, an araban, essentially free of other polysaccharides, has been isolated from a methylated sycamore cell wall polysaccharide fraction (*45*).

The structures of plant arabans have been investigated by methylation analysis (*11, 66, 68b, 70, 117, 108a*), by Smith degradation (*108a*) and by ^{13}C-NMR spectrometry (*68b*). All of the arabans that have been investigated have similar structures. The arabans are highly branched polymers; the arabinosyl residues are largely in the furanose ring form; and the glycosidic linkages are uniformly in the α-anomeric configuration. In addition, arabinose is universally the L rather than the D isomer.

The glycosyl-linkage compositions of the arabans from a number of dicots are compared in Table 1 (*11, 45, 68b, 70, 117, 108a*). It can be concluded, by the presence of an O-methyl group on carbon 5, that the terminal- and 3-linked arabinosyl residues are in the furonase ring form. The fact that all the arabinosyl linkages are susceptible to hydrolysis by relatively mild acidic conditions is evidence that all the arabinosyl residues, including the 5-linked, 3,5-linked and 2,5-linked residues, are in the furonase ring form (*70, 123*).

Nuclear magnetic resonance analysis of *Rosa glauca* araban has provided additional evidence that the arabinosyl residues are α-5-linked and in the furanose configuration. The C-1 resonance expected of α-arabinofuranosyl residues, but not the C-1 resonances expected of the β-arabinofuranosyl or α- and β-arabinopyranosyl residues, was detected by ^{13}C-NMR analysis (*68b*). The proton NMR spectrum is consistent with α- or β-furanosyl residues and β-pyranosyl residues, but not with α-pyranosyl residues (*68b*). The α-anomeric nature of these linkages is confirmed by the negative optical rotations, from −181 to −108, exhibited by such arabans (*11, 68b*).

The degree of polymerization of arabans has been estimated by converting the reducing end of the polymers to arabitol with sodium borohydride, hydrolyzing the polymers, and determining the ratio of arabitol to arabinose. This method has provided evidence that two different arabans isolated from the bark of *Rosa glauca* have degrees of polymerization of 34 and 100 (*68b*), while an araban from willow has a degree of polymerization of 90 (*70*).

Table 1. *Glycosyl-Linkage Compositions (Mole %) of the Arabans of Dicots*

Arabinosyl linkage	Soybean[a]	Lemon[b]	Mustard[c] Cotyledons	Mustard[d] seed	Rape-[e] seed	Rose[f]	Sycamore[g]	White[h] willow	Aspen[i]
T-furanosyl	39.2	30.0	36.0	39.6	34.0	44.8	31.7	40.0	31.0
T-pyranosyl	0	0	0	0	0	0	0	2.4	2.7
5-	30.0	38.4	38.0	25.4	25.7	21.0	22.5	24.0	39.0
3-	0	0	0	0	0	8.0	0	4.4	0
3,5-	14.2	15.0	21.0	28.6	31.5	13.9	18.5	8.9	11.4
2,5-	6.0	4.6	0	tr	tr	3.5	8.0	5.7	6.5
2,3,5-	10.5	12.0	4.2	6.3	8.7	8.5	7.2	13.8	9.5

[a] Isolated from soybean (*Glycine max*) meal (*11*).
[b] Isolated from lemon peel (*Citrus limon L.*) pectin (*11*).
[c] Isolated from mustard (*Sinapis alba L.*) cotyledon (*108a*).
[d] Isolated from mustard (*Sinapis alba L.*) seed (*11*).
[e] Isolated from rapeseed (*Brassica campestris*) (*117*).
[f] Isolated from the bark of rose (*Rosa glauca*) (*68b*).
[g] Isolated from a mixture of methylated polysaccharides from sycamore (*Acer pseudoplatanus*) cell walls (*45*).
[h] Isolated from the bark of white willow (*Salix alba L.*) (*70*).
[i] Isolated from the bark of aspen (*Populus tremuloides*) (*68a*).

There is not much information about the arrangement of the differently linked arabinosyl residues in arabans. The best work to date is that of Rees and Richardson (108a) who have studied an araban from mustard cotyledons using the Smith degradation. Their result ruled out the possibility of regions of long, unbranched 5-linked arabinosyl residues. The evidence suggested that branched and unbranched arabinosyl residues occur near each other in the chain.

A number of complex pectic polysaccharides have been demonstrated to contain arabinosyl residues (23, 110, 118, 121, 123). These studies have generally not been carried to the point of determining whether the arabinosyl residues of the pectic polysaccharides exist as relatively long araban chains or whether the arabinosyl residues exist as mono-, di-, or tri-oligosaccharide sidechains attached to the other pectic polysaccharides (23, 110, 118, 121, 123). One investigation, using methylation analysis, has provided evidence that the arabinosyl residues of rapeseed pectic polysaccharides are present as mono- or disaccharide sidechains (19). On the other hand, glycosyl-linkage analyses of the pectic polysaccharides of sycamore primary cell walls (123) and studies of these polysaccharides, using mild acid hydrolysis for selective cleavage of the furanosyl linkages, suggest the presence of homo-arabans (123).

It is difficult to draw even a tentative structure of the primary cell wall araban. Clearly, branched arabans are important primary cell wall components. Efforts are currently under way in this laboratory to isolate and structurally analyze the araban of sycamore cell walls. This study is augmented by the availability of two recently purified enzymes, endo-α-1,5-arabanase and an exo-arabinosidase (69, 130).

E. Galactan

Galactans have been isolated from citrus pectin (80), from white willow (126), and from beech (96). As with the arabans, no homogalactan has ever been isolated directly from primary cell walls, although the glycosyl linkages which comprise those homogalactans which have been studied are also present, in similar ratios, in primary cell walls (123).

The pectic galactans are primarily β-4-linked polymers. The 4-linkage has been established by methylation analysis (94, 126). The galactosidic linkages were shown to be in the β-anomeric configuration by the fact that these linkages are susceptible to hydrolysis by a β-1,4-endogalactanase (80) and by their low positive optical rotation (80). In addition, oligosaccharides produced from the intact galactan by partial acid hydrolysis (126) are susceptible to further hydrolysis by a β-galactosidase.

Finally, the β-configuration of some of the galactosidic linkages has also been established by chromatographic comparison to known standards of oligosaccharides derived from the galactan by partial acid hydrolysis (96).

Those galactans which have been studied have degrees of polymerization ranging from 33 (126) to 50 (94). These values were obtained by vapor pressure osmosis (126) and by comparing the ratio of terminal to internal sugars as obtained by methylation analysis (94).

Galactans have been obtained which contain 6-linked in addition to 4-linked galactosyl residues (94, 96, 126). In two of the cases studied, the 6-linked residues accounted for approximately 4% of the polymer and are therefore quantitatively minor components of the polysaccharide (94, 126). On the other hand, beech galactan is a polysaccharide with a major amount of 6-linked residues although the amount of the polysaccharide accounted for by the 6-linked galactosyl residues has not been determined (96). The fact that 6-linked and 4-linked galactosyl residues are present in a single polymer has been established by isolation of the trisaccharide: Gal-(1$\xrightarrow{\beta}$6)-Gal-(1$\xrightarrow{\beta}$4)-Gal (96).

Homogalactans have not been isolated from primary cell walls, but the presence of such galactans in the walls is inferred by the detection of large amounts of 4-linked galactosyl residues upon methylation analysis of total cell walls and of pectic fractions of cell walls (123). In addition, small oligomers of β-4-linked galactosyl residues have been isolated in relatively large amounts from sycamore cell walls after treating the walls with an endo-β-1,4-galactanase which can only hydrolyze galactans which contain four contiguous β-4-linked galactosyl residues (80).

Although the pectic polysaccharides probably do contain β-4-linked homogalactans, many of the galactosyl residues of the pectic polysaccharides are probably not part of homogalactans (10, 12, 19, 23, 118, 123, 127). The galactosyl residues on one pectic polymer have been shown to occur as β-4-linked dimers rather than as longer oligosaccharides or polymers (19). There are several pectic polysaccharides that have been demonstrated to contain 3- and 6-linked galactosyl residues (10, 23, 123). The sycamore cell walls contain appreciable amounts of terminal and 3-, 6-, 3,6-, and 2,6-linked galactosyl residues (45, 123). It is likely that these galactosyl residues are part of an arabinogalactan (123); arabinogalactans are discussed in Section IV F.

Several galactose-containing oligosaccharides have been isolated from plant polysaccharides. These include Gal-(1→2)-Xyl, GlcA-(1→6)-Gal, and GlcA-(1→4)-Gal (18) and GalA-(1→4)-Gal (127). The occurrence of these oligosaccharides in primary cell walls has not been demonstrated.

Table 2. The Glycosyl Compositions (Mole %) of a Variety of Arabinogalactans

Glycosyl Residue	Rapeseed cotyledon[a]	Rapeseed flour[b]	Larch[c]	Maple sap[d]	Extracellular tobacco[e]	Extracellular Sycamore I[f]	Extracellular Sycamore II[g,h]	Soybean Cotyledon[i]
Arabinose	48	90	14	45	46	31	34	30
Galactose	46	10	84	51	44	69	31	71
Uronic acid	6[j]	0	0	0	3[k]	0	12[k]	0
Rhamnose	0	0	tr	5	8	0	4	0

[a] Isolated from rapeseed (*Brassica campestris*) cotyledon meal (*116*).
[b] Isolated from rapeseed (*Brassica campestris*) flour (*87*).
[c] Isolated from Japanese larch (*Larix leptolepis*) (*14*).
[d] Isolated from Maple sap (*Acer saccharum*) (*1*).
[e] Isolated from the medium of suspension-cultured tobacco (*Nicotiana tabacum*) (*72*).
[f] Isolated from the medium of suspension-cultured sycamore (*Acer pseudoplatanus*) cells (*24*).
[g] Isolated from the medium of suspension-cultured sycamore (*Acer pseudoplatanus*) cells (*75*).
[h] Also contains 14% xylosyl residues which are thought to arise from a contaminating xylan.
[i] Isolated from soybean (*Glycine max*) meal (*7, 98, 99*).
[j] Shown to be glucuronosyl residues.
[k] The type of uronic acid was not determined.

There is not sufficient information at this time for writing a preliminary structure of the galactans of primary cell walls. Clearly, such galactans exist, but whether they contain glycosyl constituents other than β-4-linked galactosyl residues has not been determined.

F. Arabinogalactan

Arabinogalactans have been isolated from the tissues of a variety of dicots. However, no arabinogalactan has been isolated from a source known to contain only primary cell walls. The glycosyl compositions of the arabinogalactans isolated from rapeseed cotyledons, rapeseed flour, larch wood, maple sap, the medium of suspension-cultured tobacco cells, the medium of suspension-cultured sycamore cells, and from soybean cotyledons are summarized in Table 2.

Unlike the arabans and galactans discussed earlier, there is considerable variation in the sugar compositions of the arabinogalactans. The arabinogalactan isolated from rapeseed (*Brassica campestris*) flour (*87*) contains 90% arabinosyl residues while the arabinogalactan isolated from larch (*Larix leptolepis*) (*14*) contains 88% galactosyl residues. Three of the arabinogalactans that have been studied have been demonstrated to contain rhamnosyl residues, whereas the other four arabinogalactans do not contain rhamnosyl residues. Three of seven arabinogalactans have been demonstrated to contain uronosyl residues.

The glycosyl linkage analyses of these arabinogalactans are summarized in Table 3. They clearly show that the arabinogalactan of soybean cotyledons is very different from the other arabinogalactans. The soybean arabinogalactan has a β-4-linked galactosyl backbone with arabinosyl dimers glycosidically linked to C-3 of some of the galactosyl residues (*7, 98, 99*). The arabinosyl dimers have the structure Ara_f-$(1 \rightarrow 5)$-Ara_f. The other arabinogalactans summarized in Table 3 are more similar to each other but still vary a great deal in the ratios of the arabinosyl and galactosyl residues. The differences in the glycosyl-linkage compositions (Table 3) reflect the difference in the glycosyl compositions (Table 2). Except for the soybean arabinogalactan, all of the arabinogalactans are characterized by the presence of significant amounts of 3,6-linked galactosyl residues and terminal arabinofuranosyl residues.

The arabinogalactans are also related to one another in that they all appear to have a galactan backbone with arabinosyl sidechains. The structures of these polysaccharides have been further investigated by partial acid hydrolysis and by periodate oxidation (*14, 72, 87, 116*). The arabinosyl residues are hydrolyzed preferentially by mild acidic

Table 3. Glycosyl Linkage Compositions (Mole %) of the Arabinogalactans Listed in Table 2

Glycosyl Linkage	Rapeseed cotyledon[a, d]	Rapeseed flour	Larch	Maple sap	Extracellular Tobacco	Extracellular Sycamore I	Extracellular Sycamore II[b]	Soybean cotyledon
T-Ara$_f$	32	48	+[c]	22	23	+++	38	14
T-Ara$_p$	0	1	+	0	15	0	0	0
2-Ara$_f$	0	0	0	0	0	0	1	0
3-Ara$_f$	11	0	+	11	0	0	1	0
5-Ara$_f$	2	3	0	0	15	+	2	14
2,5-Ara$_f$	2	40	+++	0	0	0	5	0
T-Gal	9	0	+	39	8	++	4	0
3-Gal	0	0	0	0	0	+0	6	0
4-Gal	5	0	++	0	2	++	0	57
6-Gal	0	0	0	0	0	0	6	0
3,4-Gal	29	8	++++	22	33	+++	33	14
3,6-Gal	5	0	0	0	0	0	0	0
3,4,6-Gal	0	0	0	6	6	0	3	0
T-Rha								

[a] The sources and references are the same as in Table 2.
[b] The xylosyl residues have been left out of this calculation (see Table 2).
[c] Approximate linkage quantitation is represented by + through ++++.
[d] Contains 7% T-GlcA.

conditions, suggesting that the arabinosyl residues are attached to the galactan by furanosidic bonds.

The disaccharide L-Ara$_p$-(1$\overset{\beta}{\rightarrow}$3)-L-Ara$_f$ has been isolated from larch arabinogalactan (14), while the disaccharide L-Ara$_p$-(1$\overset{\beta}{\rightarrow}$5)-L-Ara$_f$ has been isolated from the arabinogalactan of the extracellular medium of suspension-cultured tobacco (Nicotiana tabacum) cells (72). More extensive hydrolysis of the arabinogalactans isolated from larch, from the extracellular medium of suspension-cultured tobacco cells, and from the extracellular medium of suspension-cultured sycamore cells (14, 24, 72) have led to the isolation from each arabinogalactan of Gal-(1$\overset{\beta}{\rightarrow}$3)-Gal and Gal-(1$\overset{\beta}{\rightarrow}$6)-Gal. Clearly, these arabinogalactans have galactosyl residues attached to one another by both 1,6- and 1,3-linkages. SMITH degradation of the arabinogalactans, isolated from larch (14), from rapeseed (87, 116), and from the extracellular medium of suspension-cultured tobacco cells (72) supports the result obtained by partial acid hydrolysis in that the periodate oxidation studies do demonstrate that the backbones of the polysaccharides are galactans and that the galactosyl residues of the galactans are glycosidically linked to one another through either or both C-3 and C-6.

The presence of arabinogalactans in the primary cell walls is supported primarily by the results of a single study (123). The existence of an arabinogalactan was suggested in studies of an endopoly-galacturonase-released pectic fraction obtained from the walls of suspension-cultured sycamore cells (123). The main arabinosyl and galactosyl containing components of these pectic polysaccharides appeared to originate from a β-4-linked galactan and a highly branched araban. However, glycosyl linkages were also detected that are characteristic of arabinogalactans. The endopolygalacturonase-released pectic polysaccharides contained terminal and 3-, 6-, and 3,6-linked galactosyl residues. These residues were detected in amounts totaling approximately 5% of the pectic fraction. In addition, terminal, 3-, 5-, and 2,5-linked arabinofuranosyl linkages were also detected in substantial amounts in the pectic polysaccharides, but these residues could have originated from a branched araban.

Partial acid hydrolysis of the sycamore primary cell wall polysaccharides released by the endopolygalacturonase did not significantly alter the amounts of the branched galactosyl residues that were detected. This result suggests that the primary walls may contain an arabinogalactan which is similar to the one isolated from larch (14); the primary wall arabinogalactan may possess a low percentage of arabinosyl sidechains. These results are also consistant with the presence in primary walls of a branched galactan that lacks arabinosyl sidechains.

It is difficult to suggest a structural model for the arabinogalactan of primary cell walls. Indeed, since no arabinogalactan has been purified from the primary cell walls of dicots, it is not certain that arabinogalactans exist in these walls. In addition, the great variability among the arabinogalactans isolated from dicots would make it difficult to make any generalized model of arabinogalactans.

G. Rhamnogalacturonan II

A previously unknown pectic polysaccharide, rhamnogalacturonan II, has been isolated from the walls of suspension-cultured sycamore cells (*44*). Hydrolysis of rhamnogalacturonan II yields the rarely observed cell wall sugars 2-O-methyl fucose, 2-O-methyl xylose and apiose. The methylated sugars, 2-O-methyl xylose and 2-O-methyl fucose, have long been recognized as trace components of pectic polymers (*8, 15, 18, 30*). Apiose has also been recognized as a component of the pectic polysaccharides of *Lemna* species (see Section IV H). However, these three sugars have never previously been recognized to be associated in a single cell wall polysaccharide, although all three sugars have been isolated from leaves of deciduous trees (*26*). The previously isolated apiose-containing pectic polysaccharide is not structurally related to rhamnogalacturonan II.

Rhamnogalacturonan II is solubilized from the walls of suspension-cultured sycamore cells by the action of endo-α-1,4-galacturonase. Rhamnogalacturonan II is separated from the other pectic polysaccharides solubilized by the enzyme by anion exchange and gel permeation chromatography. As isolated, rhamnogalacturonan II is size homogeneous, containing between 25 and 50 glycosyl residues (Table 4).

Table 4. *Glycosyl Composition of Rhamnogalacturonan II*

Glycosyl Residue	Wt. % of Recovered Carbohydrate	Number of residues in a polymer 39 residues long
Galacturonic Acid	28	10
Rhamnose	18	7
Galactose	12	5
Arabinose	13	5
Apiose	7	3
2-O-Methyl Fucose	5	2
Glucuronic Acid	3	2
2-O-Methyl Xylose	3	2
Fucose	4	2
Glucose	2	1

Methylation analysis has provided information about the glycosyl linkages of which rhamnogalacturonan II is composed. The polysaccharide is characterized by a wide variety of terminal glycosyl residues including T-galacturonosyl, T-galactosyl, T-arabinosyl, T-2-O-methyl xylosyl, T-2-O-methyl fucosyl, and T-rhamnosyl. Rhamnogalacturonan II also contains 2-linked glucuronosyl, 3'-linked apiosyl, 3-linked rhamnosyl, 3,4-linked rhamnosyl and 3,4-linked fucosyl residues. The large amount of terminal glycosyl residues suggests a highly branched molecule. It is not yet possible to draw even a partial structure for rhamnogalacturonan II.

H. Apiogalacturonan

A component of the cell walls of duckweed *(Lemna minor)* has been isolated and identified as an apiogalacturonan, with apiosyl and galacturonosyl residues as the only components *(32, 51, 63, 64)*. Apiose containing galacturonans are also reported to be present in other plant tissues *(26)*, although these polysaccharides may contain other glycosyl components. Apiogalacturonans have never been isolated from a source containing only primary cell walls; therefore, their presence in such walls is still uncertain.

A partial structure of the apiogalacturonan from *Lemna minor* has been established by partial enzymic hydrolysis, by partial acid hydrolysis, and by periodate oxidation *(63, 64)*. A dimer of apiose (apiobiose) has been isolated following partial acid hydrolysis of the apiogalacturonan. Apiobiose has the structure D-Api$_f$-(1 → 3')-D-Api$_f$ *(64)*. Most of the apiose of the polysaccharide can be accounted for by this dimer. The homogalacturonan recovered following partial acid hydrolysis of the apiogalacturonan was degraded by a crude pectinase preparation providing some evidence that the galacturonsyl residues are α-4-linked *(63)*. The apiogalacturonan possesses a very low content of methyl-esterified galacturonosyl residues. It has not been ascertained whether the apiobiosyl sidechains are attached to the galacturonosyl residues through C-2 or C-3.

V. The Hemicelluloses

A. Xyloglucan

Xyloglucan is perhaps the most thoroughly understood of the noncellulosic polysaccharides of primary cell walls *(31)*. Xyloglucans were first characterized as an amyloid component of seeds *(21, 60, 67, 79,*

Table 5. Glycosyl Compositions (Mole %) of a Variety of Xyloglucans and Amyloids

Glycosyl Residue	Tamarindus[a]	Nasturtium[b,i]	Rape Seed I[c]	Rape Seed II[d]	BEPS[e]	REPS[f]	SEPS[i,g]	Sycamore[j,h] cell wall
Glucose	48	55	64	48	46	51	46	31
Xylose	36	27	24	34	36	30	37	36
Galactose	16	18	12	10	10	10	7	14
Fucose	0	0	0	7	8	7	6	7

[a] Isolated from Tamarindus indica seeds (41, 79).
[b] Isolated from nasturtium (Tropeoleum majus) seeds (41, 67).
[c] Isolated from Brassica campestris seeds (115).
[d] Isolated from Brassica campestris seeds (21, 119).
[e] BEPS = Bean extracellular polysaccharides. Isolated from the medium of suspension-cultured true bean (Phaseolus vulgaris) cells (calculated from data presented in ref. 133).
[f] REPS = Rose extracellular polysaccharides. Isolated from the medium of suspension-cultured Rose (Rosa glauca) cells (29).
[g] SEPS = Sycamore extracellular polysaccharides. Isolated from the medium of suspension-cultured (Acer pseudoplatanus) cells (24, 31).
[h] Isolated from the cell walls of suspension-cultured sycamore Acer pseudoplatanus (calculated from the data presented in ref. 31).
[i] For these sources of xyloglucan or amyloids, the glucosyl, galactosyl and xylosyl residues were shown to be in the D-configuration.
[j] This prepration is known to be contaminated with pectic polysaccharides.

115, 119). Considerably later, xyloglucans were isolated from the medium of suspension-cultured sycamore (*24*) cells, and, finally, from the primary cell walls of suspension-cultured sycamore cells (*31*). The basic structure of xyloglucans was elucidated by KOOIMAN (*79*) who studied the amyloid of *Tamarindus indica* seeds. The xyloglucans of primary cell walls were isolated and structurally characterized before it was recognized that the xyloglucans are very similar to the amyloids (*31*). The widespread occurrence of the amyloids (*21, 67, 79, 115, 119*) and xyloglucans (*24, 29, 31*) shows that polysaccharides isolated from tissues other than primary cell walls can, at times, serve as excellent models for the cell wall polysaccharides.

The composition of the xyloglucans isolated from a variety of sources is presented in Table 5. There are differences between the xyloglucans. For example, only half of the xyloglucans have been shown to contain fucosyl residues. However, all of the cells possessing primary cell walls produce xyloglucans containing fucosyl residues. It is possible that all xyloglucans contain fucosyl residues, when they are synthesized, but during differentiation (to secondary walls), the fucosyl residues may be removed. It is also possible that xyloglucans without fucosyl residues are present in primary cell walls but have not yet been detected.

The structure of the xyloglucans has been determined by a combination of methylation analysis and chromatographic separation of the oligosaccharides produced by partial enzymic digests. The only major differences in the methylation analyses (Table 6) of the various xyloglucans result from the presence or absence of the terminal fucosyl residues. The fucosyl-containing xyloglucans yield the same partially methylated partially acetylated alditols as the non-fucosyl-containing xyloglucans except for the addition of terminal fucosyl residues to C-2 of the majority of the otherwise terminal galactosyl residues.

The terminal residues were shown by KOOIMAN (*79*) to be linked via α-glycosidic bonds to the C-6 of glucosyl residues. KOOIMAN hydrolyzed *Tamarindus indica* xyloglucan with a crude commercial mixture of enzymes called "Luizym" and recovered from the enzymic digest almost all of the xylosyl residues as $Xyl-(1 \xrightarrow{\alpha} 6)-Glc$. He methylated the disaccharide to establish that the xylosyl residues are attached to C-6 of the glucose and also showed that the glycosidic linkage was the α anomer by demonstrating that the disaccharide possesses a highly positive optical rotation. KOOIMAN (*79*) also established that the α-linked disaccharide possessed a different melting point from that of the naturally occurring β-linked disaccharide, $Xyl-(1 \xrightarrow{\beta} 6)-Glc$. ASPINALL *et al.* (*21*) isolated $Xyl-(1 \xrightarrow{\alpha} 6)-Glc$ by acetolysis of rapeseed hull xyloglucan, characterizing the disaccharide by methylation analysis and by optical rotation.

Table 6. *Glycosyl-Linkage Compositions (Mole %) of the Xyloglucans and Amyloids Isolated from a Variety of Sources*[a]

Glycosyl Residue	Tamarindus indica[e]	Nasturtium	Rape Seed I[d]	Rape Seed II	BEPS	REPS	SEPS[c]	Sycamore Cell Walls
4-Glc	16	+[f]	20	13	11	17	13	13
4,6-Glc	33	+	32	39	33	30	32	29
T-Xyl	16	+	20	25	24	28	28	29
2-Xyl[b]	16	+	4	6	7	21	8	7
T-Gal	16	+	12	3	2	2	2	3
2-Gal	0	0	0	6	7	+	6	7
T-Fuc	0	0	0	6	8	+	5	7

[a] Xyloglucan and amyloid sources and references are the same as in Table 5.
[b] This figure for 2-linked Xyl may also contain some 4-linked Xyl in all the preprations except *Tamarindus indica* and nasturtium.
[c] Calculated from ref. *119*; original data from ref. *132*.
[d] Also contained 4% T-Glc and 8% 6-Glc.
[e] Also contained between one and two percent of both T-Ara$_f$ and 2,4,6-Glc.
[f] + = detected but not quantited.

It is assumed, because of KOOIMAN's results with the amyloids, that all of the xylosyl residues of the cell wall xyloglucans are linked to C-6 of glucosyl residues. At least a portion of the xylosyl residues of the sycamore wall and extracellular xyloglucans are linked to C-6 of glucosyl residues (*31*). This was established by analysis of enzyme-produced hepta- and nonasaccharides (see below). The anomeric configuration of the xylosidic bonds of the primary cell wall xyloglucans have not been determined, but, by analogy to the amyloids, these bonds may be assumed to be in the α configuration.

The galactosyl residues of xyloglucans are linked through a β-galactosidic bond to the C-2 of xylosyl residues. This was established by the isolation of Gal-(1 $\overset{\beta}{\rightarrow}$ 2)-Xyl following partial acid hydrolysis or acetolysis of the xyloglucans isolated from *Tamarindus indica* seed (*79*), from nasturtium *(Tropeoleum majus)* seed (*67*) and from rapeseed *(Brassica campestris)* hulls (*21*). The existence of the Gal-(1 → 2)-Xyl linkage in sycamore extracellular xyloglucan has been established by isolation and characterization (see below) of pentasaccharide "d" in Fig. 6 (*129*). At least a portion of the glucosyl residues of rapeseed hull xyloglucan has been shown to be β-linked by the isolation of cellobiose after partial acid hydrolysis (*21*). Similarly, at least a portion of the glucosyl residues of soybean and true bean cell walls and extracellular polysaccharides has been established to be β-4-linked by their susceptibility to hydrolysis by an endo-β-1,4-glucanase (*133*). The fact that xyloglucans hydrogen-bond to cellulose is strong additional evidence that the backbone of the xyloglucans is a β-4-linked glucan.

The terminal fucosyl linkages of sycamore extracellular xyloglucan were shown to be linked to the C-2 of the galactosyl residues by taking advantage of the acid lability of the fucosidic bonds. The fucosidic linkages are hydrolyzed when the xyloglucan is treated at pH 2 and at 110° for 1 hr. Following this treatment, most of the terminal fucosyl and 2-linked galactosyl residues disappear, while an equivalent amount of terminal galactosyl residues are formed (*31*). The anomeric configuration of this fucosidic linkage is not known, although the linkage is hydrolyzed by a mixture of enzymes known to contain an α-1,2-fucosidase (*27, 129*).

The structures of the xyloglucans isolated from the walls and culture medium of suspension-cultured sycamore cells have been further studied by digestion of the xyloglucans ,with a purified endo-β-1,4-glucanase (*31*). The endoglucanase-produced oligosaccharides have been fractionated by Bio-Gel P-2 chromatography into 4 quantitatively major components: a void peak, and oligomers composed of 22, 9 and 7 glycosyl residues. (The details of the chemical characterization of these oligosaccharides will not be repeated here but are found in Ref. *31*.)

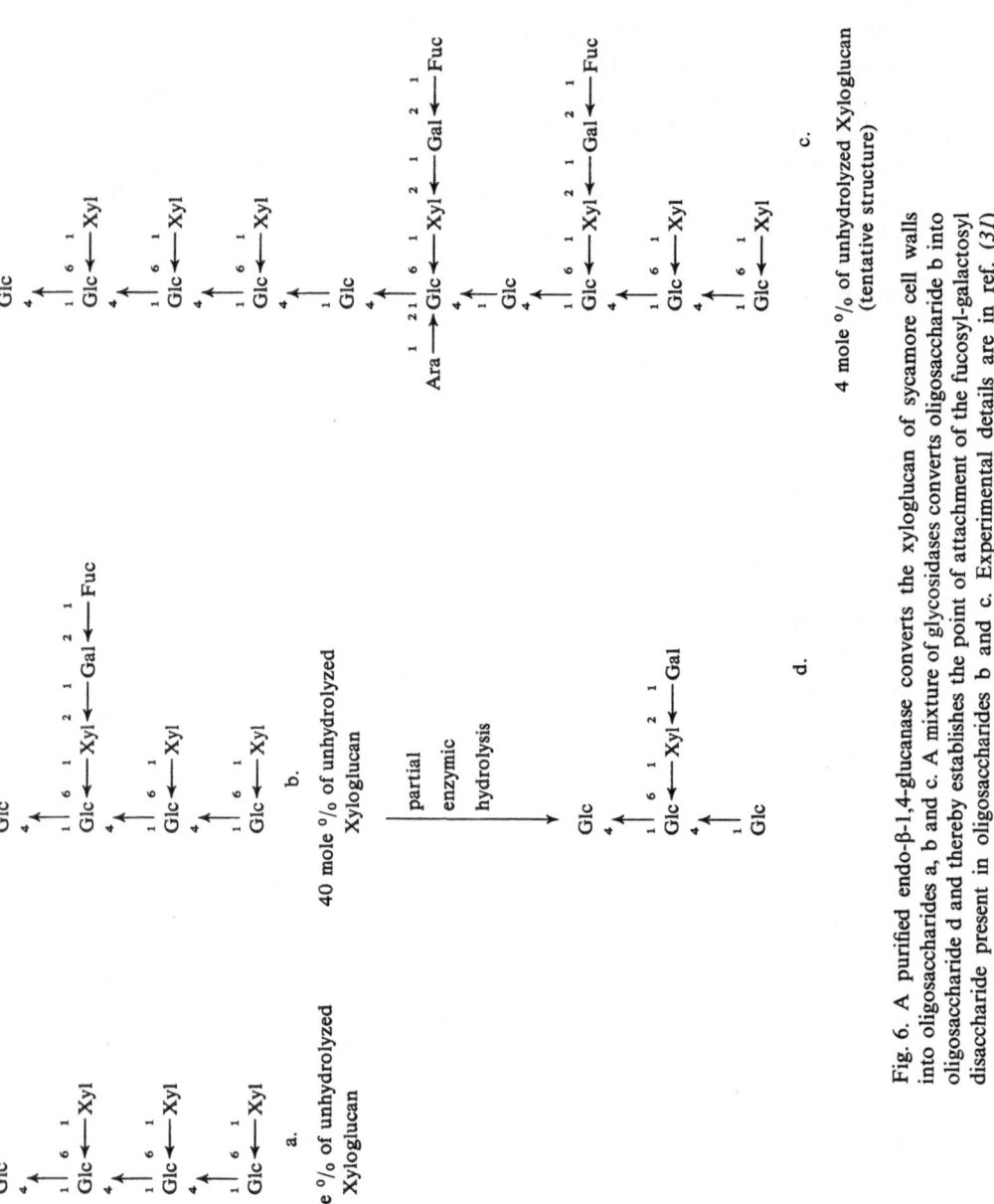

Fig. 6. A purified endo-β-1,4-glucanase converts the xyloglucan of sycamore cell walls into oligosaccharides a, b and c. A mixture of glycosidases converts oligosaccharide b into oligosaccharide d and thereby establishes the point of attachment of the fucosyl-galactosyl disaccharide present in oligosaccharides b and c. Experimental details are in ref. (31)

The structures and mole % of the total xyloglucan accounted for by each of these peaks are summarized in Fig. 6. The location of the Fuc-(1→2)-Gal disaccharide in the nonasaccharide was not originally determined (*31*). It has since been determined by partial enzymic hydrolysis of the nonasaccharide and isolation of the resulting definitive pentasaccharide labeled "d" in Fig. 6 (*129*).

The molecular weight of sycamore cell culture medium xyloglucan has been estimated as 7600 which represents about 50 glycosyl residues. This number is consistent with that found by KOOIMAN for *Tamarindus indica* seed xyloglucan (*79*).

The relative mole percents of the xyloglucan oligosaccharides a, b, and c (Fig. 6) suggests that these occur in the xyloglucan chains in a ratio of 10 : 10 : 1, respectively. The smallest possible xyloglucan that can be constructed from 10 repeats each of oligosaccharides "a" and "b" and a single copy of oligosaccharide "c" would contain 182 glycosyl residues. Since this value is much larger than the experimentally measured 50 glycosyl residues, it suggests that there exists at least two different xyloglucan species. For example, there could be one species made of a dimer of oligosaccharide c and another species made of 3 molecules each of oligosaccharides a and b. If this were true, the ratio of the two species would be one molecule of the former to approximately 7 molecules of the latter. Clearly, the exact nature of the xyloglucan is not known. Indeed, the following additional uncertainties concerning the structure of xyloglucans remain to be elucidated.

1) The Ara$_f$-(1→2)-Glu linkage shown in oligosaccharide "c" (Fig. 6) is based solely on the finding of equal molar amounts of terminal arabinofuranosyl and 2,4,6-linked glucosyl residues.

2) Some glucosyl residues could be attached to C-6 of other glucosyl residues with an equivalent amount of xylosyl residues attached to C-4 of glucosyl residues. This possibility has been ruled out in the case of *Tamarindus indica* seed xyloglucan, but has not been ruled out in sycamore cell wall xyloglucan.

3) The anomeric linkages have not been determined for sycamore cell wall xyloglucan. They are assumed to be consistent with the glycosidic linkages of the other xyloglucans, that is, all the glucosyl and galactosyl residues are β-linked, while the xylosyl residues are α-linked. The fucosyl linkage is also thought to be the α-anomer.

B. Xylan

Xylans, with a β-4-linked backbone, have generally been considered to be quantitatively major constituents of the secondary cell walls of dicots (*131*). The secondary cell wall xylans of different plants often differ in the nature of the sidechains glycosidically-linked to the xylan backbone.

The sidechain reported most frequently is a terminal 4-O-methyl-glucuronosyl residue (22, 28, 40, 53, 120, 125). Xylans characterized by the presence of these monomethylated mono-uronosyl sidechains have been isolated from rapeseed cotyledon meal (Brassica campestris), from midrib of tobacco (Nicotiana tabacum), from aspen wood (Populus tremuloides) and from the bark of white willow (Salix alba L.). The uronosyl residues are always linked to the C-2 of xylosyl residues. One uronosyl residue is linked to one in every 7 to 10 xylosyl residues (22, 120, 125). The xylan isolated from soybean (Glycine max) hull cell wall contains only one unmethylated glucuronosyl residue attached to about one in every 30 xylosyl residues (17).

Some purified xylans contain sugars other than xylose and glucuronic acid. For example, 4-linked glucosyl residues have been shown to be interspersed with the 4-linked xylosyl residues in the xylan backbone (28, 65). Two such (4-O-methyl-D-glucuronosyl) glucoxylans have been isolated from the sapwood of the lateral roots of the sugar maple (Acer saccharum Marsh). These glucoxylans have molar ratios of D-glucosyl to D-xylosyl to 4-O-methyl-D-glucuronosyl residues of 3 : 36 : 1 and 0.5 : 25 : 1 (28). Three neutral glucoxylans have been isolated from barberry (Berberis vulgaris) leaves; two of these gluco-xylans have small proportions (< 7%) of terminal, non-reducing galactosyl and arabinopyranosyl residues in the main chain (65). Galacturonosyl and rhamnosyl residues have also been reported in a xylan purified from birch (Betula verrucosa) meal (114). Three-linked rhamnosyl residues (about 1%) have been reported to be interspersed with 4-linked xylosyl residues in the backbone of a xylan purified from lucerne (Medicago sativa) stems (22).

A xylan has been isolated from the stalks of tobacco (Nicotiana tabacum) by alkaline extraction (52). Methylation analysis, acid hydrolysis and hydrolysis with an endo-β-1,4-xylanase confirm that the tobacco xylan is a linear, unbranched chain of β-4-linked xylosyl residues.

The first xylan, or, more accurately, a glucuronoarabinoxylan, to be clearly established as a constituent of the primary cell walls of a dicot has now been characterized (46). The glucuronoarabinoxylan constitutes 5% of the walls of suspension-cultured sycamore cells (46). The glucurono-arabinoxylan was extracted with alkali from walls which had been pre-extracted with endopolygalacturonase to remove most of the pectic polysaccharides. The alkali soluble polysaccharides were fractionated by anion exchange chromatography. A xylose-rich fraction was further purified by gel filtration on Bio-Gel P-100 and then on Agarose 1.5 m. The first gel filtration column removed contaminating pectic polysaccharides and the second column removed a glucan. The xylan-containing fractions from the Agarose column were shown to possess a constant sugar composition across the polysaccharide peak. The glycosyl composition of the glucuronoarabinoxylan is presented in Table 7. The glucuronoarabinoxylan has been shown, by methylation analysis, to contain terminal and 4-, 2,4- and 3,4-linked xylosyl residues. The polysaccharide also contains terminal and 2-linked arabinofuranosyl residues as well as terminal 4-O-methyl glucuronosyl and terminal glucuronosyl residues. Future experiments, including β-elimination of the uronosyl residues (see Section X), should yield a definitive covalent structure of this cell wall polysaccharide.

Table 7. *The Glycosyl Composition*
(Mole %) of Sycamore Cell Wall
Glucuronoarabinoxylan

Glycosyl Residue	Mole %
Xylose	69
Arabinose	17
Glucuronic Acid	10
4-O-Methyl Glucuronic Acid	2

VI. Non-Cellulosic Glucan

Non-cellulosic β-linked glucans have frequently been obtained from the tissues of monocots (*37, 38, 55, 102, 134*). There is one report of the isolation of a β-glucan from the cell walls of the hypocotyls of a dicot, namely mung beans *(Phaseolus aureus) (36)*. The glucan was extracted by hot water from cell walls prepared from 3-day-old hypocotyls. Extracts of the cell walls of older hypocotyls were deficient in this glucan. Methylation analysis and periodate oxidation studies *(36)* of the hypocotyl glucan indicate that the glucan contains 3-linked and 4-linked glucopyranosyl residues in the molar ratio of 1.0 : 1.7. Partial hydrolysis of the glucan released oligosaccharides containing both β-3- and β-4-linked glucosyl residues as well as other oligosaccharides containing only β-3- or β-4-linked glucosyl residues.

The hypocotyl tissue from which the β-glucan was obtained contains both primary and secondary cell walls. Thus, it is not known whether such glucans exist in primary cell walls of dicots. If such a glucan does exist in primary cell walls, the glucan would be difficult to classify. The glucan does not appear to be associated with galacturonosyl residues. Therefore, the glucan is not, by definition, a pectic polymer. The glucan, solubilized by hot water, is also not strongly hydrogen-bonded to cellulose. Therefore, the glucan is not, by definition, a hemicellulose.

VII. Cellulose

Cellulose is the most abundant polysaccharide in nature and is probably the most studied cell wall polymer. Most of the structural studies of cellulose have been carried out with material from secondary cell walls; little data is available for primary cell wall cellulose. The chemistry of cellulose has been reviewed numerous times (56, 78, 107). The present review considers only the major aspects of cellulose structure and assumes that the cellulose of primary walls is similar to that of secondary walls.

Cellulose is composed of long, linear chains of β-4-linked glucosyl residues (56, 107). The glucan chains aggregate by hydrogen-bonding along their lengths to form thin, flattened, rod-like structures often referred to as cellulose fibers (56, 57, 77, 78, 107). The most recent estimate of the cross sectional dimensions of these fibers (by electron microscopy) is 4.5×8.5 nm (107). A single such fiber has been estimated to consist of 60—70 glucan chains (107). However, such measurements are difficult to make accurately, in part because cellulose fibers may be enlarged by the adherence of other cell wall polysaccharides (107).

The degree of polymerization of the glucan chains within the cellulose fibers has been measured but the resulting estimates may not reflect the degree of polymerization *in vivo*. The degree of polymerization of the cellulosic glucan chains can only be estimated following solubilization of the glucans. The solubilization procedures are likely to break the glucan chains. The best available estimate of the degree of polymerization of the glucan chains of primary cell wall cellulose comes from Marx-Figini and Schulz (91, 92) working with cotton. These workers used viscometric methods to study derivitized glucan chains and determined a degree of polymerization of 6000—7000 (92). However, the literature (107 and references cited therein) contains a variety of chain length values for the cellulosic glucans of a variety of tissues.

It may be that the glucan chains of cellulose have no natural ends, that is, once a chain is initiated, it never ends except when a fiber is physically separated from its synthetic enzymes. This idea is supported by the electron microscopic observation that the cellulose fibers do not appear to have natural termination points. It may also be that the fibers have unlimited length but that the individual glucans have a finite length; the ends of the glucan chains may overlap to result in fibers of indeterminate length.

The aggregated glucans within a fiber are so ordered that they are, in fact, crystalline (56, 57, 77, 78, 107, 112). X-ray diffraction studies indicate that all of the glucan chains within a cellulose fiber have a parallel orientation, that is, the reducing ends of the glucan chains face in the same direction (57, 112). The X-ray diffraction studies were performed on the highly crystalline cellulose of the cell walls of the alga *Valonia ventricosa*. It seems likely, but it has not yet been established, that the glucan chains of primary cell wall cellulose also have a parallel orientation.

Purified cellulose invariably contains, in addition to a preponderance of glucosyl residues, minor amounts of other glycosyl residues (100, 107). One must consider the possibility that the non-glucosyl residues are normal constituents of the glucan chains, perhaps representing glucan chain termination points. Alternatively, the non-glucosyl residues may originate from noncovalently although tightly bound hemicelluloses. The latter possibility is supported by the evidence which indicates that the cellulose fibers of dicot primary cell walls are completely covered by hemicelluloses hydrogen-bonded to the fiber surface (31, 128). As the cellulose fibers of secondary cell walls have a considerably greater cross section than the cellulose fibers of primary cell walls (100, 107), it may be that the primary cell wall fibers aggregate to form secondary cell wall fibers. The non-glucosyl residues of cellulose may in fact originate from hemicellulose trapped between the aggregating cellulose fibers.

VIII. Cell Wall Protein

A. Hydroxyproline-Rich Proteins

The primary cell walls of dicots contain between 5 and 10% protein (83, 107, 123). The cell wall protein is exceptionally rich in hydroxyproline (~ 20%). The wall protein also has a relatively high content of alanine, serine and threonine. A high content of these amino acids is characteristic of the structural proteins of animals (83). This characteristic amino acid composition and the inability to extract much of the protein from cell walls under non-degradative conditions (84) indicates that the protein is a structural component of the primary cell wall (83, 85).

Fragments of the hydroxyproline-rich protein obtained from the primary cell walls of dicots invariably contain arabinosyl and galactosyl residues (83, 85). The hydroxyproline-rich protein fragments used for these studies have generally been isolated from the walls of suspension-cultured sycamore (Acer pseudoplatanus) and tomato (Lycopersicon esculentum) cells. A series of hydroxyproline arabinosides have been isolated from such wall preparations. The hydroxyproline-arabinosides are obtained by 0.2 M barium hydroxide hydrolysis of the peptide linkages of cell walls or of glycopeptides obtained by digestion of the walls with a crude mixture of polysaccharide- and protein-degrading enzymes. The hydroxyproline arabinosides are a mixture of mono-, di-, tri- and tetra-arabinosides glycosidically linked to the hydroxyl group of hydroxyproline. The mixture of hydroxyproline-arabinosides have been separated chromatographically on Chromobeads B (86). The hydroxyproline tetra-arabinoside is the predominant species of the dicot primary cell wall protein. In most analyses, no unglycosylated hydroproxyline is detected (81, 82, 86).

Methylation analysis of the tetra-arabinosides isolated from primary cell walls of tomato and sycamore indicates that the arabinosyl residues are terminal and 2- and 3-linked (71, 123). Akiyama and Kato (2) have studied the hydroxyproline arabinosides obtained from suspension-cultured cells of tobacco (Nicotiana tabacum). These workers used periodate oxidation (Smith degradation), methylation analyses, NMR, and optical rotation to show that the structure of the hydroxyproline tetra-arabinoside is: $\text{Ara}_f\text{-}(1 \xrightarrow{\beta} 3)\text{-Ara}_f\text{-}(1 \xrightarrow{\beta} 2)\text{-Ara}_f\text{-}(1 \xrightarrow{\beta} 2)\text{-Ara}_f\text{-}(1 \xrightarrow{\beta} 4)\text{-}$hydroxyproline.

Single galactosyl residues are glycosidically attached to the serine hydroxyls of the hydroxyproline-rich cell wall proteins of suspension-cultured tomato cells (84, 85). This was shown by removing the arabinosides from the intact cell walls by acid hydrolysis (pH 1 for 1 hr at 100° C). The hydroxyproline-rich wall protein, with arabinosyl residues removed, is susceptible to proteolysis with trypsin. The resulting solubilized tryptides have been separated by cation exchange and gel filtration chromatography. The composition of the tryptides has been determined by amino acid analysis (85). Some of the tryptides have been sequenced by subtractive N-terminus identification and further partial acid hydrolysis (85).

Each of the hydroxyproline-rich wall protein tryptides contains a pentapeptide of serine-(hydroxyproline)$_4$, while most of the tryptides contain one or more galactosyl residues (85). One tryptide, which was found to contain 2 residues each of galactose and serine, was subjected to β-elimination by several different methods. The elimination procedures converted serine to alanine or cysteic acid with a concomitant release of free galactose (85). These results demonstrated the covalent attachment of a single galactosyl residue to each seryl residue in the tryptide. Similar evidence has also been obtained for the existence of galactosyl-serine linkages in the hydroxyproline-rich glycoprotein of carrot cell walls (39).

The covalent attachment of arabinose and galactose to the hydroxy-proline-rich proteins of primary cell walls is a generally accepted fact (83, 85). However, the available evidence suggests that the hydroxyproline-rich glycoprotein is *not* covalently attached to any of the other cell wall polymers. The evidence does not rule out the possible existence of strong, non-covalent bonding between the hydroxyproline-rich glyco-protein and the other wall polymers.

A hydroxyproline-rich glycoprotein is secreted by suspension-cul-tured sycamore cells into their culture medium (75). The carbohydrate component of the glycoprotein is an arabinogalactan. The structure of the arabinogalactan portion of this glycoprotein has been studied by methylation analysis and found to be structurally similar to a protein-free arabinogalactan which is also present in the culture medium of suspension-cultured sycamore cells (see Section IV F).

The hydroxyproline-rich proteins of the cell walls and extracellular culture medium of suspension-cultured sycamore cells both contain arabinosyl and galactosyl residues. In spite of these compositional similarities, POPE (106) has found that the arabinogalactan-containing hydroxyproline-rich protein of the culture medium is structurally dis-similar from the hydroxyproline-rich protein of cell wall. He established this dissimilarity by comparing the hydroxyproline-arabinosides obtained by barium hydroxide hydrolysis of these two glycoproteins. POPE found that 80 mole % of the hydroxyproline arabinosides isolated from the cell wall glycoprotein can be accounted for by the tetra-arabinoside, while only 4 mole % of the hydroxyproline arabinosides obtained from the culture medium glycoprotein can be accounted for by the tetra-arabinoside. Perhaps, more importantly, he failed to find any arabinogalactan associated with the cell wall hydroxyproline-rich protein, while fully 50% of the glycoprotein of the extracellular macromolecules is accounted for by arabinogalactan.

It appears that the primary cell walls of suspension-cultured sycamore cells contain two hydroxyproline-rich proteins. One of these appears to be a structural protein found only in the wall, while the second is a glycoprotein which, in culture, is present in the wall in only small amounts and is predominently found in the culture medium. Of course, even the culture medium arabinogalactan hydroxyproline-rich protein is likely to be found in the cell wall in the intact tissues, as there is no culture medium for it to be dispersed in.

A composite model of a portion of the hydroxyproline-rich structural glycoprotein of primary cell walls is depicted in Fig. 7.

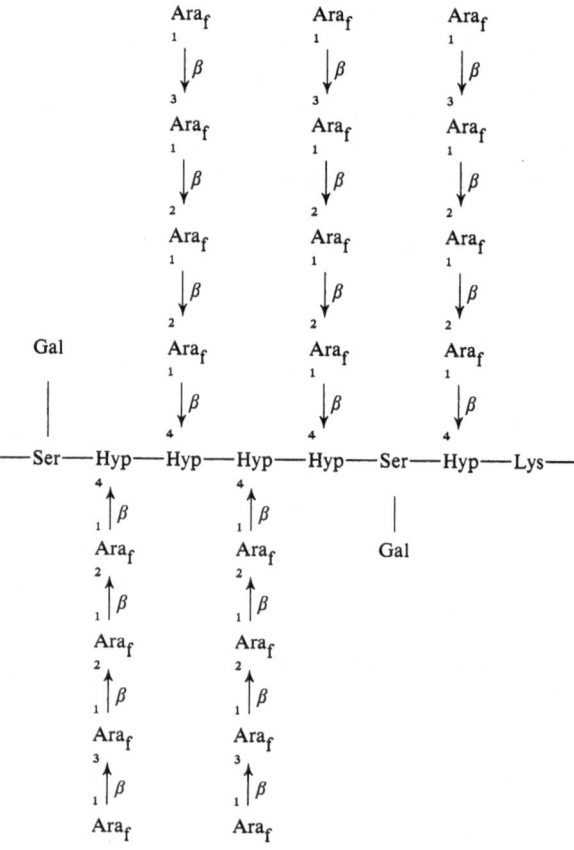

Fig. 7. A tenative model for a portion of the hydroxyproline-rich structural glycoprotein of dicot primary cell walls. This model is adapted from results described in ref. (2) and (85)

B. Hydroxyproline-Rich Glycoproteins With Lectin-Like Properties

Several of the hydroxyproline-rich glycoproteins extracted from plant tissues have carbohydrate-binding activity; these glycoproteins have the characteristics of lectins. These lectins or lectin-like glycoproteins have compositions which are similar to that of the cell wall hydroxyproline-rich glycoprotein.

A true lectin which was isolated from potato tubers and shown to bind N-acetyl glucosamine residues was the first lectin demonstrated to contain hydroxyproline (6). Allen and Neuberger (6) showed that the potato tuber lectin is composed of 50% protein and 50% carbohydrate. Hydroxyproline accounts for 16% of the amino acids of the lectin while arabinosyl residues account for 92% of the carbohydrate.

Recently, JERMYN and MAY YEOW (68) discovered that the seeds and/or tissues of a wide variety of dicots possess hydroxyproline-containing glycoproteins with lectin-like properties. It is not known whether these hydroxyproline-containing lectin-like glycoproteins are present in cell walls. Nevertheless, it is interesting that these glycoproteins are about 90% carbohydrate and only 10% protein; arabinosyl and galactosyl residues account for most of the carbohydrate of the lectins. These glycoproteins have been called "all-β" lectins as they bind to a variety of β-linked hexopyranosyl residues.

Two lectin-like protein fractions have been extracted from the cell walls of mung bean *(Phaseolus aureus)* seedlings (73, 74). These lectin-containing fractions bind specifically to galactosyl residues. It has not yet been ascertained whether these lectin-like proteins contain hydroxyproline. The existence of these galactosyl-binding proteins in the wall has led to the suggestion that the lectins may be involved in establishing a non-covalent protein-glycan network (73, 74).

IX. Interconnections Between the Primary Cell Wall Polymers of Dicots

A. Introduction

Chemical studies of primary cell walls are still largely concerned with elucidating the identity of the wall polymers as well as the structures of these polymers. Sufficient progress in characterizing the wall components has been made to allow some efforts to determine the manner in which the wall polymers interact with each other. The major effort in this direction has been the work of TALMADGE, BAUER, KEEGSTRA and ALBERSHEIM (31, 75, 123) which culminated in a preliminary model of the primary cell walls of dicots. This section will review the data concerning the interconnections of wall polymers and will propose some changes in the preliminary model.

B. The Pectic Polysaccharides are Covalently Interconnected

It has been established by many lines of evidence that the neutral pectic polysaccharides, the araban and galactan, are covalently attached to the acidic pectic polysaccharide, rhamnogalacturonan I (45, 123).

Araban, galactan, and both rhamnogalacturonan I and rhamnogalacturonan II are solubilized from isolated cell walls by a highly purified endopolygalacturonase. The endopolygalacturonase has been shown to be free of arabinase and galactanase activities (45, 123). A solubilized complex, containing the araban, galactan and rhamnogalacturonan I, co-chromatograph as a single acidic polymer on DEAE-Sephadex (45, 123). The araban and galactan would not be retained on the DEAE-Sephadex unless the neutral polysaccharides were strongly attached to the acidic rhamnogalacturonan. Further evidence for the interconnections of these polymers is provided by their co-chromatography on Agarose 5m (45). Particularly strong evidence for the covalent connection of these polymers comes from studies in which the glycosidic linkages to C-4 of the uronosyl residues of rhamnogalacturonan I have been chemically cleaved. Beta-elimination of the uronosyl residues, under a variety of conditions, results in a drastic reduction in the molecular size of both the araban and galactan. No araban or galactan has ever been extracted free of rhamnogalacturonan I from a primary cell wall.

Homogalacturonans have always been assumed to be attached to rhamnogalacturonans. One line of evidence for this has been the isolation from sycamore cell walls of oligogalacturonides containing 10 or more galacturonosyl residues in which the reducing ends of these oligogalacturonides are covalently attached to single rhamnose residues (123). Another line of evidence comes from the fact that both the homogalacturonans and the rhamnogalacturonans, with associated galactans and arabans, are released from the cell walls by the same enzyme, the endopolygalacturonase. In other words, glycosidic bonds susceptible to the same enzyme result in the solubilization of all of these polymers. Barrett and Northcote (30) have obtained evidence for the interconnection of these polymers in studies of a pectic fraction isolated from apple fruit. These workers have demonstrated that a pectic fraction rich in both neutral sugars and in galacturonic acid migrates as a single acidic component upon electrophoresis. However, a region of these pectic polymers rich in neutral sugars can be separated by electrophoresis from a region rich in galacturonic acid after the glycosidic bonds to some of the uronosyl residues have been cleaved by β-elimination.

The only evidence available at this time that rhamnogalacturonan II is covalently connected to the other pectic polysaccharides is that rhamnogalacturonan II is solubilized from the cell walls by the same endopolygalacturonase that solubilized the pectic polymers. The ability of the endopolygalacturonase to solubilize rhamnogalacturon II suggests very strongly that this polymer is connected to the wall through a series of 4-linked galacturonosyl residues. The question of whether rhamnogalacturonan II is connected to the other pectic polysaccharides is being investigated in our laboratory.

The pectic polysaccharides probably interact through non-covalent chemical bonding as well as through covalent bonding. Indeed, non-covalent interactions may well provide the most important inter-connections between the pectic polysaccharides and the other cell wall polymers. Calcium has long been demonstrated to confer rigidity to cell walls (121). The "egg-box model" of REES and his colleagues (61, 108, 108a) is an attractive model for the manner in which calcium strengthens cell walls. Calcium ions can fit between two or more chains of un-esterified polygalacturonic acid in such a fashion that the calcium ions chelate to the oxygen atoms of 4 different galacturonosyl residues on two different galacturonan chains. The result is the calcium ions packing somewhat like eggs in an egg-shaped box of polygalacturonans. This results not only in increased rigidity of the polyuronans, but also in cross-linking of the chains. The degree of interchain cross-linking due to the presence of calcium ions will be sensitive to the degree of methyl esterification of the galacturonans. Efforts have been made to correlate the degree of calcium cross-linking of galacturonans to the rate of cell wall elongation, but no such correlation has been established (121). The possibility of other types of non-covalent interactions between the pectic polysaccharides and other cell wall polymers certainly must be considered (47, 59), but there is little to indicate that such non-covalent interactions do exist in the cell wall. Efforts to determine whether such interactions within cell walls are structurally important are clearly called for.

C. The Hemicelluloses of Primary Cell Walls Bond Strongly to Cellulose Fibers

The quantitatively predominant hemicellulose of the primary cell walls of dicots is the xyloglucan. It has been proposed (31) that xylo-glucan bonds strongly, through multiple hydrogen bonds, to the surface of cellulose fibers. The following evidence supports the hypothesis that xyloglucan chains are hydrogen-bonded to the cellulose fibers: [1] the primary cell walls of dicots contain sufficient xyloglucan to form a monolayer-coating of the cellulose fibers of the walls (31, 75). [2] Space filling molecular models of xyloglucan show that xyloglucan is capable of forming multiple hydrogen bonds to cellulose (31). [3] The xylo-glucan may be extracted from a cellulose-xyloglucan complex by either alkali or 8 M urea, solvents which are known to break hydrogen

bonds, whether the complex has been formed *in vitro* or in the cell wall
(*31*). [4] The binding of the xyloglucan to the cell wall and to isolated
cellulose is reversible as would be expected for non-covalent hydrogen-
bonding (*31*). [5] Xyloglucans react quickly with and bond strongly to
isolated cellulose fibers in the absence of enzymes or chemical catalysis (*24, 31*). [6] Xyloglucans can be extracted from the cell wall and
from cellulose fibers in cell free systems by the action of an enzyme
which hydrolyzes the xyloglucan chains into small fragments, fragments which are not long enough to form stable hydrogen bond
complexes with cellulose (*31*). [7] Xyloglucan fragments which are too
short to form stable hydrogen bond complexes with cellulose can be
induced to form such complexes by reducing the water activity of the
solvent and thereby the opportunity for the fragments to hydrogen
bond with the solvent (*128*).

The bonding of xyloglucan to cellulose is clearly one of the major
interconnections of the cell wall polymers. This bonding may prevent
the cellulose fibers from adhering to each other to form enormous aggregates. Xyloglucan chains also offer the possibility of interconnecting
the cellulose fibers to other polymers of the primary cell walls.

Xyloglucan is not the only hemicellulose in dicot cell walls. The
glucuronoarabinoxylan, which has recently been characterized in the
primary walls of suspension-cultured sycamore cells, is structurally
related to the arabinoxylans and xylans, which are known to hydrogen
bond to cellulose (*31, 95, 103*). In addition, examination of the data
by Bauer *et al.* (*31*) suggests that arabinoxylans (or glucuronoarabinoxylans) were contaminating the xyloglucans extracted from primary
cell walls by urea. It seems likely that glucuronoarabinoxylans, and
perhaps xyloglucans, not only bind to cellulose in the cell wall, but also
bind to themselves. This possibility is increased by the observation
that plant arabinoxylans form aggregates in solution (*34, 46*). Dea *et al.*
(*47*) have published evidence that, in solution, arabinoxylan exists as a
mixture of random coils and aggregated linear chains. Such structures
can lead to gel formation and may, in fact, be involved in cross-
linking of the primary cell wall polymers.

D. Is the Hydroxyproline-Rich Glycoprotein of the Cell Wall Connected to the Other Polymers of the Cell Wall?

There is no evidence which demonstrates a covalent linkage between
the hydroxyproline-rich glycoprotein and the polysaccharides of cell
walls. The hydroxyproline-rich glycoprotein which is present in the
culture medium of sycamore cells grown in suspension culture has been

shown to be connected to an arabinogalactan (75) and this interconnection was suggested as a possible model for a cell wall interconnection between the protein and cell wall polysaccharides (75). However, POPE (106) has established that the hydroxyproline-rich protein of the extracellular fluid is structurally different from the hydroxyproline-rich protein of the cell wall and, thus, the extracellular polysaccharide model cannot be used as a model for the cell wall protein. The work of MONRO et al. (97) also supports a lack of a covalent linkage between the cell wall hydroxyproline-rich protein and the cell wall polysaccharide. On the other hand, there is a very real, albeit undemonstrated possibility that the hydroxyproline-rich glycoprotein is bonded through non-covalent interactions to the polysaccharides of the cell wall.

E. Are the Xyloglucan Chains Covalently Linked to the Pectic Polysaccharides?

Evidence has been provided for a covalent linkage between the xyloglucan chains and the pectic polysaccharides (31, 75, 123). The authors demonstrated that only a portion of the cell wall xyloglucan was covalently connected to the pectic polysaccharides (31, 75, 123). Later attempts by this laboratory to isolate large amounts of solubilized xyloglucan attached covalently to the pectic polysaccharides have been unsuccessful (94), although small amounts of xyloglucan in apparent covalent linkage with the pectic polysaccharides have been isolated (75, 94). The extent of the attachment of xyloglucans to the pectic polysaccharides and the structural importance of this interconnection remains unanswered.

F. The Current Cell Wall Model

There are four major types of polymers in the primary cell walls of dicots: hydroxyproline-rich glycoprotein, cellulose, hemicellulose and pectic polysaccharides. The mole percents of these polymers in suspension-cultured sycamore cell walls is presented in Table 8 (45, 75). The pectic polysaccharides are covalently interconnected. The hemicelluloses are non-covalently bonded to the cellulose fibers. There is no known attachment between the hydroxyproline-rich glycoprotein and the other cell wall polymers. The extent of covalent attachment between the hemicelluloses and the pectic polysaccharides is unknown. The fact that removal of the xyloglucans from the cell walls is enhanced by solubilization of the pectic polysaccharides (31, 75, 123) suggests that the hemicelluloses and the pectic polysaccharides do interact, whether it is by covalent or non-covalent bonding. Our present picture of the cell wall remains one in which the cellulose fibers are embedded in a layer of hemicellulose and that these fibers are interconnected by the pectic polysaccharides. It may be that the hydroxyproline-rich glycoprotein or other proteins within the wall, acting as lectins, participate in the cross-linking of the cell wall polymers (73, 74). The methods now available and being developed to determine the covalent and non-covalent interactions of the cell wall polymers should lead in the next few years to a far more detailed structural model of primary cell walls.

Table 8. *Polymer Composition of the Walls of Suspension-Cultured Sycamore Cells*

Wall Component	Wt. % of Cell Wall
A. Petic Polysaccharides	34
Rhamnogalacturonan I	7
Homogalacturonan	6
Arabinan	9
Galactan and possible Arabinogalactan	9
Rhamnogalacturonan II	3
B. Hemicelluloses	24
Xyloglucan	19
Glucuronoarabinoxylan	5
C. Cellulose	23
D. Hydroxyproline-rich Glycoprotein	19

X. The Future of Primary Cell Wall Research — New Methods

This review has stressed the need to isolate and chemically characterize the structures of the polymers of primary cell walls. Emphasis must be placed on working with cell wall polymers rather than with polymers obtained from other tissues and organelles. It is also important to obtain polymers from homogeneous preparations of primary cell walls rather than from walls isolated from tissues containing a mixture of wall types. Much of the early work on characterizing cell wall polymers was done with heterogeneous wall preparations. The availability of easily grown suspension-cultured cells offers the possibility of working with homogeneous preparations of primary cell walls. It is hoped that, the structural work of the future will be done with undifferentiated suspension-cultured cells or with other sources of homogeneous primary cell walls.

There is a continued need for new techniques to solubilize and fractionate cell wall polymers. The power of purified cell wall-degrading enzymes in solubilizing cell wall polymers has been convincingly demonstrated (123). Researchers in this field often avoid using purified enzymes because of the large amount of labor necessary to obtain the purified polysaccharide-degrading enzymes. In almost every case, it has not been a simple task to obtain an enzyme which attacks only a single type of glycosidic linkage within the wall. The increasing use of affinity chromatography should help to ease the task of those attempting to purify cell wall-degrading enzymes. This technique has been successfully used in our laboratory to purify a β-1,4-galactanase (80). However, most of the available wall-degrading enzymes have been purified by classical methods of chromatography. We believe that pure cell wall-degrading enzymes are almost essential for furthering the structural knowledge of primary cell walls.

A powerful method for structural elucidation of the pectic polysaccharides involves β-elimination. Methyl-esterified 4-linked uronosyl residues have long been known to undergo β-elimination (4, 101). The β-elimination was used by BARRETT and NORTHCOTE for structural investigations as early as 1965 (30). However, the β-elimination methods that have been available are limited in their value because the base used to catalyze the elimination reaction also degrades the glycosyl residue at the reducing end of the oligosaccharide eliminated from the uronosyl residue. The previously available β-elimination methods were also limited because they involved competition between the elimination reaction and the saponification of the methyl esters. The de-esterified uronosyl residues do not undergo the elimination reaction.

Methods have just become available which permit quantitative β-elimination of methylated polysaccharides containing methylated uronosyl residues (25, 88, 89). One of the newly characterized β-elimination methods has been extended for use with polysaccharides substituted with easily removed acetal groups instead of with methyl ether groups (42). Another β-elimination method is available which preserves the identity of the reducing sugar eliminated from the uronosyl residue (9, 20). An example of a reaction sequence using these recently developed methods is illustrated in Fig. 8. This reaction has been successfully used in our laboratory in the study of the pectic polysaccharides of suspension-cultured sycamore cells (45).

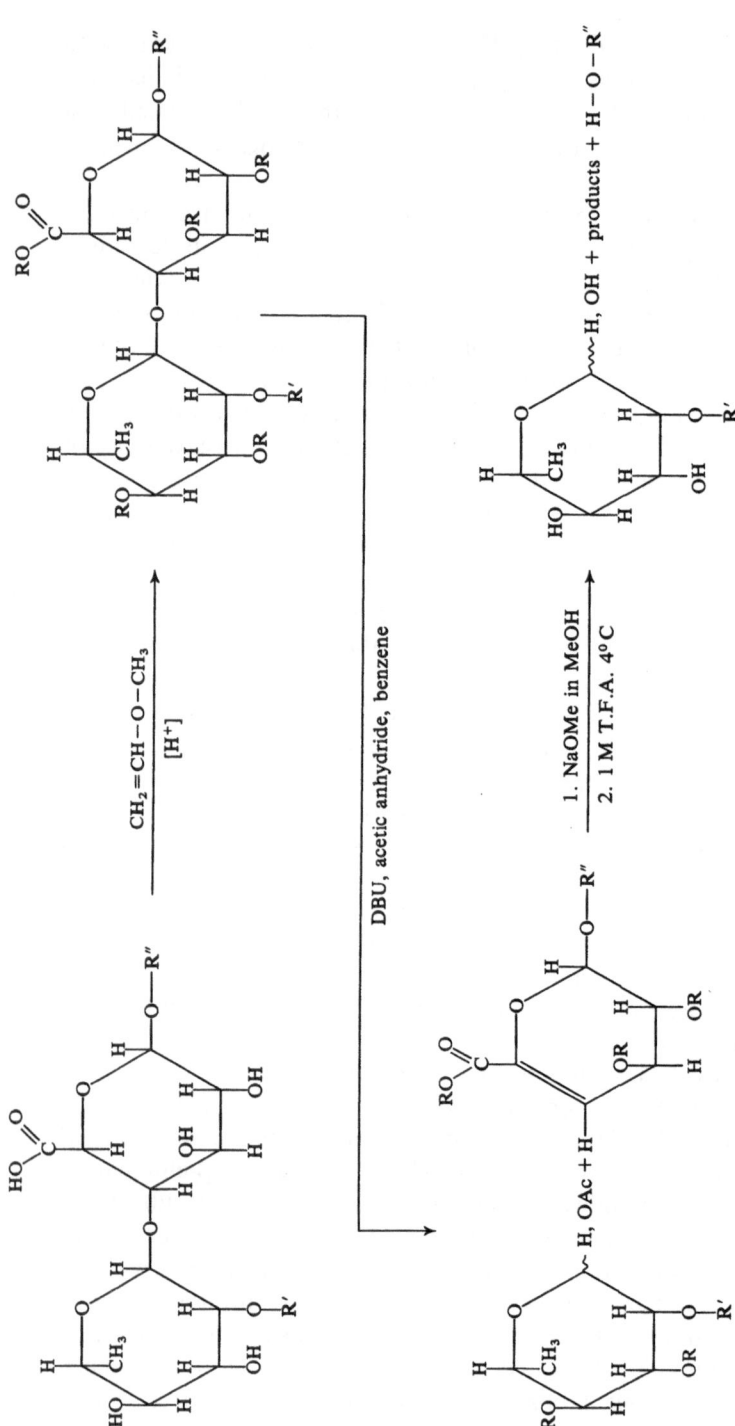

Fig. 8. A method for the β-elimination of those rhamnosyl residues attached to C-4 of galacturonosyl residues. This method (42, 20) leads to the recovery of the underivatized neutral glycosyl residues of such acidic polysaccharides

R′ and R″ are glycosyl residues

A powerful new method for the study of the structure of plant cell wall polysaccharides is high pressure liquid chromatography. Chromatographic columns and solvents are now available which allow the rapid and quantitative separation of oligosaccharides possessing a degree of polymerization up to 10 or larger. We have had success, in separating a variety of equal-length oligosaccharides. In addition, reverse phase high pressure liquid chromatographic columns are available which allow the highly efficient separation of relatively large permethylated oligosaccharides. Although it is in its infancy, the high pressure liquid chromatographic separation of methylated oligosaccharides, when combined with subsequent gas chromatographic-mass spectrometric analysis of the derived partially methylated partially acetylated alditols is opening the way for rapid and efficient sequencing of the sugars that compose oligo- and polysaccharides. Indeed, these methods offer the possibility of sequencing oligosaccharides at a rate of 1 to 2 orders of magnitude faster than has been possible until now. Most impressively, this increased rate of analysis can be carried out on amounts of oligosaccharide equivalent to at least two orders of magnitude less than were previously required for such studies.

Still another approach to the sequencing of the sugars of oligo- and polysaccharides is being pioneered by SVENSSON and his colleagues in the Department of Clinical Chemistry at the University Hospital in Lund, Sweden (*122*). These workers are developing methods using capillary column gas chromatography to separate methylated oligo-saccharides and are combining these separation techniques with direct mass spectrometric identification of the oligosaccharides present in the capillary column effluent. These methods, although technically somewhat challenging, offer the possibility of still easier sequence analysis on sub-milligram quantities of oligo- and polysaccharides. The methodology available for structural analysis of polysaccharides, including the polysaccharides of primary cell walls, is undergoing a revolution. This technological revolution will permit dramatic advances in our knowledge of cell wall structure.

References

1. ADAMS, G. A., and C. T. BISHOP: Constitution of an Arabinogalactan from Maple Sap. Can. J. Chem. **38**, 2380 (1960).

2. AKIYAMA, Y., and K. KATO: Structure of Hydroxyproline-arabinoside from Tobacco Cells. Agric. Biol. Chem. **41**, 79 (1977).

3. ALBERSHEIM, P.: The Primary Cell Wall. In: Plant Biochemistry (edited byVarner and Bonner). New York: Academic Press. 1976.

4. ALBERSHEIM, P., H. NEUKOM, and H. DEUEL: Splitting of Pectin Chain Molecules in Neutral Solutions. Arch. Biochem. Biophys. **90**, 46 (1960).

5. Albersheim, P., J. Nevins, P. D. English, and A. Karr: A Method for the Analysis of Sugars in Plant Cell-wall Polysaccharides by Gas-liquid Chromatography. Carbohyd. Res. **5**, 340 (1967).
6. Allen, A. K., and A. Neuberger: The Purification and Properties of the Lectin from Potato Tubers, a Hydroxyproline-containing Glycoprotein. Biochem J. **135**, 307 (1973).
7. Aspinall, G. O., R. Begbie, A. Hamilton, and J. N. C. Whyte: Polysaccharides of Soy-beans. Part III. Extraction and Fractionation of Polysaccharides from Cotyledon Meal. J. Chem. Soc. (C), 1065 (1967).
8. Aspinall, G. O., and A. Canas-Rodriguez: Sisal Petic Acid. J. Chem. Soc. 4020 (1958).
9. Aspinall, G. O., and A. S. Chaudhari: Base-catalyzed Degradations of Carbohydrates. X. Degradation by β-Elimination of Methylated Degraded Leiocarpan A. Can. J. Chem. **53**, 2189 (1975).
10. Aspinall, G. O., and I. W. Cottrell: Lemon-peel Pectin. II. Isolation of Homogeneous Pectins and Examination of Some Associated Polysaccharides. Can. J. Chem. **48**, 1283 (1970).
11. — — Polysaccharides of Soybeans. VI. Neutral Polysaccharides from Cotyledon Meal. Can. J. Chem. **49**, 1019 (1971).
12. Aspinall, G. O., I. W. Cottrell, S. V. Egan, I. M. Morrison, and J. N. C. Whyte: Polysaccharides of Soy-beans. Part IV. Partial Hydrolysis of the Acidic Polysaccharide Complex from Cotyledon Meal. J. Chem Soc. (C), 1071 (1967).
13. Aspinall, G. O., J. W. T. Craig, and J. L. Whyte: Lemon-peel Pectin. Part I. Fractionation and Partial Hydrolysis of Water-soluble Pectin. Carbohyd. Res. **7**, 442 (1968).
14. Aspinall, G. O., R. M. Fairweather, and T. M. Wood: Arabinogalactan A from Japanese Larch *(Larix leptolepis)*. J. Chem. Soc. (C), 2174 (1968).
15. Aspinall, G. O., and R. S. Fanshawe: Pectic Substances from Lucerne *(Medicago sativa)*. Part I. Pectic Acid. J. Chem. Soc. (C), 4215 (1961).
16. Aspinall, G. O., B. Gestetner, J. A. Molloy, and M. Uddin: Pectic Substances from Lucerne *(Medicago sativa)*. Part II. Acidic Oligosaccharides from Partial Hydrolysis of Leaf and Stem Pectic Acids. J. Chem. Soc. (C), 2554 (1968).
17. Aspinall, G. O., K. Hunt, and I. M. Morrison: Polysaccharides of Soy-beans. Part II. Fractionation of Hull Cell-wall Polysaccharides and the Structure of a Xylan. J. Chem. Soc. (C), 1945 (1966).
18. — — — Polysaccharides of Soy-beans. Part V. Acidic Polysaccharides from the Hulls. J. Chem. Soc. (C), 1080 (1967).
19. Aspinall, G. O., and K. S. Jiang: Rapeseed Hull Pectin. Carbohyd. Res. **38**, 247 (1974).
20. Aspinall, G. O., T. N. Krishnamurthy, W. Mitura, and M. Funabashi: Base-catalyzed Degradations of Carbohydrates. IX. β-Eliminations of 4-O-Substituted Hexopyranosiduronates. Can. J. Chem. **53**, 2182 (1975).
21. Aspinall, G. O., T. N. Krishnamurthy, and K. G. Rosell: A Fucogalactoxyloglucan from Rapeseed Hulls. Carbohyd. Res. **55**, 11 (1977).
22. Aspinall, G. O., and D. McGrath: The Hemicelluloses of Lucerne. J. Chem. Soc. (C), 2133 (1966).
23. Aspinall, G. O., and J. A. Molloy: Pectic Substances from Lucerne *(Medicago sativa)*. Part III. Fractionation of Polysaccharides Extracted with Water. J. Chem. Soc. (C), 2994 (1968).
24. Aspinall, G. O., J. A. Molloy, and J. W. T. Craig: Extracellular Polysaccharides from Suspension-cultured Sycamore Cells. Can. J. Biochem. **47**, 1063 (1969).
25. Aspinall, G. O., and K. G. Rosell: Base-catalyzed Degradations of Methylated

Acidic Polysaccharides: A Modified Procedure for the Determination of Sites of Attachment of Hexuronic Acid Residues. Carbohyd. Res. **57**, C23 (1977).

26. BACON, J. S. D., and M. V. CHESHIRE: Apiose and Mono-O-Methyl Sugars as Minor Constituents of the Leaves of Deciduous Trees and Various Other Species. Biochem. J. **124**, 555 (1971).

27. BAHL, O. P.: Glycosidases of *Aspergillus niger*. II. Purification and General Properties of 1,2-α-L-Fucosidase. J. Biol. Chem. **245**, 299 (1970).

28. BARDALAYE, P. C., and G. W. HAY: Structural Studies on the Hemicelluloses of the Roots of the Sugar Maple (*Acer saccharum* Marsh). Part 2. Two (4-O-Methyl-D-Glucurono-)Glucoxylans from the Sapwood of Mature Lateral Roots. Carbohyd. Res. **37**, 339 (1974).

29. BARNOUD, F., A. MOLLARD, and G. G. S. DUTTON: Une Xyloglucan β 1→4 présente dans le milieu de culture des suspensions cellulaires de *Rosa glauca*. Physiol. Vég. **15**, 153 (1977).

30. BARRETT, A. J., and D. H. NORTHCOTE: Apple Fruit Pectic Substances. Biochem. J. **94**, 617 (1965).

31. BAUER, W. D., K. W. TALMADGE, K. KEEGSTRA, and P. ALBERSHEIM: The Structure of Plant Cell Walls. II. The Hemicellulose of the Walls of Suspension-cultured Sycamore Cells. Plant Physiol. **51**, 174 (1973).

32. BECK, E.: Z. Pflanzen-Physiol. **57**, 444 (1967).

33. BJÖRNDAL, H., C. G. HELLERQUIST, B. LINDBERG, and S. SVENSSON: Gas-liquid Chromatography and Mass Spectrometry in Methylation Analysis of Polysaccharides. Angew. Chem. internat. Edit. **9**, 610 (1970).

34. BLAKE, J. D., and G. N. RICHARDS: Evidence for Molecular Aggregation in Hemicelluloses. Carbohyd. Res. **18**, 11 (1971).

35. BLUMENKRANTZ, N., and G. ASBOE-HANSEN: New Method for Quantitative Determination of Uronic Acids. Analyt. Biochem. **54**, 484 (1973).

36. BUCHALA, A. J., and G. FRANZ: A Hemicellulosic β-Glucan from the Hypocotyls of *Phaseolus aureus*. Phytochemistry **13**, 1887 (1974).

37. BUCHALA, A. J., and K. C. B. WILKIE: Non-Endospermic Hemicellulosic β-Glucans from Cereals. Die Naturwissenschaften **10**, 496 (1970).

38. — — The Ratio of β(1→3) to β(1→4) Glucosidic Linkages in Non-Endospermic Hemicellulosic β-Glucans from Oat Plant *(Avena sativa)* Tissues at Different Stages of Maturity. Phytochemistry **10**, 2287 (1971).

39. CHO, Y. P., and M. J. CHRISPEELS: Serine-O-Galactosyl Linkages in Glycopeptides from Carrot Cell Walls. Phytochemistry **15**, 165 (1976).

40. COMTAT, J., J. P. JOSELEAU, C. BOSSO, and F. BARNOUD: Characterization of Structurally Similar Neutral and Acidic Tetrasaccharides Obtained from the Enzymic Hydrolyzate of a 4-O-Methyl-D-Glucurono-D-Xylan. Carbohyd. Res. **38**, 217 (1974).

41. COURTOIS, J. E., and P. LE DIZET: Etude Comparée de la Structure de Trois Galactoxyloglucanes (amyloides) de Graines. Anales de Quimica **70**, 1067 (1974).

42. CURVALL, M., B. LINDBERG, and J. LÖNNGREN: Modification of Polysaccharides Containing Uronic Acid Residues. Carbohyd. Res. **41**, 235 (1975).

43. DANISHEFSKY, I., R. L. WHISTLER, and F. A. BETTELHEIM: Introduction to Polysaccharide Chemistry. In: The Carbohydrates IIA (edited by Pigman and Horton), pp. 375—410. New York: Academic Press. 1970.

44. DARVILL, A., M. MCNEIL, and P. ALBERSHEIM: The Structure of Plant Cell Walls VIII. A New Pectic Polysaccharide. Accepted for publication, Plant Physiology.

45. — — — Studies on the Pectic Polysaccharides of the Primary Cell Walls of Sycamore *(Acer pseudoplatanus)* Cultured Cells. Unpublished results of the authors.

46. DARVILL, J., A. DARVILL, M. MCNEIL, and P. ALBERSHEIM: Unpublished results.

47. DEA, I. C. M., E. R. MORRIS, D. A. REES, E. J. WELSH, H. A. BARNES, and J. PRICE:

Associations of Like and Unlike Polysaccharides: Mechanism and Specificity in Galactomannans, Interacting Bacterial Polysaccharides, and Related Systems. Carbohyd. Res. **57**, 249 (1977).

48. Dea, I. C. M., D. A. Rees, R. J. Beveridge, and G. N. Richards: Aggregation with Change of Conformation in Solutions of Hemicellulose Xylans. Carbohyd. Res. **29**, 363 (1973).

49. Omitted.

50. Dische, Z.: Colour Reactions of Carbohydrates. In: Methods in Carbohydrate Chemistry, Vol. I (edited by Whistler and Wolfrom), pp. 477—512. New York: Academic Press. 1962.

51. Duff, R. B.: The Occurrence of Apiose in *Lemma* (Duckweed) and other Angiosperms. Biochem. J. **94**, 768 (1965).

52. Eda, S., A. Ohnishi, and K. Kato: Xylan Isolated from the Stalk of *Nicotiana tabacum.* Agr. Biol. Chem. **40**, 359 (1976).

53. Eda, S., F. Watanabe, and K. Kato: 4-O-Methylglucuronoxylan Isolated from the Midrib of *Nicotiana tabacum.* Agric. Biol. Chem. **41**, 429 (1977).

54. English, P. D., A. Maglothin, K. Keegstra, and P. Albersheim: A Cell Wall-degrading Endopolygalacturonase Secreted by *Colletotrichum lindemuthianum.* Plant Physiol. **49**, 293 (1972).

55. Fraser, C. G., and K. C. B. Wilkie: A Hemicellulosic Glucan from Oat Leaf. Phytochemistry **10**, 199 (1971).

56. Gardner, K. H., and J. Blackwell: The Structure of Native Cellulose. Biopolymers **13**, 1975 (1974).

57. — — The Hydrogen Bonding in Native Cellulose. Biochim. Biophys. Acta **343**, 232 (1974).

58. Gilkes, N. R., and M. A. Hall: The Hormonal Control of Cell Wall Turnover in *Pisum sativum* L. New Phytol. **78**, 1 (1977).

59. Gould, S. E. B., D. A. Rees, N. G. Richardson, and I. W. Steele: Pectic Polysaccharides in the Growth of Plant Cells: Molecular Structural Factors and Their Role in the Germination of White Mustard. Nature **208**, 876 (1965).

60. Gould, S. E. B., D. A. Rees, and N. J. Wight: Polysaccharides in Germination. Xyloglucans ("Amyloids") from the Cotyledons of White Mustard. Biochem. J. **124**, 47 (1971).

61. Grant, G. T., E. R. Morris, D. A. Rees, P. Smith, and D. Thom: Biological Interactions Between Polysaccharides and Divalent Cations: The Egg-box Model. FEBS Letters **32**, 195 (1973).

62. Hall, M. A. (Editor): Plant Structure, Function and Adaptation. London and Basingstoke: The McMillan Press. 1976.

63. Hart, D. A., and P. K. Kindel: Isolation and Partial Characterization of Apiogalacturonans from the Cell Wall of *Lemna minor.* Biochem. J. **116**, 569 (1970).

64. — — A Novel Reaction Involved in the Degradation of Apiogalacturonans from *Lemna minor* and the Isolation of Apiobiose as a Product. Biochemistry **9**, 2190 (1970).

65. Henderson, G. A., and G. W. Hay: The Carbohydrates of the Leaves of Common Barberry *(Berberis vulgaris).* The Extraction, Fractionation and Structural Studies of Selected Non-Cellulosic Polysaccharides. Carbohyd. Res. **23**, 379 (1972).

66. Hirst, E. L., and J. K. N. Jones: Pectic Substances. Part VI. The Structure of the Araban from *Arachis hypogea.* J. Chem. Soc. 1221 (1947).

67. Hsu, D. S., and R. E. Reeves: The Structure of Nasturtium Amyloid. Carbohyd. Res. **5**, 202 (1967).

68. Jermyn, M. A., and Y. M. Yeow: A Class of Lectins present in the Tissues of Seed Plants. Aust. J. Plant Physiol. **2**, 501 (1975).

68a. Jiang, K. S., and T. E. Timell: Cellulose Chemistry. Technol. **6**, 499 (1972).

68b. JOSELEAU, J. P., G. CHAMBAT, M. VIGNON, and F. BARNOUD: Chemical and ¹³C-N. M. R. Studies on Two Arabinans from the Inner Bark of Young Stems of *Rosa glauca*. Carbohyd. Res. **58**, 165 (1977).

69. KAJI, A., and T. SAHEKI: Endo-arabinanase from *Bacillus subtilis* F-11. Biochim. Biophys. Acta **410**, 354 (1975).

70. KARACSONYI, S., R. TOMAN, F. JANECEK, and M. KUBACKOVA: Polysaccharides from the Bark of the White Willow (*Salix alba* L.): Structure of an Arabinan. Carbohyd. Res. **44**, 285 (1975).

71. KARR, A. L.: Isolation of an Enzyme System which will Catalyze the Glycosylation of Extensin. Plant Physiol. **50**, 275 (1972).

72. KATO, K., F. WATANABE, and S. EDA: An Arabinogalactan from Extracellular Polysaccharides of Suspension-cultured Tobacco Cells. Agric. Biol. Chem. **41**, 533 (1977).

73. KAUSS, H., and D. J. BOWLES: Some Properties of Carbohydrate-binding Proteins (Lectins) Solubilized from Cell Walls of *Phaseolus aureus*. Planta (Berl.) **130**, 169 (1976).

74. KAUSS, H., and C. GLASER: Carbohydrate-binding Proteins from Plant Cell Walls and their Possible Involvement in Extension Growth. FEBS Letters **45**, 304 (1974).

75. KEEGSTRA, K., K. W. TALMADGE, W. D. BAUER, and P. ALBERSHEIM: The Structure of Plant Cell Walls. III. A Model of the Walls of Suspension-cultured Sycamore Cells Based on the Interconnection of the Macromolecular Components. Plant Physiol. **51**, 188 (1973).

76. KIVIRIKKO, K. I., and M. LIESMAA: A Colorimetric Method for Determination of Hydroxyproline in Tissue Hydrolysates. Scand. J. Clin. Lab. Invest. **11**, 128 (1959).

77. KOLPAK, F. J., and J. BLACKWELL: Deformation of Cotton and Bacterial Cellulose Microfibrils. Textile Res. J. **45**, 568 (1975).

78. — — Determination of the Structure of Cellulose II. Macromolecules **9**, 273 (1976).

79. KOOIMAN, P.: The Constitution of *Tamarindus*-Amyloid. Recueil trav. Chim. Pays-Bas **80**, 849 (1961).

80. LABAVITCH, J. M., L. E. FREEMAN, and P. ALBERSHEIM: Structure of Plant Cell Walls. J. Biol. Chem. **251**, 5904 (1976).

81. LAMPORT, D. T. A.: Hydroxyproline-O-glycosidic Linkages of the Plant Cell Wall Glycoprotein Extensin. Nature **216**, 1322 (1967).

82. — The Isolation and Partial Characterization of Hydroxyproline-Rich Glycopeptides Obtained by Enzymic Degradation of Primary Cell Walls. Biochem. **8**, 1155 (1969).

83. — Cell Wall Metabolism. Ann. Review Plant Physiol. **21**, 235 (1970).

84. — Is the Primary Cell Wall a Protein-Glycan Network? Colloques internationaux C.N.R.S. **212**, 27 (1973).

85. LAMPORT, D. T. A., L. KATONA, and S. ROERIG: Galactosylserine in Extensin. Biochem. J. **133**, 125 (1973).

86. LAMPORT, D. T. A., D. H. MILLER: Hydroxyproline Arabinosides in the Plant Kingdom. Plant Physiol. **48**, 454 (1971).

87. LARM, O., O. THEANDER, and P. ÅMAN: Structural Studies on a Water-soluble Arabinogalactan Isolated from Rapeseed *(Brassica napus)*. Acta Chem. Scand. B **30**, 627 (1976).

88. LINDBERG, B., J. LÖNNGREN, and U. RUDÉN: Structural Studies of the Capsular Polysaccharide of *Klebsiella* Type 59. Carbohyd. Res. **42**, 83 (1975).

89. LINDBERG, B., J. LÖNNEGREN, and J. L. THOMPSON: Degradation of Polysaccharides Containing Uronic Acid Residues. Carbohyd. Res. **28**, 351 (1973).

90. LOWRY, O. H., N. J. ROSEBROUGH, L. FARR, and R. J. RANDALL: Protein Measurement with the Folin-phenol Reagent. J. Biol. Chem. **193**, 265 (1951).

91. Marx-Figini, M.: Comparison of the Biosynthesis of Cellulose *in vitro* and *in vivo* in Cotton Bolls. Nature **210**, 754 (1966).

92. Marx-Figini, M., and G. Schulz: Über die Kinetik und den Mechanismus der Biosynthese der Cellulose in den Höheren Pflanzen (nach Versuchen an den Samenhaaren der Baumwolle). Biochim. Biophys. Acta **112**, 74 (1966).

93. McNeil, M., and P. Albersheim: Chemical-ionization Mass Spectrometry of Methylated Hexitol Acetates. Carbohyd. Res. **56**, 239 (1977).

94. — — Studies on citrus pectin and sycamore cell wall. Unpublished results.

95. McNeil, M., P. Albersheim, L. Taiz, and R. L. Jones: The Structure of Plant Cell Walls. VII. Barley Aleurone Cells. Plant Physiol. **55**, 64 (1975).

96. Meier, H.: Studies on a Galactan from Tension Wood of Beech *(Fagus silvatica L.).* Acta Chem. Scand. **16**, 2275 (1962).

97. Monro, J. A., D. Penny, and R. W. Bailey: The Organization and Growth of Primary Cell Walls of Lupin Hypocotyl. Phytochem. **15**, 1193 (1976).

98. Morita, M.: Polysaccharides of Soybean Seeds. I. Polysaccharide Constituents of "Hot-Water-Extract" Fraction of Soybean Seeds and an Arabinogalactan as its Major Component. Agr. Biol. Chem. **29**, 564 (1965).

99. — Polysaccharides of Soybean Seeds. II. A Methylated Arabinogalactan Isolated from Methylated Product of "Hot-Water-Extract" Fraction of Soybean Seed Polysaccharides. Agr. Biol. Chem. **29**, 626 (1965).

100. Mühlethaler, K.: Ultrastructure and Formation of Plant Cell Walls. Ann. Review Plant Physiol. **18**, 1 (1967).

101. Neukom, H., and H. Deuel: Alkaline Degradation of Pectin. Chem. and Industry, June 1958.

102. Nevins, D. J., D. J. Huber, R. Yamamoto, and W. H. Loescher: β-Glucan of *Avena* Coleoptile Cell Walls. Plant Physiol. **60**, 617 (1977).

103. Northcote, D. H.: Chemistry of the Plant Cell Wall. In: Annual Review of Plant Physiology (edited by Machlis, Briggs and Park), Annual Review Inc., California (1972).

104. — Personal Communication.

105. Nusbaum, B., and P. Albersheim: Unpublished results.

106. Pope, D. G.: Relationships between Hydroxyproline-containing Proteins Secreted into the Cell Wall and Medium by Suspension-cultured *Acer pseudoplatanus* Cells. Plant Physiol. **59**, 894 (1977).

107. Preston, R. D.: The Physical Biology of Plant Cell Walls. London: Chapman and Hall. (1974).

108. Rees, D. A.: Shapely Polysaccharides. Biochem. J. **126**, 257 (1972).

108a. Rees, D. A., and N. G. Richardson: Polysaccharides in Germination. Occurrence, Fine Structure, and Possible Biological Role of the Pectic Araban in White Mustard Cotyledons. Biochem. **5**, 3099 (1966).

109. Rees, D. A., and E. J. Welsh: Secondary and Tertiary Structure of Polysaccharides in Solutions and Gels. Angew. Chem. Int. Ed. Engl. **16**, 214 (1977).

110. Rees, D. A., N. J. Wight: Molecular Cohesion in Plant Cell Walls. Biochem. J. **115**, 431 (1969).

111. Sandford, P. A., and H. E. Conrad: The Structure of *Aerobacter aerogenes* A3 (SI) Polysaccharide. I. A reexamination Using Improved Procedures for Methylation Analysis. Biochem. **5**, 1508 (1966).

112. Sarko, A., and Muggli, R.: Packing Analysis of Carbohydrates and Polysaccharides. III. *Valonia* Cellulose and Cellulose II. Macromolecules **7**, 486 (1974).

113. Sharon, N.: Complex Carbohydrates. Their Chemistry, Biosynthesis and Functions. Massachusetts: Addison-Wesley Publishing Co. 1975.

114. Shimizu, K., and O. Samuelson: Uronic Acid in Birch Hemicellulose. Svensk Papperstidning **76**, 150 (1973).

115. SIDDIQUI, I. R., and P. J. WOOD: Structural Investigation of Water-Soluble Rape-Seed *(Brassica campestris)* Polysaccharides. I. Rape-Seed Amyloid. Carbohyd. Res. **17**, 97 (1971).

116. — — Structural Investigation of Water-Soluble Rapeseed *(Brassica campestris)* Polysaccharides. II. An Acidic Arabinogalactan. Carbohyd. Res. **24**, 1 (1972).

117. — — Structural Investigation of Oxalate-Soluble Rapeseed *(Brassica campestris)* Polysaccharides. III. An Arabinan. Carbohyd. Res. **36**, 35 (1974).

118. — — Structural Investigation of Oxalate-Soluble Rapeseed *(Brassica campestris)* Polysaccharides. IV. Pectic Polysaccharides. Carbohyd. Res. **50**, 97 (1976).

119. — — Structural Investigation of Sodium Hydroxide-Soluble Rapessed *(Brassica campestris)* Polysaccharides. V. Fucoamyloid. Carbohyd. Res. **53**, 85 (1977).

120. — — Structural Investigation of an Acidic Xylan from Rapeseed. Carbohyd. Res. **54**, 231 (1977).

121. STODDART, R. W., A. J. BARRETT, and D. H. NORTHCOTE: Pectic Polysaccharides of Growing Plant Tissues. Biochem. J. **102**, 194 (1967).

122. SVENSSON, S.: Personal Communication.

123. TALMADGE, K. W., K. KEEGSTRA, W. D. BAUER, and P. ALBERSHEIM: The Structure of Plant Cell Walls. I. The Macromolecular Components of the Walls of Suspension-Cultured Sycamore Cells with a Detailed Analysis of the Pectic Polysaccharides. Plant Physiol. **51**, 158 (1973).

124. TAYLOR, R. L., and H. E. CONRAD: Stoichiometric Depolymerization of Polysaccharides and Glycosaminoglycuronans to Monosaccharides Following Reduction of their Carbodiimide-Activated Carboxyl Groups. Biochem. **11**, 1383 (1972).

125. TOMAN, R.: Polysaccharides from the Bark of White Willow *(Salix alba L.)*. Structure of a Xylan III. Cellulose Chem. and Tech. **7**, 351 (1973).

126. TOMAN, R., S. KARÁCSONYI, and V. KOVÁČIK: Polysaccharides from the Bark of White Willow *(Salix alba L.)*: Structure of a Galactan. Carbohyd. Res. **25**, 371 (1972).

127. TOMAN, R., Š. KARÁCSONYI, and M. KUBAČKOVÁ: Studies on the Pectin Present in the Bark of White Willow *(Salix alba L.)*: Fractionation and Acidic Depolymerization of the Water-Soluble Pectin. Carbohyd. Res. **43**, 111 (1975).

128. VALENT, B. S., and P. ALBERSHEIM: The Structure of Plant Cell Walls. V. On the Binding of Xyloglucan to Cellulose Fibers. Plant Physiol. **54**, 105 (1974).

129. VALENT, B. S., M. MCNEIL, and P. ALBERSHEIM: Unpublished results.

130. WEINSTEIN, L., and P. ALBERSHEIM: The Purification of an Endo-α-1,5-Arabinase and an α-Arabinosidase. Submitted for publication.

131. WHISTLER, R. L., and E. L. RICHARDS: Hemicelluloses. In: The Carbohydrates (edited by Pigman and Horton). New York: Academic Press. 1970.

132. WHITE, E. V., and P. S. RAO: Constitution of the Polysaccharide from Tamarind Seed. J. Amer. Chem. Soc. **75**, 1953.

133. WILDER, B. M., and P. ALBERSHEIM: The Structure of Plant Cell Walls. IV. A Structural Comparison of the Wall Hemicellulose of Cell Suspension Cultures of Sycamore *(Acer pseudoplatanus)* and of Red Kidney Bean *(Phaseolus vulgaris)*. Plant Physiol. **51**, 889 (1973).

134. WILKIE, K. C. B., and S.-L. WOO: Non-Cellulosic β-D-Glucans from Bamboo, and Interpretative Problems in the Study of all Hemicelluloses. Carbohyd. Res. **49**, 399 (1976).

135. WORTH, H. G. J.: The Chemistry and Biochemistry of Pectic Substances. Chem. Reviews **67**, 465 (1967).

136. ZITKO, V., and C. T. BISHOP: Structure of a Galacturonan from Sunflower Pectic Acid. Canadian J. Chem. **44**, 1275 (1966).

(Received January 12, 1978)

Dehydroamino Acids, α-Hydroxy-α-amino Acids and α-Mercapto-α-amino Acids

By Ulrich Schmidt, Johannes Häusler, Elisabeth Öhler, and Hans Poisel, Institut für Organische Chemie der Universität Wien, Austria

Contents

Acknowledgement. The authors express their sincere thanks to Dr. Adrian Stephen (Sandoz Forschungsinstitut, Wien) for the translation.

I. Introduction and Scope

The problem of the mechanism of the transformation of amino acids into α-keto acids claimed the attention of M. Bergmann as long ago as the early 1930s, long before the discovery of transaminations by pyridoxal-containing enzymes. Bergmann postulated dehydroamino acids as intermediates, formed by dehydrogenation of the corresponding amino acid in the peptide chain (49). These considerations led him to develop practicable syntheses of dehydroamino acid derivatives (39, 40, 46, 50), and to the study of enzymic cleavage of dehydropeptides (48, 49). The degradation of several dehydropeptides by crude pancreas preparations was discovered, and a "dehydropeptidase" isolated from kidneys, which cleaves dehydropeptides with a free carboxy group at the dehydroamino acid. Notwithstanding the publication in 1948 of a review article on these "dehydropeptidases" (149), their existence is today a controversial question. It is possible that the enzymes in question are simply normal carboxypeptidases.

During the last ten years the chemistry of dehydroamino acids and dehydropeptides has been the object of a resurgence of interest, which resulted from the isolation of numerous dehydropeptides, some of them possessing antibiotic activity, as metabolites of microorganisms. In this context, the biosynthesis of dehydropeptides has also acquired new significance. In addition to the β-elimination pathway from the corresponding serine or cysteine peptides, direct dehydrogenation of peptides in fungal metabolism appears to play a role (91, 204, 281). This latter possibility is suggested by the identification of dehydroamino-acids in several metabolites which correspond to no natural mercapto- or hydroxyamino acids. The isolation of the 7-methoxycephalosporins (1) (271), presumable formed by addition of methanol to the primary peptide dehydrogenation product — the acylimino species — lends additional support to this assumption. In addition the recently discovered "tryptophane side chain α,β-oxidase" points at a direct dehydrogenation (281).

The α,β-double bond in amino acid derivatives and peptides represents, in addition to the amino and carboxy groups, the introduction of a third highly reactive function into the molecule. It is therefore pertinent in a discussion of the α,β-dehydroamino acids to devote some attention to their primary addition products, such as derivatives of α-mercapto- and α-hydroxy-α-amino acids. Further topics relevant in this context are their relationship to β-hydroxy- and β-mercapto-α-amino acid derivatives (elimination-addition sequence), as well as syntheses and reactions of pyruvoylamino acids, which result from the hydrolysis of dehydropeptides and can possibly serve as precursors of the latter by condensation with amino acid amides. On the other hand, β,γ- and γ,δ-dehydroamino acids will be excluded from the scope of this discussion. The isolated double bonds of these compounds undergo the normal olefin reactions and display no unusual characteristics.

II. Occurrence and Biosynthesis of Dehydroamino Acids, Hydroxy Amino Acids, Mercapto Amino Acids and the Corresponding Dioxopiperazines

A. Occurrence

Numerous α,β-dehydroamino acids have in recent years been identified as constituents of fungal metabolites. Their characterization and typical structural features are summarized in Table 1. In most of these metabolites, almost all of which are cyclic compounds (cyclopeptides or cyclodepsipeptides) and which frequently possess antibiotic properties, D-amino acids also occur.

Some of the metabolites isolated from Streptomyces strains, such as ostreogrycin (112), griseoviridin (6, 7, 8, 88, 119, 120, 256), telomycin (363) and thiostreptone (61) are active against Gram positive bacteria. Nisin, which is used as a food preservative, and subtilin possess a broad antibiotic spectrum (155, 156, 160, 161, 200, 201, 277). Especially noteworthy are the tuberculostatic properties of capreomycin (*81, 137, 365*), tuberactinomycin (*191, 280, 420—427, 446*) and viomycin (*66, 67, 78, 79, 81, 82—85, 96, 97, 115, 165, 209—212*), which have also found clinical application. Berninamycin inhibits bacterial protein synthesis (*324*). In none of these dehydropeptides, however, has it been possible to identify a mode of action which can be assigned specifically to the presence of the dehydroamino-acid.

Table 1. *Occurrence of Dehydro Amino Acids in Natural Products*

	Type of Compound	Dehydro Amino Acid	X-Ray Analysis (X) Structure Elucidation (Str) Synthesis (Syn) Related Publications (Rel)
Alternariolide AM-Toxin	Cyclodepsipeptide	DH-Ala	Str.: (*290, 291, 418*) Syn: (*230*)
Althiomycin	Peptide	DH-Cys	Str: (*88A*)
Antibiotic A-2315	Cyclic Amide	DH-Ser	Rel: (*163*)
Antibiotic LL-BM 547 β	Cyclopeptide	β-Ureido-DH-Ala	Str: (*258*)
Antibiotic LL-BM 547 α (= De-β-lysyl-viomycin)	Cyclopeptide	β-Ureido-DH-Ala	Str: (*228, 229, 258*)
Antibiotic A-128-OP	Cyclopeptide	DH-Trp	Str: (*195A*)
Berninamycin A	Cyclopeptide	DH-Ala	Str: (*235*)
Bleomycin	Glycopeptide	DH-Cys	Str: (*408A, 419A*)
Capreomycin IA and IB	Cyclopeptide	β-Ureido-DH-Ala	Str: (*365*) Syn: (*365*)
Griseoviridin	Cyclic Amide	DH-Ser	X: (*88*) Rel: (*56, 256*)
Madumycin	Cyclodepsipeptide	DH-Ser	Str: (*67A*)
Micrococcin P	Cyclopeptide	DH-Aba DH-Cys	Str: (*427A*)
Nisin	Peptide with 4 Heterodetic Cycles	DH-Ala DH-Aba	Str: (*154*) Rel: (*151, 152, 153, 160*)
Nosiheptide	Cyclopeptide	DH-Ala DH-Aba DH-Cys	X: (*113, 296*) Str: (*317*)

Type of Compound	Dehydro Amino Acid	X-Ray Analysis (X) Structure Elucidation (Str) Synthesis (Syn) Related Publications (Rel)
Ostreogrycin A (= Virginiamycin) Cyclodepsipeptide	DH-Pro DH-Ser	Str: (112, 202) X: (114A)
Ostreogrycin G Cyclodepsipeptide	DH-Ser	Str: (202A)
Pencolide α-Imido Acrylic Acid Derivative	DH-Aba	Str: (55, 75, 405)
Siomycin Cyclopeptide	DH-Ala DH-Aba DH-Cys	Str: (415)
Stendomycin group Cyclodepsipeptides	DH-Aba	Str: (59, 60, 62)
Subtilin Peptide with 4 Heterodetic Cycles	DH-Ala DH-Aba	Str: (156) Rel: (152, 155, 157, 160, 200, 201, 277)
Telomycin Cyclodepsipeptide	DH-Trp	Str: (363)
Tentoxin Cyclotetrapeptide	N(Me)-DH-Phe	Str: (263, 264) Syn: (325) Rel: (224, 262)
Thiostrepton Cyclopeptide	DH-Ala DH-Aba DH-Cys	Str: (415) Rel: (11)
Tuberactinomycin O Cyclopeptide	β-Ureido-DH-Ala	X: (446) Str: (191, 427) Syn: (411) Rel: (426)
Tuberactinomycin A and N Cyclopeptide	β-Ureido-DH-Ala	Str: (421, 424, 427) Rel: (426)
Tuberactinomycin B (= Viomycin) Cyclopeptide	β-Ureido-DH-Ala	X: (83) Str: (170, 191, 209, 280, 427) Rel: (74, 213)
Versimide α-Imido Acrylic Acid Derivative	DH-Ala	Str: (76) Syn: (18, 77)

DH = α,β-Didehydro;

Aba = α-Amino-butyric acid; DH-Ser = 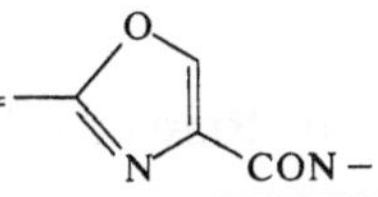   DH-Cys =

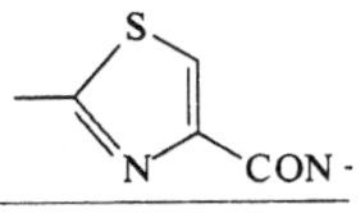

Dehydroalanine units have been identified at the active sites of the enzymes histidine ammonia lyase (*144, 434*) and L-phenylalanine ammonia lyase (*164*), either by reduction with NaB^3H_4 (*164, 434*), or by addition of nitromethane (*144*) and subsequent reduction to the amine. Following hydrolysis it was possible to detect tritiated alanine or α,γ-diamino-butyric acid respectively.

The only natural derivatives of α-hydroxy-α-amino acids found to date are the 7-methoxycephalosporins (**1**) (cephamycins) (*271*), an important class of therapeutically valuable antibiotics.

The available data on piperazinediones have been the subject of a recent detailed survey by P. G. Sammes (*341*). Since then only few additional reports on naturally occurring unsaturated cyclodipeptides (*91, 188, 189, 204, 248, 249*), 3-hydroxysubstituted cyclodipeptides (**2**) (*124, 440*) and epidithiocyclopeptides (*27, 36, 135, 260, 399*) have been published.

Epidithiopiperazinediones (**3**) are cyclodipeptides of α-mercapto-α-amino acids in the form of their disulfides. All compounds of this structural type possess high cell toxicity and have antiviral properties by virtue of their inhibition of RNA synthesis. The isolation, together with the epidithiopiperazinediones, of the analogous unsaturated piperazinediones is a significant finding with a bearing on biosynthetic pathways.

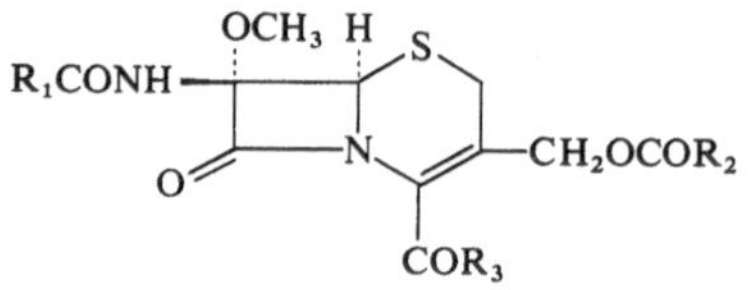

(**1**) 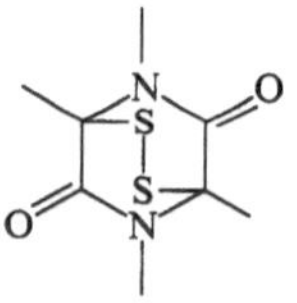   (**3**)

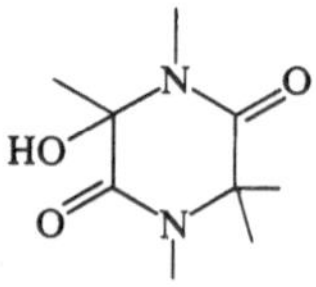

(**2**)

Scheme 1

B. Biosynthesis

The recently discovered enzyme "tryptophane side chain α,β-oxidase" dehydrogenates acetyltryptophanamide and other peptides containing tryptophane. Apart from the fact that no hydrogen peroxide is formed in this transformation, nothing is as yet known about the mechanism (*281*).

Only a few studies have been made on the stereochemistry of dehydro-amino acid formation *in vivo*. The biosynthetic elaboration of crypto-echinulin A (*91*) and mycelianamide (*203, 204*) has been the subject of feeding experiments with stereoselectively tritiated L-tryptophane and L-tyrosine respectively. *Cis*-dehydrogenation was established in both cases.

The addition of the cysteine mercapto group to a dehydrovaline unit was for long assumed to be a key step in the biosynthesis of penicillins and cephalosporins*. Subsequent workers visualized this addition as *following* the formation of the β-lactam ring (*105*). Tritium labelling experiments showed the earlier assumption of a "thioaldehyde" inter-mediate (dehydrocysteine) to be incorrect (*14*). Several groups have in-vestigated the steric course of the incorporation of valine into penicillin (*1, 22, 214*). If this biosynthesis proceeds *via* dehydrogenation of valine to dehydrovaline and subsequent addition of cysteine −SH**, two pos-sibilities are open: either *cis*-dehydrogenation followed by *trans*-addition, or *trans*-dehydrogenation and *cis*-addition. Most recent studies with labelled precursors have revealed that the postulated cysteinyldehydro-valine intermediate plays most probably no part in penicillin biosyn-thesis (*123, 187*), and that cephalosporin biosynthesis also occurs by a path independent of a dehydrovaline intermediate (*215*). On the other hand the isolation of the 7-methoxycephalosporins led to the assumption that this class of natural products is formed by methanol addition to a dehydro intermediate (acyliminocephalosporanic acid) (*271*).

The numerous compounds classed as peptide alkaloids (*429*) in-corporate a β-aryloxy-α-amino acid [e. g. frangulanine (**4**)] in their ansa bridge, conceivably formed by addition of a phenolic group to a dehydroamino acid***. In harmony with this suggestion is the isolation of a member of this class with an open chain structure (**5**) containing a dehydrovaline unit and a free phenolic hydroxy group (*247*).

Most of the fungal metabolites in which dehydroamino acid units occur (*80*) are cyclic peptides and also contain D-amino acids as structural elements. Ingenious speculations have been published concern-ing the significance of these observations with regard to biosynthesis

---

* For a recent summary of work on β-lactam biosynthesis see ref. (*2*).

** The assumption was made that this addition takes place after β-lactam formation (*105*). For biomimetic experiments see ref. (*24, 105, 274, 275*).

*** Attempted base catalyzed addition of phenol to dehydroamino acid derivatives *in vitro* has not been successful. A radically catalyzed addition (of a phenoxy radical) is at least conceivable.

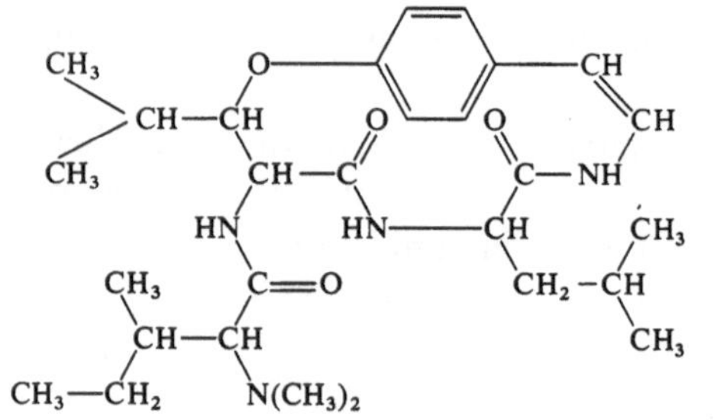

CH₃
  >CH—CH
CH₃   O   O   CH
         CH—C   C—NH   CH
       HN   HN———CH   CH₃
    CH₃   C=O   CH₂–CH
       CH—CH   CH₃
  CH₃—CH₂   N(CH₃)₂

(4) Frangulanine

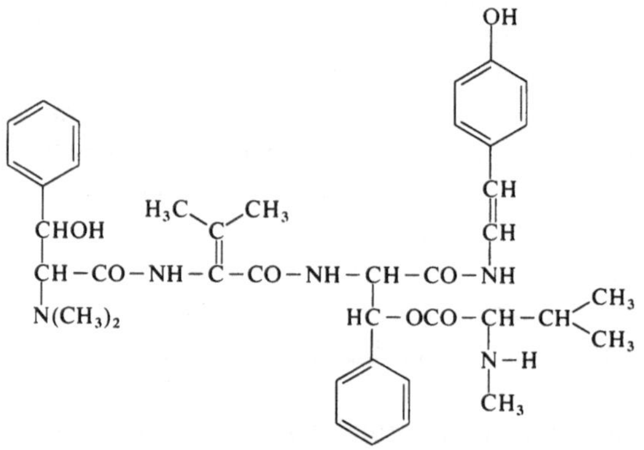

OH

CHOH   H₃C   CH₃   CH
CH—CO—NH—C—CO—NH—CH—CO—NH   CH
N(CH₃)₂   HC–OCO–CH–CH   CH₃
                N–H   CH₃
                CH₃

(5) Lasiodine A

Scheme 2

(255)*. As the biosynthesis of cyclic peptide antibiotica proceeds independently of the normal RNA-controlled peptide synthesis, one may presume that their elaboration takes place by way of an "enlargement" of cyclopeptides by insertion of L-amino acids, starting with piperazinediones.

The D-amino acid subunits of a cyclopeptide are not incorporated as such, but are formed by inversion of configuration within the peptide chain. It seems plausible to assume that this inversion takes place in the piperazinedione ring** or in the cyclopeptide *via* a dehydroamino acid intermediate.

---

    * Literature on the following data under ref. (255).
    ** Facile inversion of amino acid subunits in the piperazinedione ring is frequently observed. Furthermore, the stability of the *trans*-piperazinediones is in many cases greater, making possible the formation of an L-D-cyclopeptide from an L-L-cyclopeptide under thermodynamic control.

Although little is known about the formation of dehydroamino acid units in peptides and cyclopeptides, a number of possible pathways (Chart 1) are open to discussion:

A, B) The observation that the dehydroamino acids most frequently encountered are dehydroalanine and dehydroaminobutyric acid gives rise to the supposition that these are principally formed from serine, cysteine and threonine units by β-elimination reactions. A large number of β-eliminations have been carried out *in vitro* on S-methylcysteine, O-tosylserine and O-phosphorylserine units incorporated in a peptide chain (cf. Section IV.A.2).

O, D) A hypothetical pathway is the formation of an acylimino acid unit by N-hydroxylation of the amino acid within the peptide chain followed by loss of water. α-Acylimino carboxylic acid derivatives readily rearrange to acylenamino carboxylic acid derivatives (*314*). N-Hydroxy compounds, such as mycelianamide and pulcherrimic acid [for a review on cyclic hydroxamic acids see ref. (*26*)], have been found within the class of dehydrocyclopeptides. Although N-hydroxypyrazinones are common, it is not known whether these compounds are biogenetically linked with the piperazinediones. It is appropriate to point out that N-hydroxylation of amides is a familiar degradative pathway in the metabolic processes of higher organisms (*412*). The facile elimination of water from N-acylhydroxylamino acids to form acylenamino acids (*377*) should also be noted in this context.

M) The possibility that water may be eliminated from α-hydroxy-α-amino acids must also be mentioned. α-Hydroxy-α-amino acids can be formed by the addition of amides to α-keto acids, a reaction of which many examples *in vitro*, both intermolecular and intramolecular, are known (cf. Sections IV.A.6 and V.). The conversion of α-hydroxy-α-amino acids into dehydroamino acids or α-mercapto-α-amino acids is a further step of which examples *in vitro* can be cited (cf. Section V.).

K) A final route which cannot a priori be excluded as a possible biosynthetic pathway is the direct α-oxidation of amino acid derivatives to α-hydroxy-α-amino acid derivatives followed by elimination of water. *In vitro* oxidation of piperazinediones to α-hydroperoxides occurs very readily (cf. Section V.). The isolation of numerous dehydropiperazinediones and corresponding mercaptopiperazinediones containing proline (or "benzoproline", i.e. dihydroindole) units (*341*) necessitates the consideration of this route as a possible biosynthetic pathway. The hydroperoxypiperazinediones can readily be converted into dehydro- and mercaptopiperazinediones.

Chart 1 summarizes the possible biological relationship between dehydroamino acids and amino acids, α- and β-hydroxy-α-amino acids, α- and β-mercapto-α-amino acids, and α-keto-acids and their aminals. All these reactions may have their parallels *in vitro*.

Chart 1. Possible Biological Relationship Between Dehydroamino Acids and Amino Acids, α- and β-Hydroxy-α-amino Acids, α- and β-Mercapto-α-amino Acids, α-Keto-acids and Their Aminals

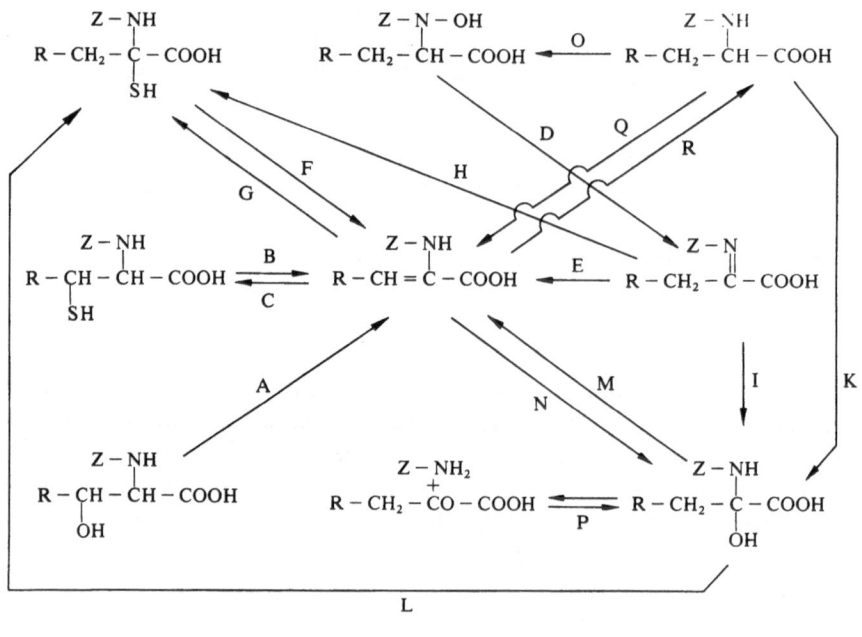

cf. section:

A	IV.A.2.1.	K	V.B.
B	IV.A.2.2.	L	V.C.
C	VI.A.2.1.	M	IV.A.3, IV.B.3
D	IV.A.4.	N	VI.A.2.2, VI.A.2.3
E	IV.F.	O	II.B.
F	Ref. (283)	P	IV.A.6, VI.A.2.4.
G	VI.A. 2.2	Q	IV.C., IV.B.4, IIB
H	Ref. (314)	R	VI.A.2.5.
I	VI.B.		

III. Determination of Structure and Configuration of Dehydroamino Acid Derivatives

The structures of complex dehydrocyclopeptides have in many cases been established by X-ray crystallography (cf. Table 1). Hydrolytic degradation and amino acid analysis have also been frequently applied, the dehydroamino acid being isolated and identified as the α-keto acid (cf. Section VI.A.2.3). Reports are also to be found in the literature of mercaptan addition to the dehydropeptide followed by hydrolysis and identification of the β-substituted α-amino acid formed by addition to the double bond (62, 154, 155).

Assignments of the configuration of acylenamino acid derivatives depend mainly on ¹H-NMR spectroscopic measurements, in particular because the influence of the carboxy and acylamino substituents on the chemical shifts of β-alkyl and vinyl protons. Thus BROWN and SMALE (75) have shown that the vinyl proton in 2-benzamido- and 2-phthalimido-crotonic acid esters absorbs at lower field when it and the acylamino group are mutually cis (E isomer). The configurational assignments deduced in this study from the chemical shifts of the β-methyl protons must however be corrected (see below).

AUSTEL and STEGLICH (20) based their assignments of configuration at the double bond on a study of the chemical shifts measured for the protons of the 4¹-substituents of alkylidene-2-trifluoromethyl-Δ²-oxazolinones before and after ring opening to trifluoroacetylamino carboxylic acids. They found that the protons of β-alkyl groups are more strongly deshielded when the latter are cis to the carboxy group than when they are trans [cf. (334)]. This method has been exploited by H. POISEL

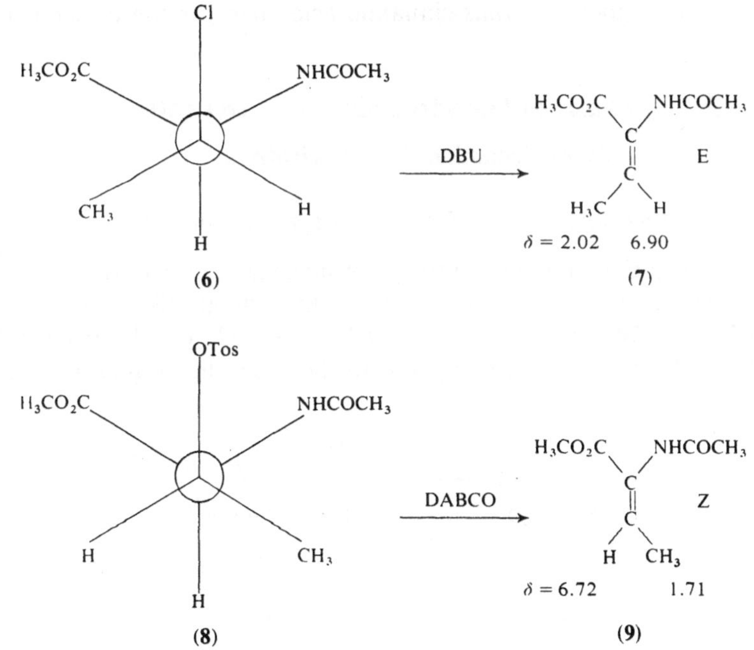

Scheme 3

and U. Schmidt (*310*), and by Olsen *et al.* (*393*) in determining the stereochemistry of a series of acyl-enamino acid esters. Their results agree with those derived by Olsen from a systematic study of elimination reactions (*395*). Thus elimination of HCl from erythro-α-acylamino-β-chloro-D,L-butyrate (**6**) by brief treatment with diazabicycloundecene yielded predominantly the *E*-isomer (**7**) expected from a *trans* E_2-elimination mechanism. Prolonged reaction times, or use of diazabicyclooctane as base, caused partial isomerization of the initially formed *E*-isomer and mixtures of *Z*- and *E*-isomers (**7**) and (**9**) were isolated. Elimination of toluenesulfonic acid from N-acyl-O-tosyl-D,L-threonine methyl ester (**8**) led to the sole formation of the stable *Z*-isomer (**9**), an observation which agrees with the results of earlier work by H. Poisel and U. Schmidt (*310, 314*) who found that, under both acidic and basic conditions, the *Z*-isomer in a series of acylenamino acid esters is the thermodynamically more stable product.

It should be noted that the NMR spectroscopy criteria detailed above are not the only available methods applicable for the determination of configuration. U. Schöllkopf and coworkers (*360*) have correlated the configuration of isomeric β-substituted formylaminoacrylic acid esters with the chemical shifts of the alkoxy groups. UV data have been used to assign configurations to the isomeric dehydrophenylalanine derivatives (*325*), an approach based on the ε-values for cinnamic acid, which are higher for *trans*-cinnamic acid than for the *cis*-isomer.

IV. Synthesis of Dehydroamino Acid Derivatives

A. Acylenamino Acid Derivatives

1. Acylenamino Acid Derivatives via Oxazolinones

One of the earliest known syntheses of acylenamino acid derivatives (**11**) consists of the ring opening of 4-alkylidene-(aralkylidene-)-Δ^2-oxazolin-5-ones (**10**) by attack of a nucleophile HY on the carbonyl group. As nucleophiles also readily add to the exocyclic double bond of

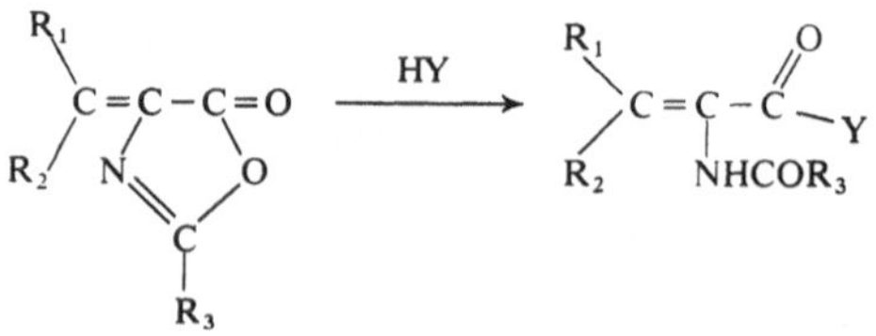

Scheme 4

such oxazolinones with ultimate formation of β-substituted amino acid derivatives, the success of this synthesis depends on the addition step being reversible. For example, reaction with OH^-, OR^- or NHR_1R_2 leads to the acylenamino acid derivatives, while powerful nucleophiles such as thiolate or azide react further.

The chemistry of unsaturated oxazolinones has been the subject of several detailed reviews (*25, 95, 106, 107, 126, 127, 322, 398*) so that attention here will be focussed primarily on those more recent studies not dealt with in the literature cited. Unsaturated oxazolinones for the synthesis of dehydropeptides can be obtained by several different routes:

α) by Erlenmeyer synthesis,
β) by Bergmann-Stern synthesis,
γ) by dehydration of β-hydroxy-α-acylamino acids,
δ) by dehydrogenation of saturated oxazolinones.

α) The classical Erlenmeyer synthesis — the condensation of acylglycines with aldehydes and ketones in the presence of acetic anhydride/sodium acetate — is above all suitable for the preparation of unsaturated oxazolinones with an aryl substituent in position 2 or 4 (*95*). STEGLICH has compiled a summary of the reaction conditions employed in this condensation (*398*). The development of dehydropeptide synthesis from unsaturated oxazolinones, in particular from those accessible by Erlenmeyer synthesis, is primarily due to BERGMANN and coworkers (*29, 41—44, 51, 53, 114*). In contrast to saturated oxazolinones which are readily hydrolysed by water or alcohol, acid or basic catalysts are necessary for the hydrolysis of unsaturated (stabilized) oxazolinones. In most cases a smooth reaction is observed with amines and the sodium salts of amino acids. BERGMANN *et al.* applied this principle to the preparation of a whole series of dehydrophenylalanine peptides, e.g. (**12**), with Gly, Leu, Glu, Tyr, Arg, Ser, Phe, Phenylserine and Cys as C-terminal amino acid. They also succeeded (*114*) in condensing peptides with a C-terminal glycine unit with aldehydes (benzaldehyde, p-hydroxybenzaldehyde) under the conditions of the Erlenmeyer synthesis to form oxazolinones of unsaturated peptides, which could be opened by alkaline hydrolysis to yield doubly unsaturated peptides, e.g. (**13**). Appropriate application of this principle permitted the construction of tetra- and pentapeptides with 2—4 dehydroamino acid units, although it is probable that the products obtained were mixtures of stereoisomers because of the possibility of *cis-trans*-isomerism at the α,β-double bond (*322*). More recently, PIERONI *et al.* (*306*) have prepared numerous peptides with one or two dehydrophenylalanine units in the same way, and determined their structure by modern spectroscopic methods.

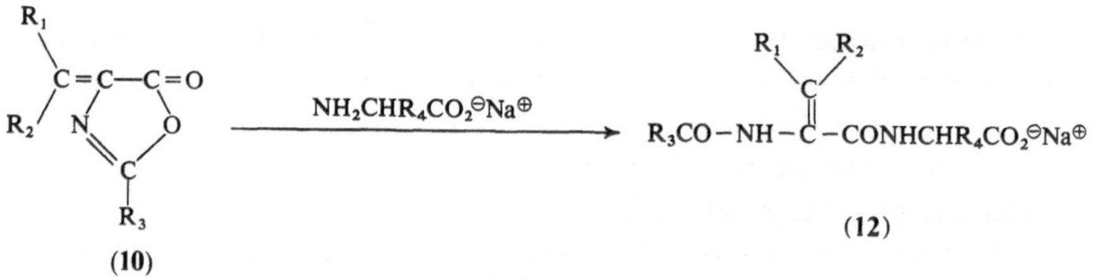

$$R_3CO-NH-\overset{\overset{\textstyle R_1 \quad R_2}{\textstyle \diagup \!\! \diagdown}}{\underset{}{C}}-CONHCHR_4CO_2^{\ominus}Na^{\oplus}$$

(10) → NH₂CHR₄CO₂⁻Na⁺ → (12)

Scheme 5

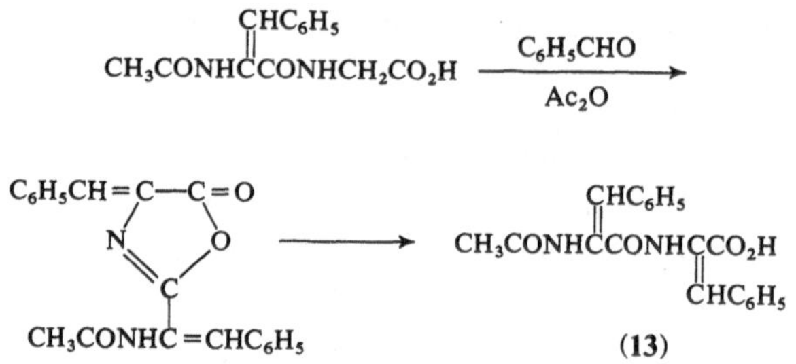

$$\text{CH}_3\text{CONH}\overset{\overset{\textstyle \text{CHC}_6\text{H}_5}{\textstyle \|}}{\text{C}}\text{CONHCH}_2\text{CO}_2\text{H} \xrightarrow[\text{Ac}_2\text{O}]{\text{C}_6\text{H}_5\text{CHO}}$$

$$\text{CH}_3\text{CONH}\overset{\overset{\textstyle \text{CHC}_6\text{H}_5}{\textstyle \|}}{\text{C}}\text{CONH}\overset{}{\text{C}}\text{CO}_2\text{H}$$

C₆H₅CH=C—C=O / N—O / CH₃CONHC=CHC₆H₅

(13)

Scheme 6

β) In 1926, Bergmann and Stern described an alternative route to unsaturated oxazolinones (40). Treatment of α-haloacylamino acids with acetic anhydride, usually in the presence of bases such as pyridine or sodium acetate, affords Δ^2- or Δ^3-oxazolinones (14) and (15), depending on the stabilization of the double bond system by the substituents R [cf. (396, 398)]. The 2-methyl-4-alkylidene (aralkylidene) oxazolinones obtained by Bergmann et al. (40, 114) and by Greenstein and Price (319) from the chloroacetyl derivatives of Phe, Tyr, Leu, Norleu, Val and Aba, from which the corresponding acetyldehydroamino acids could be obtained by hydrolysis, were most probably Δ^2-oxazolinones. In contrast, the product obtained from α-bromopropionyl alanine, which can very readily be cleaved to pyruvic acid, is probably an example of a Δ^3-oxazolinone.

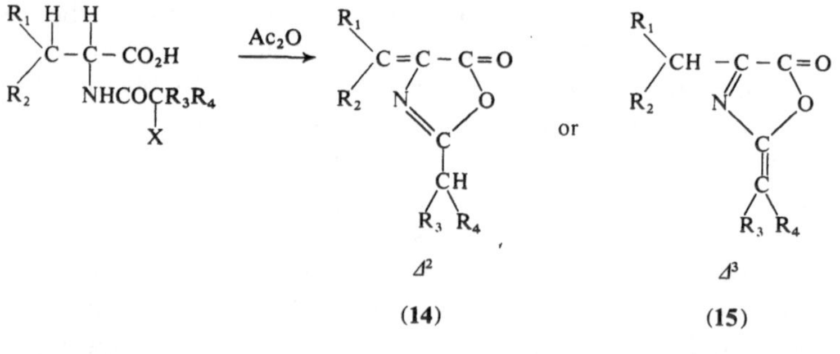

Δ^2 (14)   or Δ^3 (15)

Scheme 7

RIORDAN and STAMMER (*333, 334*) chose α,β-dibromoacylamino acids as starting materials in order to favour the Δ^2-oxazolinone structure by the additional double bond in the 2-substituent.

In the reaction of N-chloroacetylphenylserine with acetic anhydride, oxazolinone formation was followed by the elimination, not of HCl, but of H_2O, to yield 4-benzylidene-2-chloroacetyl-Δ^2-oxazolinone (*52*). Hydrolysis, followed by treatment with ammonia, afforded the first example (**16**) of a dehydropeptide in which the dehydroamino acid was not at the N-terminal end.

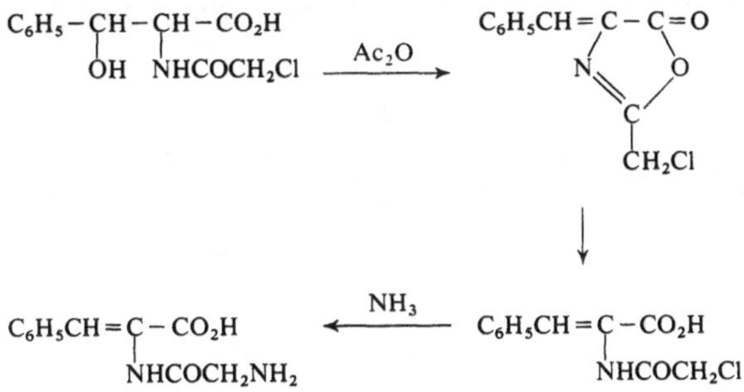

$$C_6H_5-\underset{\underset{\displaystyle OH}{|}}{C}H-\underset{\underset{\displaystyle NHCOCH_2Cl}{|}}{C}H-CO_2H \xrightarrow{Ac_2O}$$

$$C_6H_5CH=\overset{\displaystyle }{C}-\overset{\displaystyle C}=O$$

with N and O ring closed by C–CH$_2$Cl

$$C_6H_5CH=\underset{\underset{\displaystyle NHCOCH_2NH_2}{|}}{C}-CO_2H \xleftarrow{NH_3} C_6H_5CH=\underset{\underset{\displaystyle NHCOCH_2Cl}{|}}{C}-CO_2H$$

(**16**)

Scheme 8

A more general route to chloroacetyldehydroamino acids was described by KURITA et al. (*226*), who applied the Bergmann-Stern synthesis to dichloroacetylamino acids.

γ) Oxazolinone formation from β-hydroxyamino acids leads to unsaturated oxazolinones owing to simultaneous elimination of the β-substituent. ERLENMEYER (*117*) described the formation of 2-methyl-4-benzylidene-oxazolinone by the reaction of phenylserine with acetic anhydride. CARTER et al. (*92, 93, 94*) obtained 2-phenyl-4-ethylidene-oxazolin-5-one from N-benzoylthreonine, N-benzoylallothreonine, and their O-methyl, O-acetyl and O-benzoyl derivatives by treatment with benzoylchloride in pyridine. Trifluoroacetic anhydride converts threonine into 4-ethylidene-2-trifluoromethyl-2-oxazolin-5-one (*20*).

δ) Several attempts have been made, especially in recent years, to oxidize saturated to unsaturated oxazolinones. Oxazolinone formation activates the α-carbon atom of the corresponding amino acid, so that it can, for example, by easily brominated (*100, 315, 417*). MORIN and GORDON (*267*) attempted the selective oxidation of the C-terminal amino acid of a peptide chain by first converting it into an oxazolinone. Treatment of the oxazolinones obtained from N-benzoyl- or N-carbobenzoxydipeptides with Pb(OAc)₄ or Hg(OAc)₂ resulted however in attack on both amino acid units and the formation of mixtures of products.

Following publication by Weygand and coworkers (432, 433) of their studies of the bromination of 2-trifluoroacetyl-Δ^3-oxazolinones at the β-carbon atom of the corresponding amino acid, Breitholle and Stammer (68, 69) described the elimination of HBr from these bromo-derivatives (17) to yield unsaturated Δ^2-oxazolinones (18). The latter yielded N-trifluoroacetyldehydroamino acid anilides (19) or peptides on treatment with aniline or amino acid esters respectively. Müller and Steglich (270) used this method independently for the synthesis of N-trifluoroacetyl-dehydrodipeptides (20) containing C-terminal alanine.

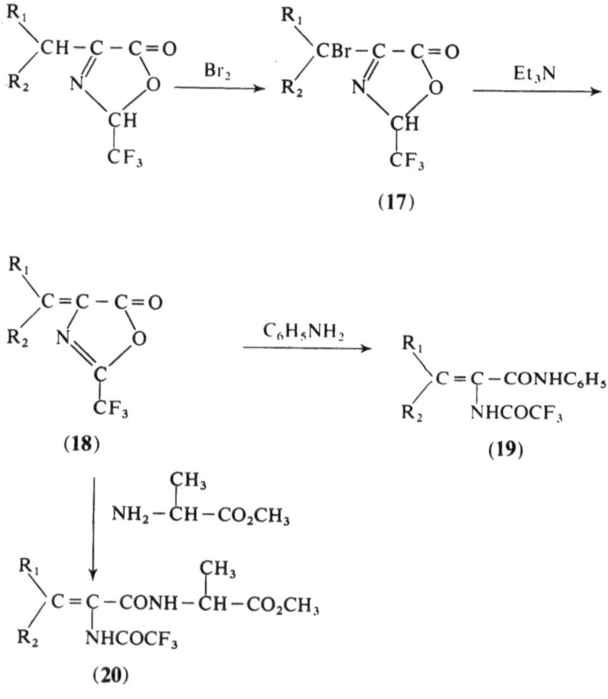

Scheme 9

Riordan and Stammer (335, 336) employed o-chloranil to oxidize saturated Δ^2-oxazolinones, and obtained dioxinones. Treatment of the latter with nucleophiles such as CH_3O^-, $PhNH_2$ or $PhCH_2SH$ afforded α-substituted amino acid derivatives, possibly by addition to the acylimine in equilibrium with the quinone adduct. Use of a non-nucleophilic base coupled with scavenging of the phenolic hydroxy groups by etherification made possible the isolation of acylenamino acid derivatives.

2. Acylenamino Acid Derivatives by β-Elimination Reactions

Amino acids with appropriate β-substituents have frequently been employed as α,β-dehydroamino acid precursors, particularly in the form of peptide chain subunits. Elimination reactions from derivatives of β-hydroxy- and β-mercapto-α-amino acids are the examples most often encountered. Of less importance are amino acid derivatives with β-halogen substituents (*38, 132, 337, 386, 394, 395*), as these require more drastic elimination conditions. Mannich bases of monoalkyl acylaminomalonates have also found use as precursors of dehydroalanine derivatives (*177, 438*).

The major advantage of this approach to the synthesis of unsaturated peptides lies in the fact that the elimination can be carried out as the final step. Peptides can thus be constructed with optically active serine and cysteine derivatives, and the formation of diastereoisomers is avoided.

2.1. β-Elimination Reactions of β-Hydroxy-α-amino Acid Derivatives

Racemic versimide and a series of analogous imidoacrylic acid, crotonic acid and dimethylacrylic acid esters have been prepared by direct dehydration of the corresponding β-hydroxy compounds with potassium hydrogen sulphate in boiling DMF (*77*).

β-Eliminations from serine *phosphates* have been observed on dephosphorylation of phosphoproteins (casein, phosvitin) (*10, 261, 299, 361*). β-Elimination from serine phosphate esters by means of dilute aqueous alkali or diethylamine in aprotic solvents has occasionally been used for the synthesis of dehydroamino acid units (*302, 331, 332*).

*O-Benzoyl*serine compounds have also found only 'occasional application in β-elimination reactions: in the alkaline methanolysis of N-benzyloxycarbonyl-S-benzoylcysteinyl-O-benzoylserine ester, ZERVAS and FERDERIGOS observed the formation of cyclolanthionine derivatives, formed by inter- or intramolecular Michael addition of the SH-function to the dehydroalanine unit formed by simultaneous elimination of benzoic acid (*448, 449*).

In contrast, β-elimination from esters of serine-O-sulfonic acid has been a popular and successful method for the preparation of dehydroamino acid units, particularly within peptides (*230, 286, 302, 347, 395, 430*). For this purpose, the peptide containing serine with unprotected OH-group is synthesized by conventional methods, treated with tosyl chloride in pyridine to form the toluene sulphonate ester, and then subjected to elimination with aqueous alkali or with diethylamine in water-free solvents. Although the dehydroamino acid compound is often obtained in good yield (Path A), the following complications are some-times encountered (cf. Chart 2).

Path B: Oxazoline formation by intramolecular nucleophilic substi-tution of the sulfonyl group by the oxygen atom of an N-acyl group (*35, 143, 273, 289*). Compounds with an urethane protective group are never observed to undergo this reaction. Polar solvents favour oxazoline formation, strong bases β-elimination. The course of the reaction is also influenced by the C_α-H acidity of the substrate, for which reason ring closure to the oxazoline is particularly common with *amides* of β-tosyloxy-α-acylamino acids.

Path C: Aziridine formation. β-Tosyl derivatives of serine and threo-nine compounds protected on nitrogen by trityl or tosyl groups, can cyclize to aziridines in strongly basic media (*272, 288, 289, 387*). In the case of serine derivatives, this reaction is observed only when either the C_α-H acidity is reduced (amides) or the nitrogen atom retains high nucleophilicity (N-trityl derivatives). The tendency of threonine deri-vatives to form aziridines is much stronger, and less dependent on these factors. Thus, treatment of N,O-ditosylthreonine *ester* with bases yields primarily the aziridine (*272*)*.

Path D: Hydantoin formation. In certain circumstances hydantoins can be formed from β-tosyloxy-N-benzyloxycarbonylamino acid amides in basic media (*273*). This reaction can probably be avoided by choice of less drastic conditions.

Path E: Nucleophilic substitution of the β-tosyloxy group. Direct nucleo-philic substitution can occur on treatment of O-tosylserine derivatives with strongly nucleophilic bases (*303, 304, 305, 413, 450*). To the best of our knowledge, only a single example of an analogous reaction with a threonine derivative has been reported (*266*). The products obtained can in principle be formed either by direct nucleophilic substitution or by an elimination-addition sequence. This question can only be settled by a study of the stereochemistry of the products, as the elimination-addition sequence destroys the optical activity of the serine or threonine unit. β-Elimination is favoured both by a higher C_α-H acidity in the sub-strate (COOR > CONH > COO⁻) and by a greater basic strength of the attacking nucleophile (*143, 300, 391, 450*). For example, N-benzyloxycarbonyl-O-tosylserine esters react with sodium thiolacetate

---

* Okawa *et al.* (*272*) reported the cyclization of two β-tosyloxy-α-*acylamino* acid derivatives to aziridines. This report, which stands in contradiction to the results of other authors, has since been corrected (*273, 289*). Careful analysis of the spectroscopic data showed the reaction products to be oxazolines.

or with sodium benzyl selenide with conservation of optical activity (*413, 450*), but with the more strongly basic anions benzylmercaptide or tritylmercaptide with complete racemization (*302, 303, 447, 450*). The more weakly acidic O-tosylserylglycine esters, however, suffer no racemization on reaction with benzyl mercaptide or trityl mercaptide (*450*). In contrast, both O-tosylserine esters (*63*) and β-halopropionic acid derivatives (*250, 268, 431*) react with alkali metal salts of malonic esters by way of an elimination-addition mechanism.

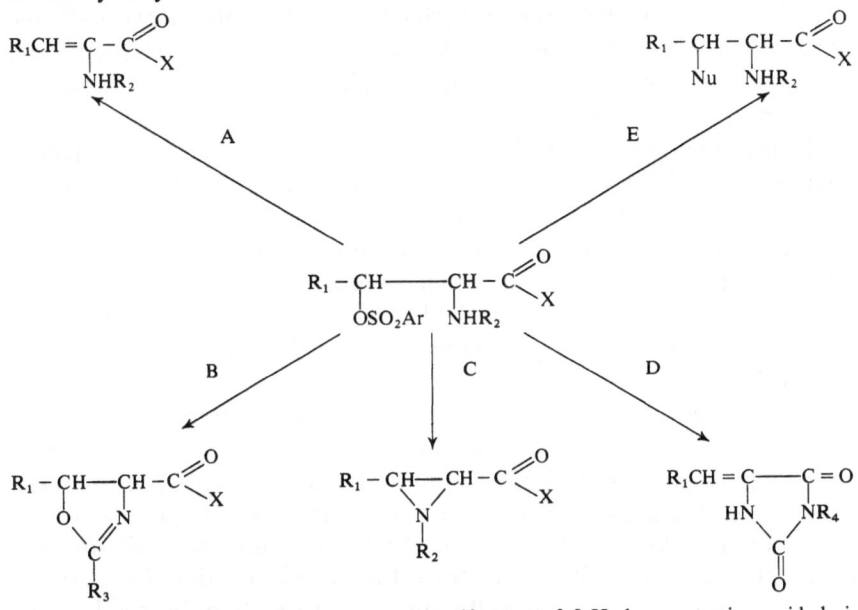

Chart 2. Possible reactions between O-Aryl sulfonates of β-Hydroxy-α-amino acid derivatives and bases or nucleophiles

Path B R_1 = H, CH₃
          R_2 = Z–Phg, Z–Phe, RCO, Bz–Gly, Z–Gly ...
          X   = NHR

Path C R_1 = H, CH₃
          R_2 = Tos, Trit
          X   = O–Alk, NHR

Path D R_1 = H
          R_2 = Z
          X   = NHR

Path E R_1 = H, CH₃
          R_2 = Z, Boc, Z–Gly, Z–Glu (OBzl), ...
          X   = O–Alk, NHR
          Nu = S–COR, SeBzl, S–Trit, S–CH₂–COO$^\ominus$, S–(CH₂)₂NH₂, S–CH₂C₆H₅

2.2 β-Elimination Reactions from Cysteine Derivatives

Many reports exist of β-eliminations from N-acylcysteine derivatives on treatment with silver (I) oxide or silver (I) carbonate (*58, 147*),

HgO (*340*), iron salts (*28*) or dicyclohexylcarbodiimide (*205*). Reactions of S-alkyl- or S-acylcysteine derivatives in basic media are often accompanied by β-elimination side reactions (*180, 245, 246, 302*). Elimination is favoured in polar solvents and is also dependent on the C_α-H acidity of the substrate. Accordingly, esters of S-alkylcysteines undergo elimination more readily than derivatives of other types.

PATCHORNIK and coworkers studied the elimination reactions of S-2,4-dinitrophenyl-cysteine compounds (*390, 391*). The method proved suitable for the conversion of cysteine peptides into dehydroalanine peptides, in which the double bond can then be exploited as the site of attack in the specific (hydrolytic or oxidative) cleavage of the peptide (*298—300*).

Dinitrophenylation of free SH-groups is conducted in aqueous medium at pH 5—6, conditions under which amino groups do not react. Dinitrothiophenol is eliminated by treatment with 0,1 N NaOH, which is accompanied by a certain risk of racemization of other amino acids. Other side reactions which may occur are described in the original literature which should also be consulted for information on β-eliminations from β-isothiocyanatoalanine derivatives (*406*) and cystine peptides (*141, 407*).

The simplest and most satisfactory method for preparing dehydropeptides is, to the best of our knowledge, the β-elimination from sulfonium salts of β-alkylthio-α-amino acid units (*286, 287, 297, 327, 391*). A peptide is assembled which incorporates a β-alkylthio-α-amino acid subunit, which is then converted into the sulfonium salt followed by treatment with weak base to form the dehydropeptide. PATCHORNIK and SOKOLOVSKY suggested sulfonium salt formation by methylation with methyl bromide in formic acid (*297, 391*). After evaporation of the formic acid, treatment of the sulfonium salt with aqueous bicarbonate solution affords the dehydropeptide directly. RICH alkylates the thioether group with methyl fluorosulfonate and then eliminates with triethylamine in a water-free solvent (*327, 329*), a method he has used for the preparation of peptides containing dehydroalanine, dehydroaminobutyric acid or dehydrovaline units. For the synthesis of peptides containing the latter two dehydroamino acids, peptide substrates incorporating racemic β-methylthio-α-amino butyric and valeric acids respectively were prepared. No side reactions, such as ring closure to the aziridine or oxazoline, were observed in these eliminations. The procedure is not suitable for compounds containing other groups which can be alkylated, such as lysine, methionine or histidine. Peptides with acid labile protective groups or which can undergo side reactions under the influence of formic acid (formylation, N-O-shift), should be methylated with methyl fluorosulfonate.

An elimination reaction analogous to those of O-tosylserine derivatives has been reported for an S-tosylcysteine derivative (274).

Thermolysis of β-alkylsulfinyl-α-amino acid derivatives is an alternative reaction which has been applied to the synthesis of dehydroamino acid compounds (325, 326). Esters and peptides with dehydroalanine, dehydroaminobutyric acid, dehydrovaline and dehydrophenylalanine units have in many cases been obtained in excellent yields. The method is especially suitable for peptides with base sensitive groups. The elimination appears to take place with particular facility in compounds in which the amino acid unit bearing the sulfoxide group is present as a tertiary amide (326). Dehydrophenylalanine and dehydroaminobutyric acid derivatives are formed as mixtures of Z- and E-isomers on sulfoxide elimination.

As might be expected, analogous selenoxides undergo elimination with still greater facility. These oxides are so unstable that only the dehydroamino acid compounds resulting from elimination can be isolated (428).

Thermal cleavage of a sulfone has been exploited in a synthesis of racemic versimide (18), while GROSS (159) mentions a base catalyzed elimination from an S-methylcysteinesulfone derivative to yield the dehydroalanine compound.

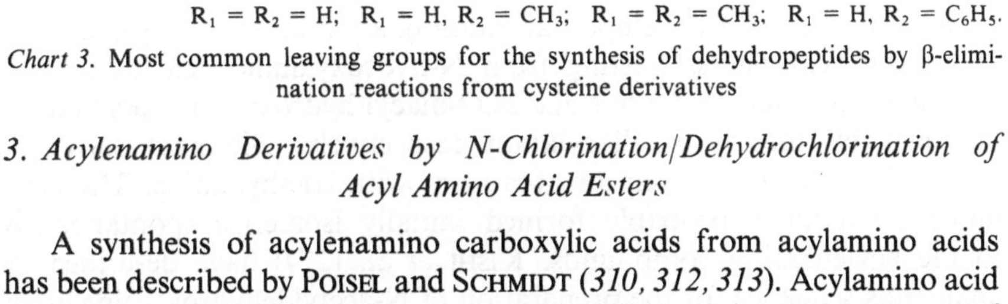

$$Pep_1NH - \underset{\underset{R_1}{\overset{|}{\underset{}{C - X}}}{\overset{}{}}}{CH} - COPep_2 \longrightarrow Pep_1NH - \underset{\underset{R_1}{\overset{|}{C}}{\overset{}{R_2}}}{C} - COPep_2$$

$$X = -\overset{\oplus}{S}(CH_3)_2; \quad S-\langle\bigcirc\rangle-NO_2; \quad S(O)R;$$

$$\overset{|}{NO_2}$$

$$R_1 = R_2 = H; \quad R_1 = H, R_2 = CH_3; \quad R_1 = R_2 = CH_3; \quad R_1 = H, R_2 = C_6H_5.$$

Chart 3. Most common leaving groups for the synthesis of dehydropeptides by β-elimination reactions from cysteine derivatives

3. Acylenamino Derivatives by N-Chlorination/Dehydrochlorination of Acyl Amino Acid Esters

A synthesis of acylenamino carboxylic acids from acylamino acids has been described by POISEL and SCHMIDT (310, 312, 313). Acylamino acid esters are N-chlorinated with t-butylhypochlorite in the presence of catalytic quantities of base. A full equivalent of base can then be used to eliminate HCl, with the initial formation of acylimino esters as unstable primary products which then either add solvent (alcohol) to form α-acylamino-α-alkoxy acid derivatives (22), or (in the absence of protic solvents) rearrange to the acylenamino acid esters (21). The α-acylamino-α-methoxy carboxylic acid (22) esters can also be converted into acylenamino acid esters (**21**) by acid or base catalysed elimination. The esters can be easily hydrolysed to the corresponding carboxylic acids (**23**), which can then serve as components in dehydropeptide synthesis.

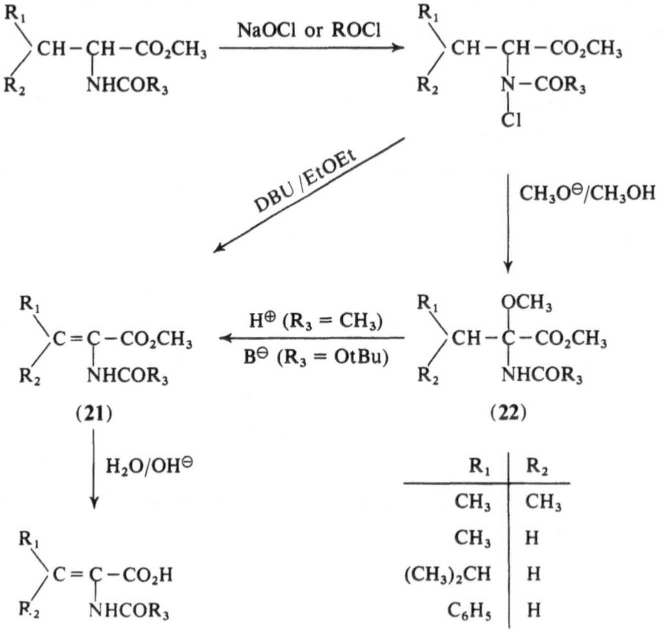

R_1	R_2
CH_3	CH_3
CH_3	H
$(CH_3)_2CH$	H
C_6H_5	H

Scheme 10

4. Acylenamino Carboxylic Acid Derivatives From N-Hydroxyamino Acid Derivatives

SHIN *et al.* (*377*) developed an additional route to N-acylenamino-carboxylic acid esters, starting from N-hydroxyamino acid esters: N-acyl-hydroxyamino acid esters and N,O-diacyl-hydroxyamino acid esters, obtained by mono- or diacylation respectively, eliminate water or carboxylic acid respectively on treatment with triethylamine. The acyl-imino compounds probably formed initially isomerize spontaneously to the acylenamino compounds. KISHI *et al.* (*274*) have described an analogous sequence for the preparation of N-acetyldehydroalanine ester.

5. Acylenamino Acid Esters from Carbonyl Compounds and Isocyanoacetic Ester or Isothiocyanatoacetic Ester

SCHÖLLKOPF *et al.* (*360*) reacted α-metallated isocyanoacetic ester with aldehydes and ketones in aprotic solvents to form formyl-enamino acid esters (**24**) *via* intermediate oxazoline carboxylic acid esters.

Applied to unsymmetrical ketones, the synthesis yields Z/E mixtures. The NMR spectra of the components allow the appropriate configurations to be assigned. The reaction has also been exploited by other authors (73), for example to introduce branched chains into ketoses.

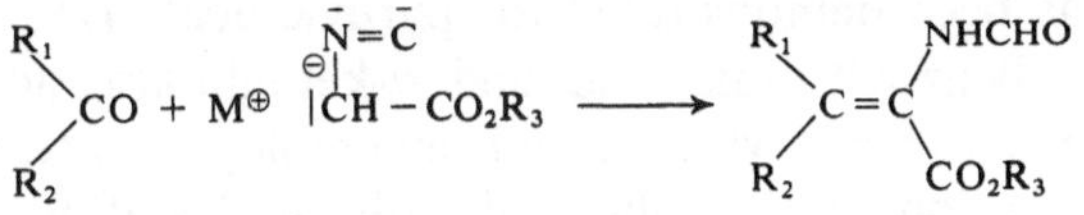

$$\begin{array}{c} R_1 \\ R_2 \end{array}\!\!\!\!CO \;+\; M^\oplus \;\;\overset{\overline{N=\overline{C}}}{|CH-CO_2R_3} \;\longrightarrow\; \begin{array}{c} R_1 \\ R_2 \end{array}\!\!\!\!C=C\!\!\!\!\begin{array}{c} NHCHO \\ CO_2R_3 \end{array}$$

(24)

Scheme 11

HOPPE (181—185) obtained acylenamino acid esters from condensations with isothiocyanates. Base catalysed condensation of α-isothiocyanatoacetic ester with carbonyl compounds affords initially 2-thioxo-1,3-oxazolidine-4-carboxylic acid esters (25) which yield 3-acyl derivatives (26) on reaction with acid halides. Alternatively, alkylation with alkyl halides yields 2-alkylthio derivatives (28). (26) and (28) can be cleaved to acylenamino acid esters (27 and 29) by treatment with strong bases such as KOtBu followed by acidification. A concerted mechanism has been suggested for the cleavage of (26), as the configuration at the double bond of the ring opened product depends on the stereochemistry of the compound (26). In contrast, ring opening of (28) is not stereospecific.

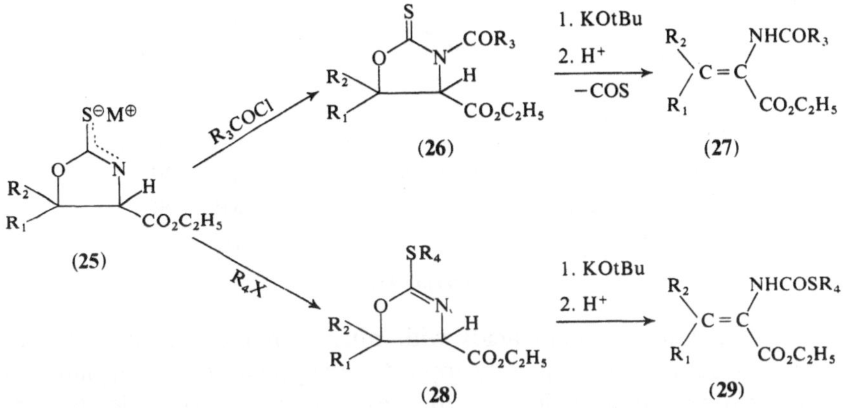

Scheme 12

6. Acylenamino Acids by Condensation of α-Keto Acids With Amides or Nitriles

In 1930 BERGMANN and GRAFE (45) observed the formation of α,α-diacetaminopropionic acid (32) and acetaminoacrylic acid (31) on warming a mixture of pyruvic acid and acetamide under reduced pressure. (32) could be converted into (31) by treatment with acetic acid (46, 337) (Scheme 13). The α-hydroxy-α-acylamino acid (30) is almost certainly the primary product, which then either reacts directly with a second amide molecule to form the bis(acylamino) derivative (32) or loses water to yield the acylaminoacrylic acid (31), to which addition of amide can occur under suitable conditions. The reversibility of this latter step has been demonstrated for pyruvic acid (179, 364)*. Phenylpyruvic acid, benzoylformic acid and α-ketoglutaric acid are further examples of α-keto acids which have been employed as starting materials (364). Condensation of glyoxylic acid with amides affords hemiaminals (33, 451). Amide components used include carbaminic acid ester and chloroacetamide (46, 227, 252). Formamide reacted with pyruvic acid to yield an isolable hemiaminal (382). A convenient summary of the course and scope of this reaction has been published by Greenstein and Winitz (150). The Bergmann-Grafe condensation has also been exploited by more recent workers, some of whom have contributed improvements in experimental procedures (15, 195, 196, 259, 321, 358, 372, 375, 402, 435).

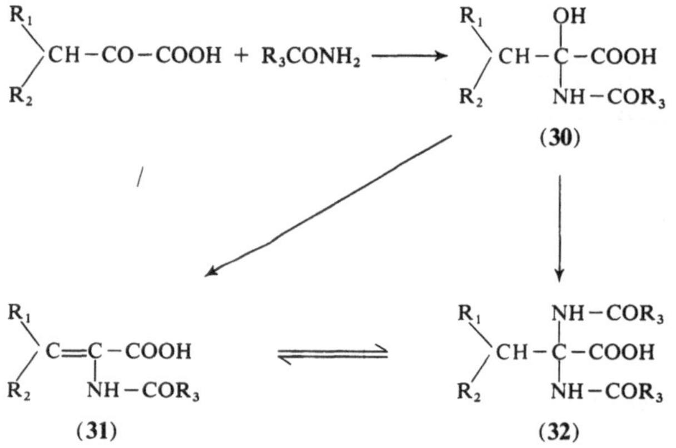

Scheme 13

The reaction of α-keto acids with nitriles in the presence of acid catalysts has also been frequently applied to the preparation of acylaminoacrylic acid derivatives (150, 318, 369). α-Acylaminoacrylic acids are obtained directly from α-halonitriles and excess keto acid under rigorous exclusion of moisture. If the reaction is conducted in the presence of water and excess nitrile, α,α-bis-acetylamino carboxylic acids are formed in very good yields, and can be converted into the acylaminoacrylic acids in a subsequent step. The summary by Greenstein and Winitz (150) should be consulted for more detailed information.

---

* Gallina (136) has described the corresponding addition of acetamide to N-acetyldehydroproline to form N,N'-diacetyl-α-aminoproline.

7. Acylenamino Carboxylic Acid Derivatives by Chain Lengthening of β-Halo-α-acylaminoacrylic Acid Derivatives

OLSEN et al. (330) describe the reaction of Z-β-halo-α-acylaminoacrylic acid derivatives with lithium dialkyl copper reagents. Replacement of the vinyl halogen by alkyl takes place with complete or predominant retention of configuration at the double bond.

B. Unsaturated Piperazinediones

Dehydropiperazinediones have frequently been found in company with their saturated analogues as metabolites of microorganisms. P. G. SAMMES is the author of a recent review of natural piperazinediones (341): to the best of our knowledge, only a few scattered reports of the isolation of dehydrodioxopiperazines have appeared since the publication of this review (91, 188, 189, 248, 249). Several synthetic approaches are available.

1. Preparation of Dehydropiperazinediones by Condensation of Piperazinediones With Carbonyl Compounds

2,5-Piperazinedione condenses easily with benzaldehyde in the presence of acetic anhydride/sodium acetate to yield 3,6-dibenzylidene-2,5-piperazinedione (343, 344). Analogous preparations have been conducted with heterocyclic and unsaturated aliphatic aldehydes (19), but not with saturated aliphatic aldehydes. The synthesis of alkylidenepiperazinediones can be realized by condensation of N,N'-diacetylpiperazinediones with aliphatic aldehydes in the presence of potassium t-butoxide (138, 140), conditions also applicable to the synthesis of monoarylidenepiperazinediones. A second condensation is however only possible with aromatic aldehydes.

2. Dehydropiperazinediones by Ring Closure of Dehydrodipeptides

M. BERGMANN was presumably the first to exploit this possibility (54). He prepared N-benzoyldehydrodipeptides (33) by azlactone synthesis, and used ketene or acetic anhydride to effect cyclization to N-benzoyldehydropiperazinediones (34).

Dehydrodipeptides (35) have also been prepared from free enamino acid esters and cyclized after hydrazinolysis of the protective group (366, 368) giving (37). Condensation with carbonyl compounds makes possible the synthesis of unsymmetrical tetradehydropiperazinediones (36).

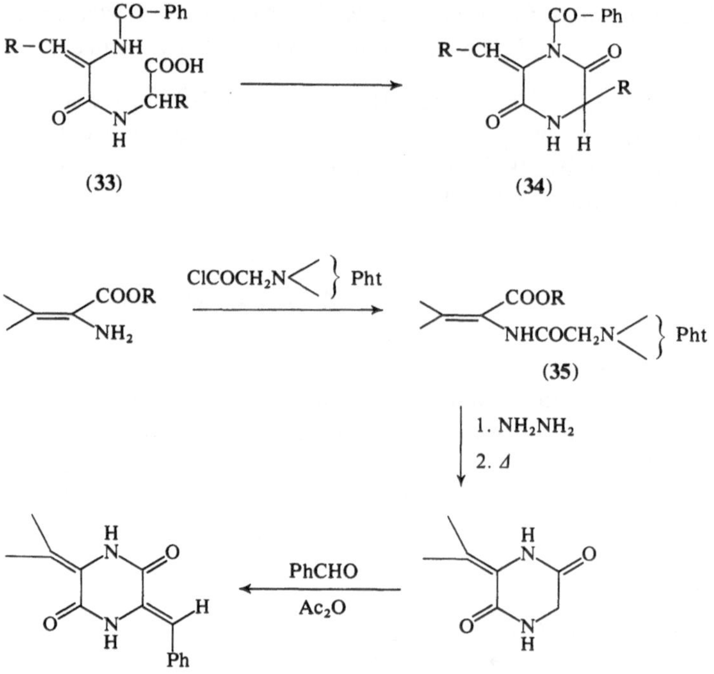

(33)   (34)

(35)

1. NH_2NH_2

2. Δ

(36)   (37)

Scheme 14

A further method used to prepare dehydropiperazinediones is the condensation of α-keto esters with chloroacetamide or chloroacetonitrile leading to (38), followed by treatment with ammonia (*367, 372, 375*). For a similar cyclization reaction with concomitant dehydration see ref. (*374*). Subsequent condensation with aldehydes affords unsymmetrical tetradehydropiperazinediones, exemplified by the synthesis of albonoursin (36) (*367, 369*). Arylidenepiperazinediones from condensation reactions always have the Z-configuration, in contrast to the alkylidenepiperazine-diones, which are normally obtained as Z/E mixtures (*57, 140, 381*).

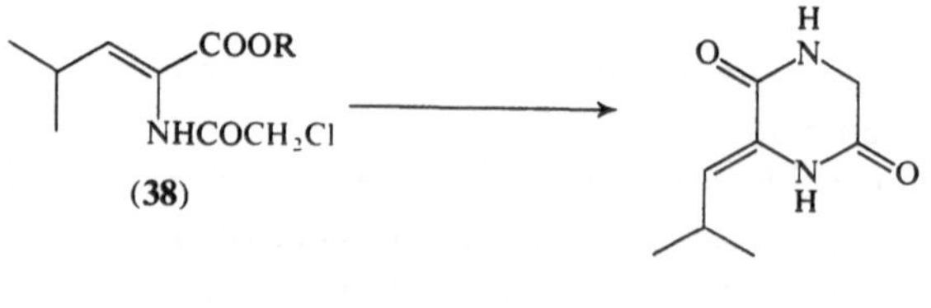

(38)

Scheme 15

3. Dehydropiperazinediones by Elimination Reactions

Piperazinediones which contain serine as a subunit sometimes eliminate water spontaneously in the course of their formation by ring closure (37). Picrorocellin (39A), which is a cyclic dipeptide containing

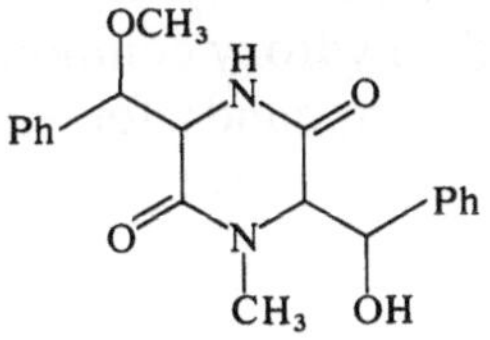

(39A) 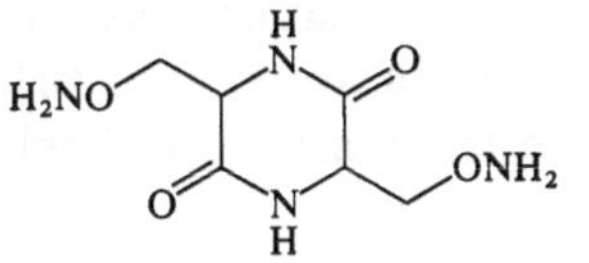   (39B)

Scheme 16

Table 2. *Examples of Piperazinediones Yielding Unsaturated Derivatives by Elimination Reactions*

	R^1	R^2	R^3	R^4	Ref.
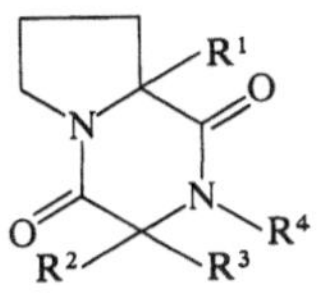	H	OH	CH₃	H	(86, 167)
	H	OH	CH₃	CH₃	(86, 167, 251)
	H	NH₂	CH₃	CH₃	(286)
	OAc	OAc	CH₃	CH₃	(285)
	OCH₃	OH	CH₃	CH₃	(311)
	H	OH	CH₂CHMe₂	H	(86)
	H	OH	CHMeEt	H	(86)
	OH	H	H	H	(168)

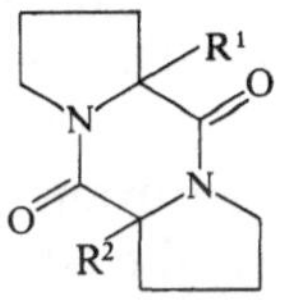	R^1	R^2	Ref.
	H	SCH₃	(283)
	SCH₃	SCH₃	(283)
	OH	OH	(283)
	OAc	OAc	(284)

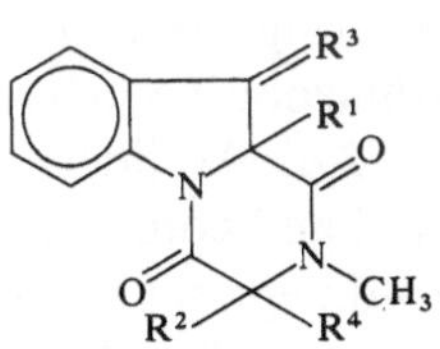	R^1	R^2	R^3	R^4	Ref.
	H	OH	H, H	CH₃	(251)
	Cl	OH	CH₃, CH₃	CH₃	(294, 295)

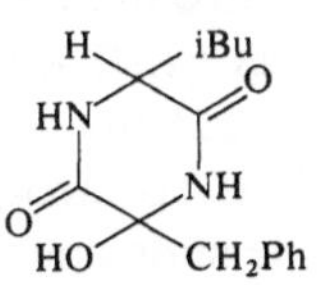	Ref. (102)

two phenylserine units, readily forms a dibenzylidenepiperazinedione on heating (*131*). Elimination of hydroxylamine from the dimer of cycloserine (**39B**) to form dimethylenepiperazinedione has been observed in basic media (*173*).

Elimination reactions have more often been carried out with piperazinediones containing suitable leaving groups (OH, acetoxy, SCH₃) in positions 3 and/or 6. In some cases (*102, 168*) the hydroxy compound is not isolated, but loses water in situ. Starting compounds for elimination reactions are summarized in Table 2.

4. Dehydropiperazinediones by Dehydrogenation Reactions

P. G. Sammes has dehydrogenated the iminoethers of piperazinediones with 2,3-dichloro-5,6-dicyanobenzoquinone (*243*). Cyclo-L-Pro-L-Pro can be oxidized with this reagent directly to pyrocoll (*243*). The use of sulfur as oxidizing agent has been successful only with N-acetylated piperazinediones (*242*).

C. Arylidene-enamino Acid Compounds

R. Grigg and J. Kemp (*148*) prepared benzylidene compounds of dehydrovaline ester, dehydroleucine ester, dehydroisoleucine ester and dehydrophenylalanine ester by dehydrogenation of the corresponding saturated benzylidene compounds with diethyl azodicarboxylate. This dehydrogenation is assumed to be an "ene reaction".

U. Schmidt and coworkers (*287, 352*) condensed dehydrovaline ester with aromatic aldehydes and pyridoxal to yield Schiff bases directly, which were employed in biomimetic experiments (VI. A. 2.1).

The simplest and most satisfactory method of preparing arylidene-enamino acid esters is the β-elimination from 2-aryl-4-thiazolidinecarboxylic acid esters (**40**) on treatment with silver carbonate (*287 A*). Arylidene-enamino acid N-hydroxysuccinimide esters were prepared and combined with amino acid esters to yield Schiff bases of dehydro-dipeptides. This elimination reaction of thiazolidine carboxylic acid esters has been applied to prepare the extremely unstable benzylidene dehydroalanine ester, although it was impossible to isolate it. Nevertheless its existence could be proved without a doubt when the reaction was carried out at 0° in trideuterioacetonitrile solution and was followed by NMR spectrometry. Within one half hour at room temperature the unstable substance started to decompose.

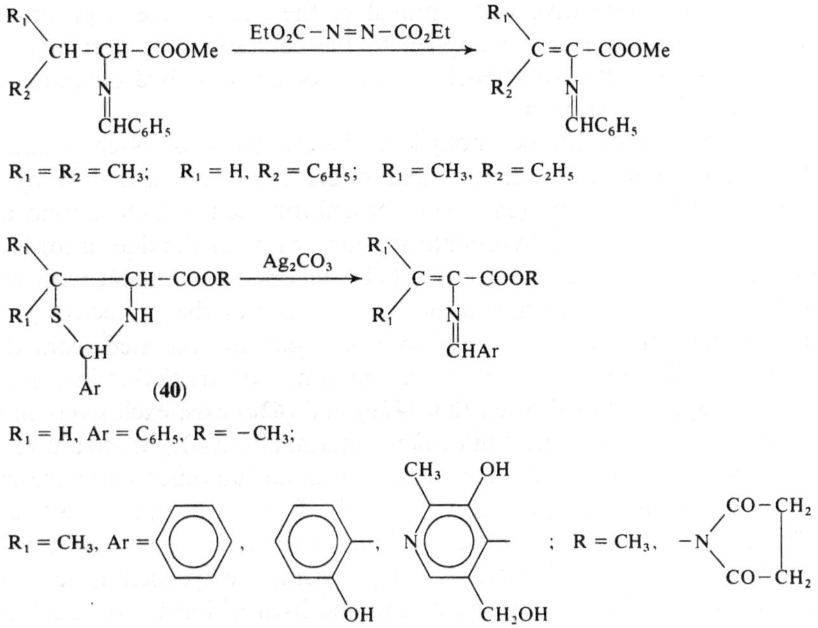

$$\underset{R_2}{\overset{R_1}{\diagdown}}CH-\underset{\underset{\overset{\|}{CHC_6H_5}}{N}}{CH}-COOMe \xrightarrow{EtO_2C-N=N-CO_2Et} \underset{R_2}{\overset{R_1}{\diagdown}}C=\underset{\underset{\overset{\|}{CHC_6H_5}}{N}}{C}-COOMe$$

$R_1 = R_2 = CH_3;$ $R_1 = H, R_2 = C_6H_5;$ $R_1 = CH_3, R_2 = C_2H_5$

$$\underset{R_1'}{\overset{R_1}{\diagdown}}\underset{\underset{\underset{Ar}{CH}}{S}}{C}\!\!-\!\!-\!\!-\underset{NH}{CH}-COOR \xrightarrow{Ag_2CO_3} \underset{R_1'}{\overset{R_1}{\diagdown}}C=\underset{\underset{\overset{\|}{CHAr}}{N}}{C}-COOR$$

(40)

$R_1 = H, Ar = C_6H_5, R = -CH_3;$

$R_1 = CH_3, Ar =$ (phenyl), (2-hydroxyphenyl), (2-methyl-3-hydroxy-4-hydroxymethyl-pyridin-5-yl) ; $R = CH_3,$ $-N\!\!\overset{\overset{CO-CH_2}{|}}{\underset{\underset{CO-CH_2}{|}}{}}$

Scheme 16 A

D. α-Enaminocarboxylic Acid Compounds

The established methods for preparing α-enaminocarboxylic acid derivatives are the reduction of unsaturated nitro or azido esters, and the oxidation of amino acid derivatives.

TATSUOKA et al. (410) and later SHIN et al. (366, 368, 373) obtained α,β-dehydrovaline ester by reduction of the α,β-unsaturated nitro ester with Al-Hg. Only β-disubstituted nitro compounds yield enamines in this reaction, α-oximino esters being obtained from the reduction of β-monosubstituted esters. The analogous reduction of β-methoxy-α-nitro esters to β-methoxy-α-amino esters, which lose methanol on warming to yield enamino esters, is described in ref. (370).

SHIN et al. (380) have recently shown that α,β-dehydro-α-azido esters, obtainable by a four-step synthesis, are smoothly reduced by Al-Hg to enamino esters. The structures of the products were confirmed by NMR-spectrometry.

BREITHOLLE and STAMMER (68, 69) attempted to prepare free enamino acid anilides from the corresponding trifluoroacetyl compounds, which are readily obtainable via oxazolinones. Only in the case of the dehydrophenylalanine derivative was removal of the trifluoroacetyl group successful, not however when applied to the corresponding derivatives of dehydrovaline, dehydroisoleucine, dehydroleucine, dehydroalanine and dehydroaminobutyric acid.

A route to α-enaminocarboxylic acid esters based on N-chlorination/dehydrochlorination of amino acid esters has been described by H. Poisel and U. Schmidt (312, 313). N-Chlorination of BOC-amino acid esters followed by dehydrochlorination with methoxide affords α-methoxy-α-BOC-amino acid esters (41), which, on treatment with HCl, undergo elimination of methanol and cleavage of the protective group in a single step. The free enamino esters (42) are liberated from their hydrochlorides by treatment with ammonia or triethylamine. NMR spectroscopic studies showed that (42a) and (42c) exist exclusively in the enamino tautomeric form, while (42b) contains about 40% imino tautomer.

Shin was unable to observe enamino-imino tautomerism in the case of dehydrovaline compounds (253, 373). E. Öhler and U. Schmidt (287, 350) were able to convert α-iminocarboxylic acid esters into α-enaminocarboxylic acid esters (42) by preparing the hydrochlorides of the former, which slowly rearranged into the hydrochlorides of the latter. This reaction is the simplest means of preparing α-enaminocarboxylic acid esters, as the α-iminocarboxylic acid esters obtained by N-chlorination/dehydrochlorination of amino acid esters can be rearranged directly, without intermediate purification.

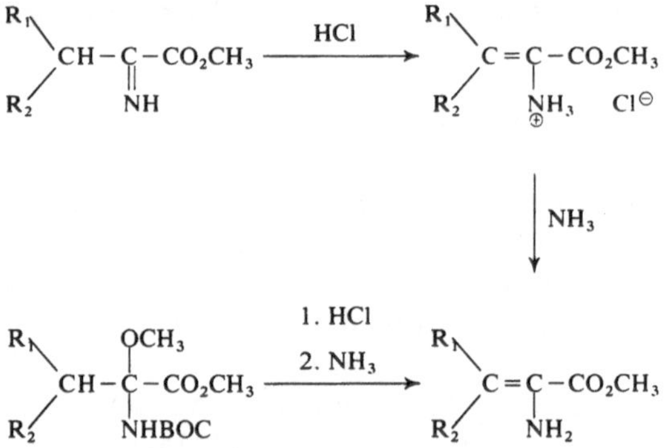

	R_1	R_2
a	CH_3	CH_3
b	H	$CH(CH_3)_2$
c	H	C_6H_5

Scheme 17

E. α-Iminocarboxylic Acid Derivatives

Most syntheses of iminocarboxylic acid derivatives start from α-keto acids or α-amino acids. Thus APPEL and HAUSS (12) described the formation of α-iminopropionic acid ester by treatment of pyruvic acid ester with triphenylphosphine imine. This product was however not isolated, but hydrogenated directly to the corresponding alanine ester.

C. SHIN et al. (253, 373) applied the same reaction to an ester of dimethylpyruvic acid and were able to isolate and characterize the α-iminovaleric acid ester containing no detectable amount of tautomeric enamine. Other reactions of N-triphenylphosphine imine with various α-keto esters have been found to yield tautomeric mixtures (371).

N-Alkyliminoacetic esters have been obtained by MASSEN et al. (254) by addition of phosphonate carbanions to tert. nitrosoalkanes.

FIAUD and KAGAN (125) condensed glyoxylic acid menthyl ester with benzylamine or α-substituted benzylamines to form the corresponding N-benzyliminoacetic acid menthyl esters.

N-Chlorination/dehydrochlorination of amino acid esters (310) is the simplest route to α-iminocarboxylic acid esters. With the exception of the leucine derivative, which was obtained as a tautomeric mixture, no enamine could be detected in the products by NMR spectroscopy. Phenylalanine ester yielded the enaminoacid ester exclusively.

F. α-Acyliminocarboxylic Acid Derivatives

α-Acylimino acids, highly reactive tautomers of acylenamino acids, have been postulated by many authors as intermediates in the oxidation of α-acylamino acids and in substitution reactions on α-substituted N-acylamino acid derivatives (292). It is normally impossible to isolate them, as they stabilize themselves either by the addition of a nucleophile, if available, or by rearrangement to the tautomeric form.

BALDWIN et al. (23) and shortly afterwards FIRESTONE and CHRISTENSEN (128) prepared 6-methoxypenicillins (44) by chlorination/dehydro-

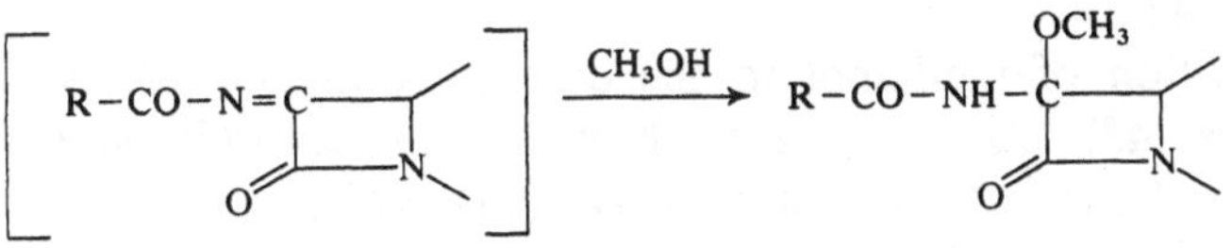

(43)   (44)

Scheme 18

chlorination of penicillin derivatives followed by addition of methanol to the intermediate acylimino compounds (**43**).

This reaction has been studied in some detail by H. Poisel and U. Schmidt (*310, 312, 314*), using several acylamino acid esters. α-Methoxy derivatives (**22**) were obtained from the N-chloro compounds on elimination of HCl with CH_3O^-/CH_3OH, while treatment with non-nucleophilic bases in aprotic solvents afforded acylenamino acid esters (**21**). Acylimino acids were undoubtedly primary elimination products, although it proved impossible to detect them.

The preparation and characterization of α-acetyliminovaleric acid methyl ester (**46**) was achieved by acetylation of the imino acid ester (**45**). Addition of methanol yielding the α-methoxy-α-acetaminovaleric acid ester and rearrangement to acetyl-α,β-dehydrovaline ester could be carried out without difficulty.

α-Acetylimino-β,β-dibromocarboxylic acid esters were observed by Shin *et al.* (*376, 378, 379*) as products of N-bromination and subsequent rearrangement of β-bromo-α-acetylamino acrylic esters.

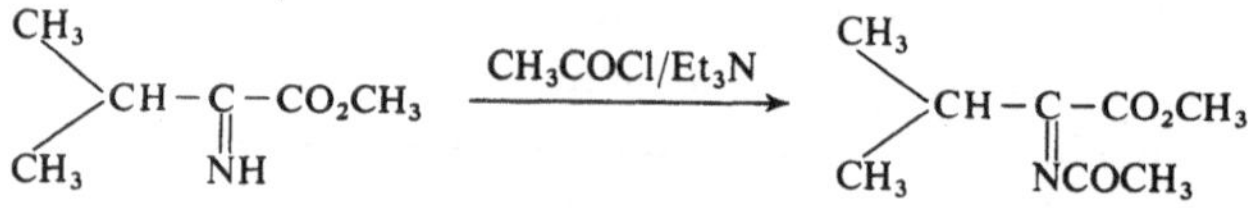

$$\begin{array}{ccc} CH_3 & & CH_3 \\ \diagdown & \xrightarrow{CH_3COCl/Et_3N} & \diagdown \\ CH-\underset{\underset{NH}{\parallel}}{C}-CO_2CH_3 & & CH-\underset{\underset{NCOCH_3}{\parallel}}{C}-CO_2CH_3 \\ CH_3 & & CH_3 \\ (\mathbf{45}) & & (\mathbf{46}) \end{array}$$

Scheme 19

V. Synthesis of α-Hydroxy- and α-Mercapto-α-amino Acids, α,α-Diamino Acids and the Corresponding Peptides and Piperazinediones

The most important α-alkoxy-α-amino acid compounds are the cephamycins (7-methoxycephalosporins) (**1**) (*5, 271*), which are of considerable therapeutic significance by virtue of their activity against Gram negative bacteria. Several methods have been developed for the conversion of penicillins and cephalosporins into their 6-(7-)methoxy derivative (**44**).

1. Addition of methanol to the corresponding dehydrocompounds [acylimines (**43**) (*23, 128, 225*), chlorovinylimines (**47**) (*400*), sulfenimines (**49**) (*146 A, 223*), and imines (**48**) (*441, 442*)].

2. Metallation of Schiff bases (e. g. arylidene-aminopenicillanic acid esters), electrophilic reaction of the carbanion (**50**) with a dialkyldisulfide followed by substitution of the alkylmercapto group of (**51**) by the

methoxy group [e. g. with methanol/Hg(OAc)$_2$] (*13, 192, 323, 392*) alternatively via an intermediate fluoro or bromo derivative (*90, 385*).

3. Via penicillin ketenimines (**52**) (*339*).

4. By nucleophilic substitution in 6-bromo-6-azido-penicillanic acid (*89*).

6-Aminopenicillin derivatives became recently available by direct substitution of penicillates with N-chloro-N-sodio-carbamates (*70—72*).

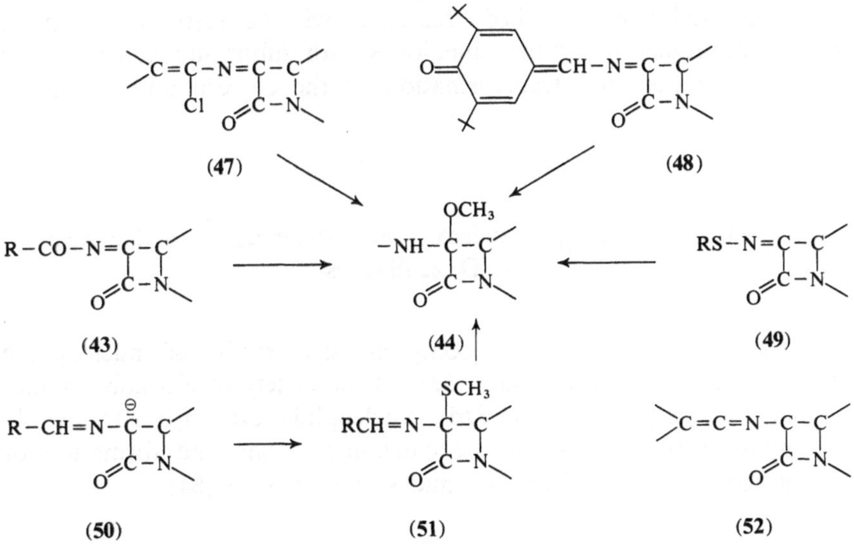

(47)   (48)

(43)   (44)   (49)

(50)   (51)   (52)

Scheme 20

For further details on the chemistry of the penicillin-cephalosporin system, some of the more recent reviews should be consulted (*65, 269, 342*).

The preparation of α-alkoxy- and α-mercapto-(α-alkylmercapto)-α-amino acids by addition reactions to acylenamino acids or acylimino acids is discussed in detail in sections VI. A. 2.2. and VI. B. Syntheses of epi-dithiopiperazinediones will be dealt with only briefly, as a detailed review of the piperazinediones has recently appeared in this series (*341*).

The epidithiopiperazinedione unit has been found to be a characteristic structural element in numerous fungal metabolites. Sulfur functions can be introduced into the piperazinedione nucleus in several ways:

1. Direct nucleophilic substitution reactions between e. g. thiolacetate or thiolate and halogenated piperazinediones or piperazinediones with hydroxy-, acetoxy- or sulfone substituents (*283, 284, 307, 416*).

2. Metallation of piperazinediones and electrophilic substitution by sulfur or disulfides (*282, 308*).

3. More complicated systems can be synthesized by preparation of dimercaptopiperazinediones by method 1, followed by mercaptalization, metallation of the mercaptals, electrophilic substitution and finally generation of the disulfide from the mercaptal via the monosulfoxide (*206—208*).

4. S_2Cl_2 has been used to form a disulfide bridge by reaction with alkali 2,5-dioxopiperazine-3,6-dicarboxylates and by addition reaction to unsaturated piperazinediones respectively (*345, 104*).

In the following, detailed treatment will be restricted to more modern approaches to piperazinediones with sulfur substituents based on the formation and transformation of the corresponding hydroxy compounds.

A. α,α-Diamino Acids, α-Hydroxy- and α-Mercapto-α-amino Acids via Oxazolinones

SHEMYAKIN has made a comprehensive study of nucleophilic reactions of 4-halooxazolinones (**53**) with a variety of alcohols, amines and mercaptans (*100*, see also *316*). Nucleophilic exchange is frequently accompanied by opening of the oxazolinone ring and formation of substituted α-amino acid esters, amides or thiolesters (**54**).

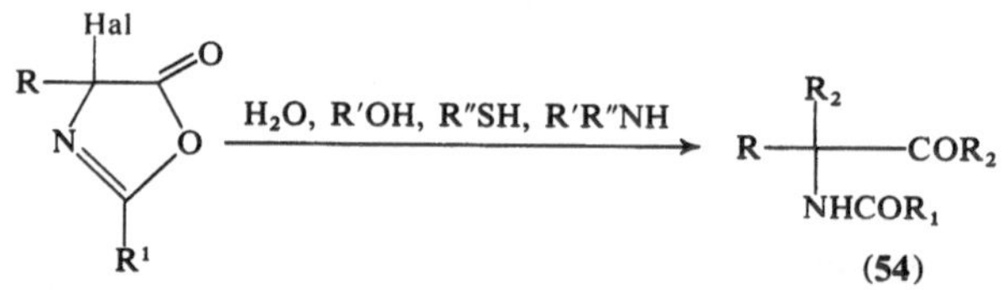

Scheme 21

Alternatively, similar amino acid derivatives (**54**) can be obtained by oxidation of the oxazolines with o-chloranil and reaction of the adducts with nucleophiles (*335*).

Not only acylimino acids (VI. B.) but also 2-alkylidene- and 2-arylidene-Δ^3-oxazolin-5-ones (**55**) (pseudooxazolones) add nucleophiles. Simultaneous ring opening is usually observed (*195, 237, 238, 239, 397*) yielding the corresponding amino acid derivatives. Reaction with ammonia or amines resp. alcohols yields α,α-diamino carboxylic acid esters (**56**) or α-alkoxy-α-acylaminocarboxylic acid esters (**58**) respectively. Mercaptans add so rapidly that mercaptooxazolinones (**57**) can be isolated and then opened with other nucleophiles.

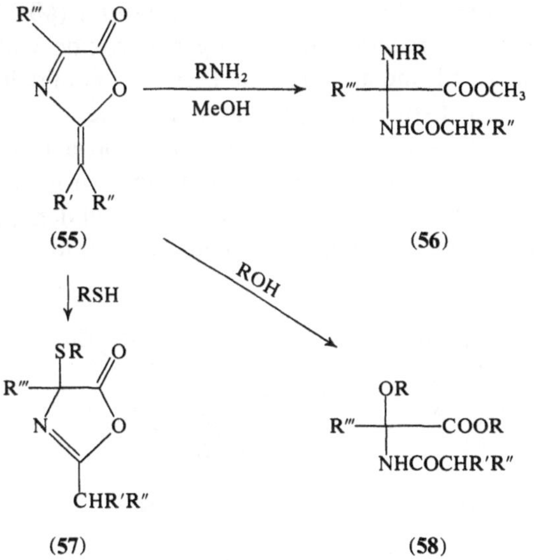

Scheme 22

B. Direct Oxidation of Amino Acid Derivatives and Piperazinediones

Little is known about the direct oxidation of amino acid derivatives. Although an α-acetoxy derivative can be prepared from hippuric acid by treatment with lead tetraacetate (99) this reaction is not applicable to N-phthaloyl- or N-benzyloxycarbonyl-glycine. Anodic oxidation of acetamidomalonic acid monoester in alcohol yielded the α-alkoxy-α-acetamino acetic acid, while in acetic acid the corresponding α-acetoxy compound was obtained (186). The oxidation of optically active N-salicylidene-alanine cobalt (III) complexes has been studied as a model reaction for enzymatic deamination of α-amino acids, and found to yield the corresponding diastereomeric α-hydroxy alanine complexes (118).

With regard to the biosynthesis of the epidithiopiperazinediones, U. SCHMIDT and J. HÄUSLER studied the photochemical oxidation of piperazinediones in some detail (168, 349), a reaction which takes place with exceptional ease in many cases. Piperazinediones which are not kept sealed in an oxygen-free atmosphere contain, if exposed to light, considerable amounts of hydroperoxides within a few weeks. Photochemical oxidation of piperazinediones in the presence of catalytic quantities of benzophenone yields mono- and bis-hydroperoxides in which the configuration of the original compounds is largely retained. As an example, the main product of photochemical oxidation of L-prolyl-L-proline anhydride (59) is the cis-bishydroperoxy compound (60). On the other hand, the nucleophilic reaction of hydrogen peroxide with L-prolyl-L-proline anhydride disulfone afforded the thermodynamically more stable bishydroperoxide of L-prolyl-D-proline anhydride (283). Direct conversion of L-prolyl-L-proline anhydride into bis-acetoxy-L-prolyl-L-proline anhydride has been achieved by treatment with lead tetraacetate (284). The same reagent oxidizes 1-methyl-3-benzylidenepiperazine-2,5-dione at the double bond to yield the vicinal diacetate (244).

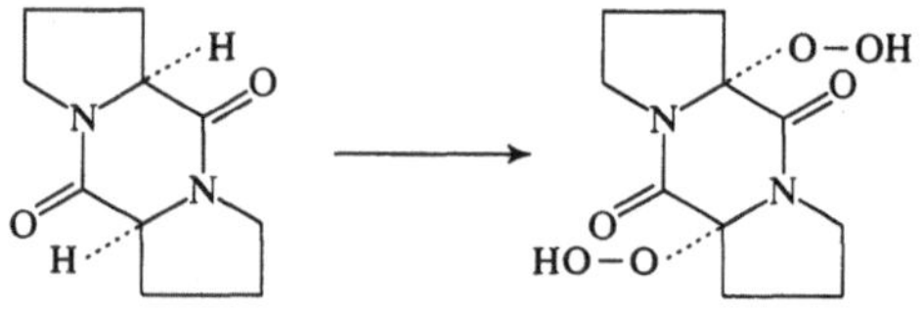

(59)   (60)

Scheme 23

C. Interconversion of α-Hydroxy- and α-Mercapto-α-amino Acids, Hydroxy- and Mercaptopiperazinediones, Pyruvoylamino Acid Amides and Dehydropeptides

α,α-Diaminoacids (61), α-hydroxy-α-amino acids (62), α-mercapto-α-amino acids (63) and α-imino acids are aminals, hemiaminals, thio-aminals and imines respectively of α-keto acids. Acyl-enamino acids (64) are the enamides of α-keto acids.

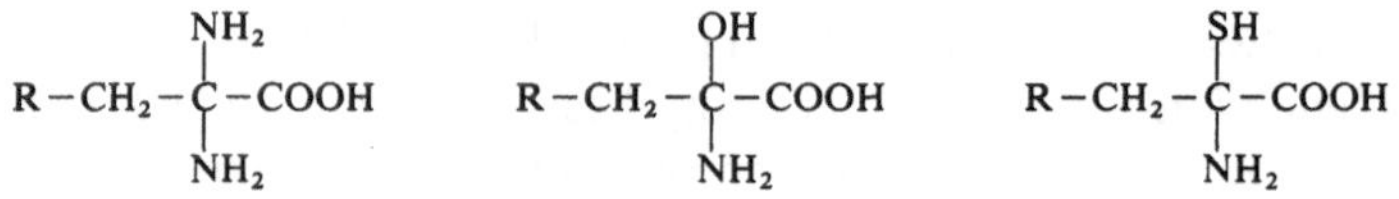

(61)   (62)   (63)

$$R-CH=C-COOH$$
$$\quad\quad\quad |$$
$$\quad\quad NH-CO-R$$

(64)

Scheme 24

α-Methoxy-α-acetamino acid esters and α-imino acid esters, which can be easily prepared from the racemic amino acids, have been used to prepare the corresponding α-keto acid esters (314A).

These formal interrelationships, which are certainly also relevant with regard to the biosynthesis of α-alkoxy-α-amino acid derivatives, dehydropeptides and epidithiopiperazinediones, can be clearly demonstrated by in vitro experiments. It has been shown by J. HÄUSLER and U. SCHMIDT (*169*) that pyruvoylamino acid amides, for which convenient synthetic routes have been worked out*, cyclize under physiological conditions to hydroxy-piperazinediones**, which can readily undergo loss of water to unsaturated piperazinediones or can be converted into mercaptopiperazinediones. The steric course of this reaction can be studied with particular ease when applied to the ring closure of N-pyruvoyl-L-proline-N-methylamide (**65**); hydroxy-L-prolyl-L-alanine anhydride (**66**) is formed in a kinetically controlled reaction. Particularly this isomer may arise from the ring closure reaction of the s-transoid conformation of the α-ketoacyl compound (**65**). (**66**) rearranges in weakly acid solution to the thermodynamically more stable hydroxy-L-prolyl-D-alanine anhydride. Treatment with ethanethiol yields the thioether (**67**) while the dehydrocyclodipeptide (**68**) is formed on dehydration. Analogous reactions have also been described by later authors (*86, 251, 295a, 295b, 445*)***.

Feasible biological relationships between α-keto-ω-acylamino acid amides and the epidithiopiperazinediones were demonstrated by E. ÖHLER and U. SCHMIDT (*285*). In weakly acid aqueous solution the acetal (**69**) underwent hydrolysis followed by "double ring closure" to the dihydroxy-piperazinedione (**70**). The latter could be converted by treatment with hydrogen sulfide into the cis-dimercaptan, which yielded the epidisulfide

---

* α-Ketoacylamides respectively α-ketoacylamino acids are available by the following routes:

1. By acylation of amines with "active" derivatives of α-keto acids: pyruvic acid chloride (*167*), p-nitrophenyl pyruvate (*167, 169*), carbodimide method (*86, 166, 178*), mixed anhydrides with phosphorous acid dichloride (*436*) and by activation with the cyclic esters of oxalic acid with 4,6-diphenyl-thieno-[3,4-d]-[1,3]-dioxolone-dioxide (H. SCHMIDT and W. STEGLICH, *346A*).

2. By acylation of amines with hydroxymaleic anhydride (*129, 167, 437*).

3. From adducts of carboxylic acid chlorides or N,N-dialkylformamide chlorides with isonitriles (*190, 419*).

4. By oxidation of lactic acid amides (*9*).

5. Via 4-acylamino-oxazolinones (*47*).

6. By thermal rearrangement of 2,3-bis(alkylimino)oxetanes (*338*).

7. By thermal decomposition of O-acylated 2-(N-alkyl-hydroxyl-amino)-carboxamides (*265*).

8. By hydrolysis of trifluoroacetyl dehydrodipeptides (*270*).

9. By addition of pyruvoyl chloride to α-iminocarboxylic acid derivatives (*295a, 295b, 311*).

** For ring closure of an N-glyoxyl amino acid amide see (*162*).

*** For questions of priority see also (*87, 351*).

(71) on dehydrogenation. (69) is the acetal of the α,α'-diketo compound corresponding to N^{ω}-alanyl-ornithineamide.

Numerous examples of analogous transformations of α-hydroxy-α-acylamino acids into α-mercapto-α-acylamino acids have been reported (*195, 294b, 451*).

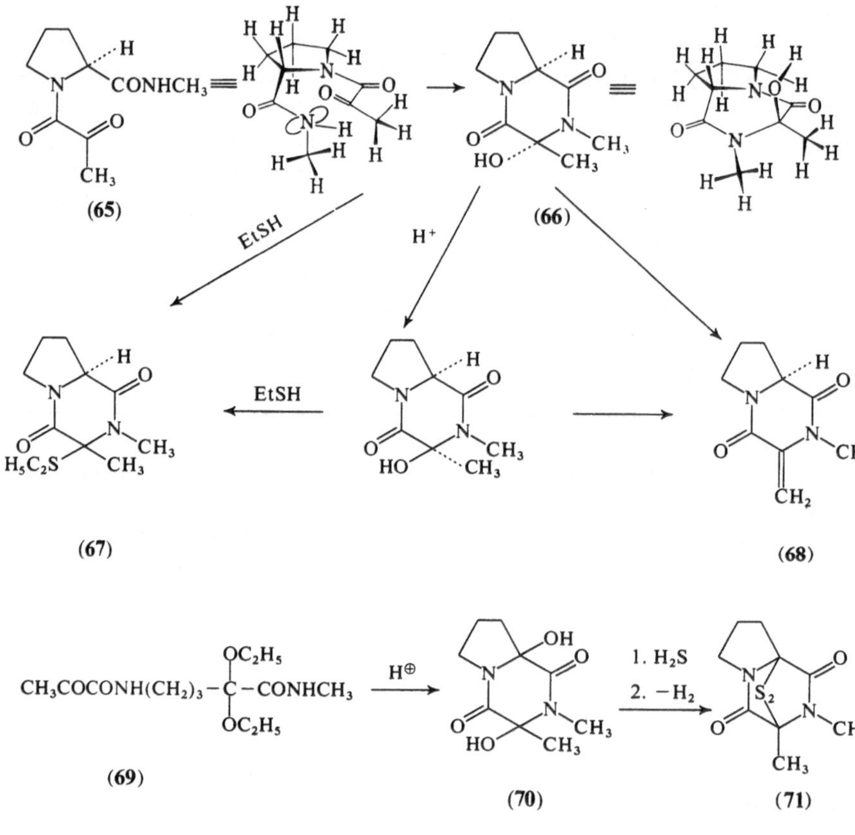

Scheme 25

The transformation of α-acylamino-α-amino carboxylic esters into α-acylamino-α-hydroxycarboxylic esters was achieved by Lucente *et al.* (*240*).

VI. Reactions of Dehydroamino Acids

A. Enaminocarboxylic Acid Derivatives

1. Reactions of the Amino and Carboxy Groups

Only a few esters of dehydroamino acids are stable and these only as hydrochlorides, so that all the reactions at the amino group studied to date are electrophilic. As the enamino group is less nucleophilic than an amino group, enamino acid esters can only be acylated by the most reactive acid derivatives, such as acid chlorides (*368, 373*) and mixed anhydrides (*69, 257, 313*). Attempted peptide formation from acylamino acids and enamino acid esters by the DCCD or DCCD/hydroxysuccinimide methods is unsuccessful. The formation of Schiff bases at the enamino group proceeds rather slowly, but can be achieved in the case of the stable dehydrovaline methyl ester by direct condensation with aromatic aldehydes (*287, 352*).

Peptide formation by acylation of amino acid esters with N-acyl-dehydroamino acids can be carried out by employing activated derivatives of the latter. Suitable derivatives are the N-carboxy-α-amino acid anhydrides (*226*), or mixed anhydrides (*257, 314*). Coupling with DCCD/N-hydroxysuccinimide is also applicable (*314B*).

For the construction of peptides with an N-methyldehydroamino acid unit, use can be made of the observation by RICH (*328*) that selective N-methylation of N-acyl enamino acid units can be carried out with methyl iodide/potassium carbonate in dimethylformamide.

2. Reactions of the Double Bond

The double bond of N-acylenamino acids and their esters represents the "*locus minoris resistentiae*". As a result of conjugation with the carboxyl group, the double bond is a good acceptor for nucleophilic addends under basic catalysis. The acylamino group makes electrophilic additions possible. It is thus possible to obtain α- or β-mercapto α-amino acid derivatives by selecting the appropriate reaction conditions for the addition of mercaptans.

Besides polar additions, radical additions such as the photochemical (*287*) or peroxide-catalysed (*121, 122*) mercaptan addition have been investigated. Catalytic hydrogenation has been the object of particularly thorough study. In cases where optical induction is possible, addition reactions afford a means of synthesizing optically active amino acids.

Table 3. *Addition Reactions to N-Acyl-α,β-dehydro Amino Acid Derivatives*

Substrate	Addend	References
N-Acyl dehydroalanine derivatives	aliphatic amines	*(16, 17, 64, 116, 134, 320)*
	imidazole	*(435)*
	N$^\alpha$-acyl-lysine	*(64, 438)*
	potassium-phthalimide	*(176)*
	1,3-dicarbonyl compounds:	
	malonic acid esters	*(63, 250, 268, 431, 435)*
	β-keto esters	*(174)*
	1,3-diketones	*(176)*
	β-sulfonyl esters (intramol, addition)	*(443)*
	nitroalkanes	*(143, 435)*
	potassium cyanide	*(175)*
	indole derivatives	*(389)*
	aliphatic thiols (base-catalysis)	*(31, 116, 154, 155, 176, 268, 287, 359)*
	aliphatic thiols (radical addition)	*(287)*
	aromatic thiols	*(31, 116, 279)*
	thiolacetic acid	
	(base catalysis)	*(30, 354)*
	(radical addition)	*(121, 122)*
	cysteine derivatives	*(340, 353, 355, 358, 388)*
	β-methyl cysteine	*(32)*
	penicillamine derivatives	*(15, 34, 231, 232, 233, 384, 402, 404)*
	homocysteine	*(356, 357)*
	mercapto acetamide	*(151)*
	β-mercapto pivaloic acid	*(58)*
	intramol. addition of thiols	*(69, 448, 449)*
	sulfinic acids	*(31, 176)*
	thiocyanogen chloride	*(236)*
	sulfur dichloride	*(236)*
N-Arylidene dehydroalanine derivatives	aliphatic thiols	*(287A)*
N-Acyl-dehydro-amino-butyric acid derivatives	intramol. addition of thiols	*(69)*
	thioglycolic acid	*(62)*
N-Acyldehydrovaline derivatives	aliphatic thiols	*(346, 401, 403)*
	intramol. addition of thiols	*(21, 98, 103, 439)*
N-Arylidene dehydrovaline derivatives	aliphatic thiols	*(287)*
N-Acyldehydroleucine derivatives	intramol. addition of thiols	*(69)*
N-Acyldehydrophenyl-alanine derivatives	aliphatic thiols	*(116)*
	aromatic thiols	*(116, 278, 279)*
	intramol. addition of thiols	*(69, 98, 383)*

2.1. Additions Leading to β-Substituted Amino Acids (see Table 3)

The addition of CH-acidic compounds, hydrogen cyanide and amines to acylaminoacrylic acid compounds has occasionally been employed for the synthesis of more complex amino acids.

Reduction with sodium borohydride and tritiated sodium borohydride (*28, 61, 144, 164, 434*) has been used in the structural elucidation of natural products with dehydroamino acid units, as has the addition of mercaptans or nitromethane. Addition of S-nucleophiles is the method of choice for the preparation of sulfur-containing amino acids such as cysteine, penicillamine, lanthionine and cystathionine. Such additions have also been often carried out in biomimetic experiments in the study of penicillin biosynthesis (see Table 3).

E. ÖHLER, E. PRANTZ and U. SCHMIDT (*287, 352*) have studied the addition of mercaptans to the arylidene-dehydrovaline esters (**72**) with reference to cysteine biosynthesis, which according to most recent ideas can proceed in two ways, catalysed by pyridoxal-containing enzymes (*111, 414,* see also *101*).

a) In the case of the "β-replacing lyases", direct substitution of OH by SH is assumed.

b) The "β-eliminating lyases" catalyse the elimination of water from pyridoxylidene serine to form pyridoxylidene amino acrylic acid, followed by the addition of SH to give cysteine.

In the Schiff bases of dehydrovaline ester which represent simple models for studying the events of path b), a remarkably increased ease of addition to the salicylidene derivative (**72b**) as compared with the benzylidene derivative (**72a**), the α,β-dehydrovaline ester and the N-acyl-dehydrovaline ester was found. Additional introduction of electron acceptors (p-nitro compound **72c**) into the azomethine, and the transition to the less electron-rich pyridine system (pyridoxylidene compound **72d**) result in a further, but not significant increase in the ease of addition.

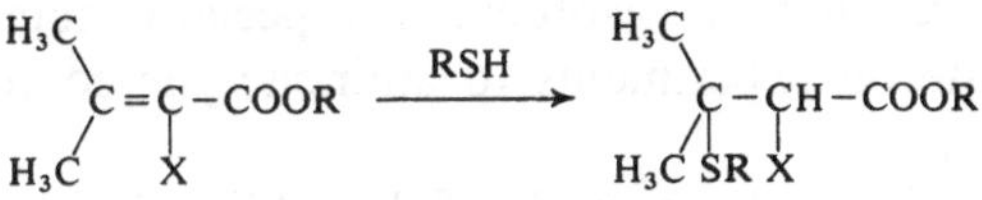

$$\begin{array}{c}
H_3C \\
\diagdown \\
\quad\;\; C=C-COOR \\
\diagup \quad\; | \\
H_3C \quad\;\; X
\end{array}
\xrightarrow{\;RSH\;}
\begin{array}{c}
H_3C \\
\diagdown \\
\quad\;\; C-CH-COOR \\
\diagup\; | \quad\; | \\
H_3C\; SR \;\; X
\end{array}$$

(**72 a–d**)

a	benzylideneamino
b	salicylideneamino
c	p-nitro-salicylideneamino
d	pyridoxylideneamino

Scheme 26

Pronounced optical induction was observed in the addition of mercaptan to N-acyldehydroalanyl-L-proline amides (73), acyl-D-cysteinyl-L-proline amides (75) being formed almost exclusively (287, 348). This stereoselective course is observed only when a strongly basic catalyst is used, and only when the proline component of the unsaturated dipeptide is present as amide. The stereoselectivity is considerably less when piperidine is the catalyst, and vanishes completely on photochemical addition. Alkylation of both nitrogen atoms is without effect on the course of reaction, while addition to acyldehydroalanylproline ester proceeds with virtually no stereoselectivity. These findings suggest a carbanion intermediate (74) fixed spatially in such a way that it can be approached by the proton from one direction only. One possible explanation seems to be the fixation of the carbanion within a carbanion-immonium ion pair (74).

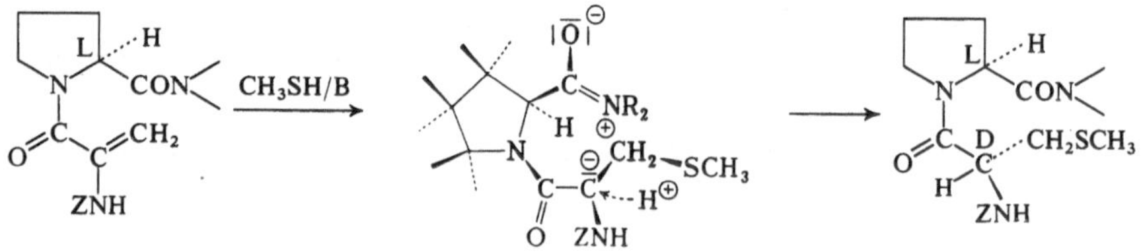

(73)   (74)   (75)

Scheme 27

The biosynthesis of the heterodetic rings in the antibiotics nisin and subtilin corresponds exactly to the steric course of this cysteine formation from dehydroalanine within a peptide.

The transition from serine to cysteine units within a peptide should also be discussed in this context. The occurence of L-seryl-L-proline units in the hormones kallidin, eledoisin and β-melanotropin, and of L-cysteinyl-L-proline units in β-corticotropin releasing factor, oxytocin, isotocin, vasopressin and vasotocin is especially noteworthy. Further biomimetic addition experiments to aminoacrylic compounds are summarized in Table 3.

A kinetic study has been made of the addition of aminothiols, thiols and mercaptoamino acids to acylaminoacrylic acid (388). As an ester group stabilizes a carbanion much more effectively, N-acetyldehydroalanine ester adds a good deal more rapidly than N-acetyldehydroalanine, a difference which is particularly noticeable in the addition of sterically hindered mercaptans such as penicillamine.

To the more inert double bond of unsaturated piperazinediones methane sulfenyl chloride or sulfur chloride was added, the β-thio compounds being formed in every case (444).

Acylamino acid derivatives with a leaving group in the β-position, such as N-acyl-O-tosylserine, undergo substitution reactions with nucleophiles. According to the leaving group, nucleophile and reaction conditions, the reaction can proceed via a nucleophilic substitution with retention of configuration or elimination/addition with racemization (see section IV. A. 2.1).

2.2. Additions Leading to α-Substituted Amino Acids

The addition of halogen to acylaminoacrylic esters with subsequent loss of hydrogen halide to form β-halo-α-acylamino-acrylic esters has been the subject of numerous reports (198, 274, 330, 376). β-Bromo-α-acylamino-acrylic ester is also obtainable by rearrangement of N-bromo-α-acylamino-acrylic ester (378). Further reaction of the β-halo-α-acyl-aminoacrylic esters with bromine or N-bromosuccinimide and subsequent treatment with alcohol affords β,β-dihalo-α-alkoxy-α-acylaminopropionic acid esters (274, 379), important intermediates for biomimetic β-lactam syntheses (275).

Kinetically controlled addition of hydrogen halide to acylamino-acrylic ester leads to α-halo-α-acylamino-propionic acid ester (76), which gradually rearranges in the reaction solution into the β-halo-derivative (236, 241, 301). The α-halo compound can be captured with mercaptans, alcohols or water, forming the appropriate α-substituted α-acylamino-propionic acid esters (77) (197, 199).

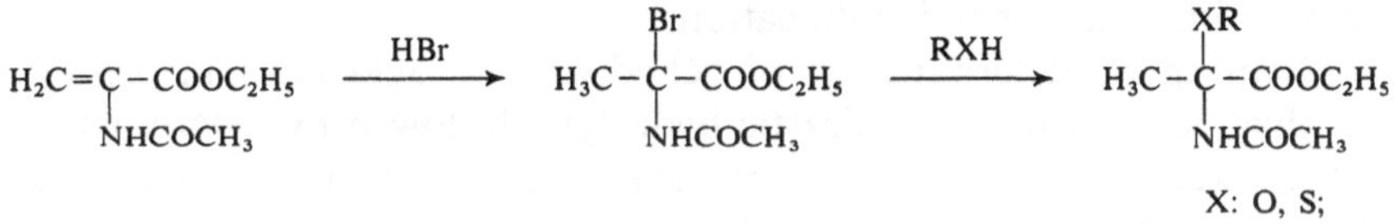

$$H_2C=\underset{\underset{NHCOCH_3}{|}}{C}-COOC_2H_5 \xrightarrow{HBr} H_3C-\underset{\underset{NHCOCH_3}{|}}{\overset{\overset{Br}{|}}{C}}-COOC_2H_5 \xrightarrow{RXH} H_3C-\underset{\underset{NHCOCH_3}{|}}{\overset{\overset{XR}{|}}{C}}-COOC_2H_5$$

X: O, S;

(76)   (77)

Scheme 28

Alkoxymercuration of acylaminoacrylic acids followed by NaBH₄-reduction affords α-alkoxy-α-acylaminopropionic acids (139).

Amides can be added to acylaminoacrylic ester even without acid catalysis to form the α,α-diacylaminopropionic acid ester (150, 364). An analogous addition has been described in the reaction of N-acyl-dehydro-proline with acetamide (136) (see Section IV. A. 6). Analogous intra-molecular additions, leading to the formation of five- and six-membered rings, proceed even under basic catalysis (see Section VI. A. 2.4).

Finally, the first step in the acid hydrolysis of an acylenamino acid to acylamide and α-keto acid is the electrophilic addition of a proton.

As this hydrolysis takes place relatively easily, a dehydroamino acid unit represents a "weak link" in a peptide chain (see Section VI. A. 2.3.).

The double bond in unsaturated dioxopiperazines is more inert than that of acylenamino acids towards addends. The acid catalysed addition of mercaptans and thiolacetic acid to various dehydrodioxopiperazines has been described by MACHIN and SAMMES (242). The kinetically controlled reaction leads to the formation of the α-product, admittedly in very moderate yield in the case of addition of thiolacetic acid.

2.3. Hydrolysis

Hydrolysis of peptides with dehydroamino acid units (78) affords two fragments, a peptide amide (79) and an α-ketoacyl peptide (80). This specific hydrolytic cleavage is not restricted to dehydropeptides, but can also be applied to peptides and proteins at the serine and cysteine units, thus serving as an aid to structural elucidation (141, 299, 300). For this purpose the SH or OH functions are converted into suitable leaving groups and subsequently eliminated (see IV. A. 2.). The dehydro-peptide bonds are then cleaved at the $N-C_\alpha$ bond of the acrylic acid system either hydrolytically (brief heating to 100° at pH 2) or in an oxidative step by treatment with bromine at room temperature or performic acid at 0°. The nitrogen of the dehydroamino acid is then the new terminal amide function of one of the fragments, the remainder of the molecule appearing as an α-ketoacyl group of the other fragment. The latter can be removed by oxidation (H_2O_2/OH^-), making the new N-terminal amino acid accessible to identification.

Quantitative determination of dehydroamino acid units in peptides can thus be carried out by drastic hydrolysis followed by determination of the α-keto acid (300), as a supplementary or alternative method to the bromine addition, which latter reaction can be carried out only in the absence of other reactive groups.

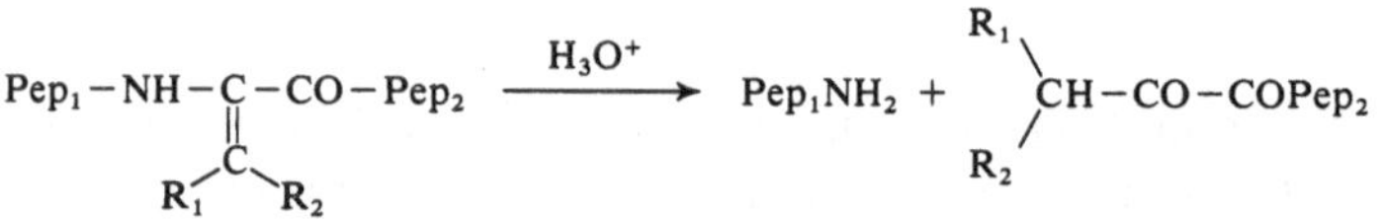

$$\text{Pep}_1-\text{NH}-\underset{\underset{R_1^{\diagup}\text{C}\diagdown R_2}{\|}}{\text{C}}-\text{CO}-\text{Pep}_2 \xrightarrow{\text{H}_3\text{O}^+} \text{Pep}_1\text{NH}_2 \ + \ \underset{R_2^{\diagup}}{\overset{R_1\diagdown}{\text{C}}}\text{H}-\text{CO}-\text{COPep}_2$$

$$(78) \qquad\qquad\qquad (79) \qquad\quad (80)$$

Scheme 29

GROSS and coworkers, exploited the facile hydrolysis of unsaturated peptides, in the solid phase synthesis of peptide amides (158, 159). The molecule which is to be elaborated is linked to the carrier through a dehydroalanine carboxyl group. After completion of the synthesis mild conditions suffice to effect hydrolysis to the peptide amide and pyruvoyl resin.

2.4. Intramolecular Additions

All base catalysed additions to the double bond of acylaminoacrylic acid derivatives yield β-substituted amino acid compounds as products of nucleophilic reactions. Only electrophilic additions catalysed by very strong acids afford α-substituted amino acids. Additions of amides to derivatives of α,β-unsaturated amino acids in neutral media, which lead to α,α-diacylamino acid derivatives have also been described (*136, 150, 179, 364*) (see IV.A.6.). Further analogous reactions are the base catalysed cyclizations of peptides with a dehydroamino acid subunit by the intramolecular addition of a neighbouring amide function to the α-C atom of a dehydroamino acid (*286, 347*). Peptides with dehydro-aminoacid units can form imidazolones (**81**) or acylaminodioxopipera-zines (**82**), depending on which of the neighbouring amide nitrogens adds to the double bond. A dioxopiperazine is only formed when the amide bond adjacent to the double bond can readily assume the s-cis conformation required for ring closure, in other words, when the dehydroamino acid is linked through the carboxyl group with prolinamide or with an N-methylaminoacid amide ($R_1 = CH_3$).

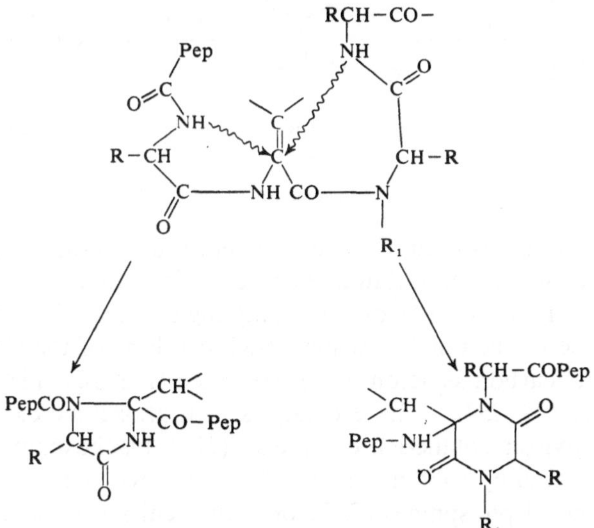

Scheme 30

Ring closure to the imidazolone is only observed when dioxopiperazine formation is not possible on structural or conformational grounds.
The six-membered ring is formed with high stereoselectivity under kinetic
control. The compound formed is usually the isomer with higher energy,
and can be rearranged into the thermodynamically more stable one under
acid or basic conditions.

2.5. Catalytic Hydrogenation

In the heterogenous catalytic hydrogenation of peptides and cyclopeptides with dehydroamino acid subunits, it has frequently been observed
that the neighbouring amino acid induces asymmetry. The influence of
valine (276), phenylalanine (110) and α-phenylethylamine (362) may be
mentioned in this context. The most pronounced effects were observed
by H. Poisel and U. Schmidt (309) in the catalytic hydrogenation of
arylidene-piperazinediones (83) (R$_2$ = Ar) containing an L-proline unit.
L,L-cyclodipeptides (84) formed can be hydrolysed to L-proline and
L-amino acid, thus representing a method of preparing aromatic amino
acids and aromatic N-methyl amino acids with an optical yield greater
than 90%. B. W. Bycroft and G. R. Lee have exploited the hydrogenation
of analogous proline-containing alkylidenedioxopiperazines (83) (R$_2$ =
Alkyl) for the synthesis of optically active aliphatic amino acids (86, 87).

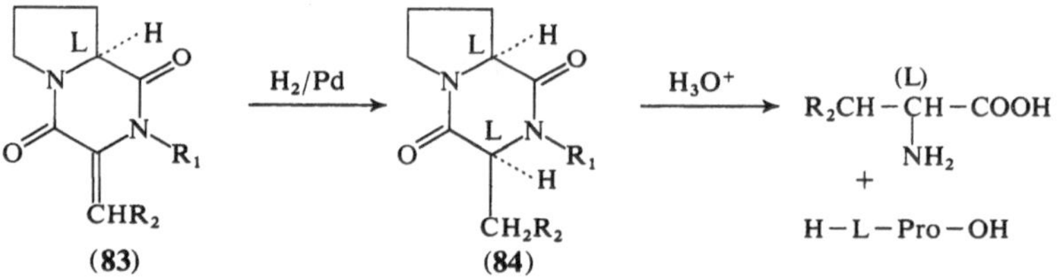

Scheme 31

Close attention has been devoted in recent years to the homogeneous
catalytic reduction of N-acylaminoacrylic acids with the aid of chiral
Rh-complexes. In some cases exceptionally high optical yields have been
achieved in these reactions*. Phosphine-Rh catalysts of the DIOP type,
i. e. with chiral carbon skeleton, have been used (108, 109, 133, 142, 145,
146, 172, 193, 194, 234), as have catalysts with phosphine oxide ligands
(409), ferrocenyl-phosphine-Rh complexes (171), bisphosphine-Rh complexes with a chiral pyrrolidine ring (3, 4), systems with chiral phosphines
(216—221) and bisphosphines (222), or with a chiral P- and C-skeleton
(130).

---

* For a recent summary of work on asymmetric synthesis see ref. (530).

Chiral Rh-complexes have recently been successfully employed for the heterogenous hydrogenation of acylaminoacrylic acids, by coupling them to a polymeric carrier (*408*).

B. Addition Reactions to α-Acylimino- and α-Iminocarboxylic Acid Derivatives

Addition reactions to the highly reactive α-acylimino carboxylic acid derivatives yielding α-substituted α-acylaminocarboxylic acid derivatives have been discussed in section IV. A. 3. and IV. F.

Addition reactions of alcohols and thiols to the less reactive α-iminocarboxylic acid derivatives remain incomplete, leading to an equilibrium which depends on the structure of the imino-compound and the addend (*314*). OTTENHEIJM observed quantitative addition of mercapto-substituted carboxylic acids to 2-ethoxycarbonyl-3,3-dimethylindolenine (*294a*), because the primary addition product was stabilized by cyclization to a thiazolidone derivative (see also *294c*). More favorable are the addition reactions of hydrogen cyanide (*311*) or carboxylic acid chlorides (*293, 311*). In the addition products of the latter (α-chloro-α-acylaminocarboxylic acid derivatives) the halogen can be displaced by other nucleophiles (thiols, alcohols, H_2O) forming the corresponding α-substituted α-amino acid derivatives (*293, 311*). This reaction combined with the ring closure of α-ketoacylamino acid amides (see V. C.) has been used to synthesize 3,6-dimercaptopiperazinediones. For example the addition product of

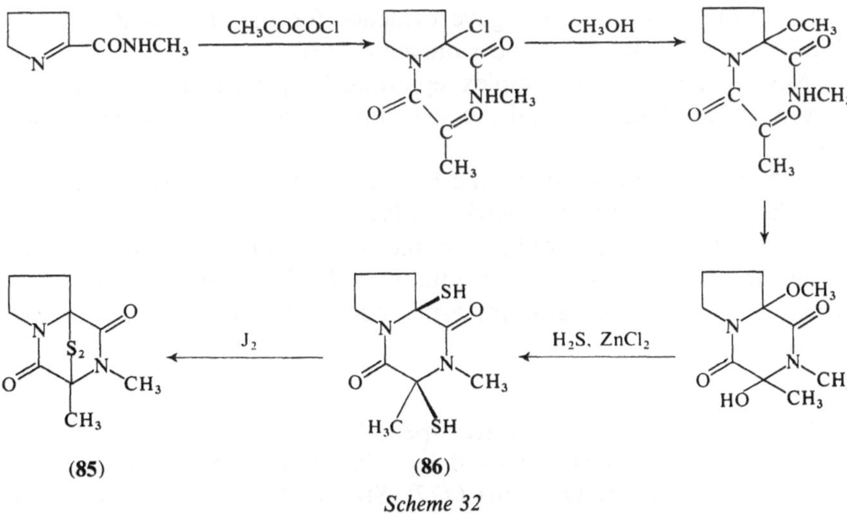

(85)   (86)

Scheme 32

pyruvoyl chloride to 1,2-dehydroproline-methylamide has been transformed to the cis-dimercaptopiperazinedione (**86**) and to the epidisulfide
(**85**). Later on a corresponding reaction sequence with an indolenine
carboxylic acid amide has been carried out (*295a, 295b, 351*).

Addendum

A considerable number of papers on related subjects has been
published since the manuscript of this review article was submitted to
publication.

ad II (in addition to Table 1):

Shimohigashi and coworkers described syntheses in the field of the
Alternaria Mali toxins and related cyclodepsipeptides (*452, 453, 454, 454A,
455*). For mass spectroscopy studies on these compounds see ref. (*456*).

The structure of capreomycin was revised (*457*) and the antibiotic itself
was synthesized (*458*) by Nomoto et al.

Gross succeeded in the synthesis of the A-ring of nisin (*459*), and
Rich continued his studies on synthesis (*462*) and conformational analysis (*460, 461*) of tentoxin and its analogues. In the tuberactinomycin
series several semisyntheses were published (*463, 464, 465*) and a total
synthesis of tuberactinomycin O was achieved (*466*).

The structures of tallysomycins A and B, a new antitumor antibiotic complex related to bleomycin, were determined by Konishi et al.
(*467*). The structures of the peptide antibiotic components of thiopeptin
containing DH-Ala and DH-Cys units were elucidated (*494*) and nitro-
gen-15 n.m.r. investigations gave evidence for the structural relationship between thiostrepton and siomycin-A (*495*).

New toxins with a complex epidithio-2,5-piperazinedione nucleus
were reported (*468, 496*) and the structure of chetomin was established
(*497*).

A biomimetic synthesis of the bisthiazole moiety of bleomycin was
published by Hecht and coworkers (*469, 496A*).

Configurational and conformational studies in the mikamycin family
(griseoviridin, ostreogrycin A, madumycin) (*471*) and the revision of the
structures of the peptide antibiotics micrococcin P_1 and P_2 were published
by Bycroft (*470*).

ad IV:

The synthesis of unsaturated peptides by dehydrosulfenation of
cysteine derivatives was improved by using phosphites and phosphines
as sulfenic acid trapping agents (*472*). Shin and coworkers synthesized

further alkyl 2-acylamino-2-alkenoates by condensation reactions of nitriles with α-haloamides (cf. chapter IV.A.6) and cyclized them to unsaturated piperazine-2,5-diones (cf. chapter IV,B,2) *(473)*.

The formation of dehydropeptides by oxidation reactions was observed *(499, 500)*. Further β-elimination reactions forming dehydro-amino acids were published *(501)*; for an intramolecular elimination reaction in the field of β-lactam antibiotics see Ref. *(509)*. KLOSTER-MEYER and WATANABE observed the formation of dehydroalanine on heating β-lactoglobulin A *(502)*. The synthesis of cycloalk-1-enylglycines with the method of SCHÖLLKOPF was described *(503)* by SUZUKI et al. Biomimetic elimination reactions with pyridoxylidene amino acids were investigated *(504, 505)*. KEITH prepared β-γ-unsaturated amino acids by rearrangement of the α-β-unsaturated compounds *(506)*. GREENLEE prepared α-vinyl-α-amino acids by α-alkylation of the anion of benzylidene α,β-dehydroaminobutyric acid ester *(507)*. CHIMIAK and KOLASA synthesized N-hydroxypeptides *(508)*, the dehydration of which may be of biological importance. An allenic amino acid derivative in the field of β-lactams was formed by elimination reaction of a β-tosyloxy-dehydro-amino acid ester *(510)*.

ad V:

7-α-Methoxy-cephalosporins and 6-α-methoxypenicillins were synthesized via iminophosphorane intermediates *(474)* and ketenimines *(475)*, respectively.

ad V.B:

TAJIMA obtained α-acylamino-α-alkoxy esters from α-acylamino acid esters reacting with DCCD/ROH *(476)*. The intermediate azlactones underwent autoxidation which was followed by DCCD-supported hydroxy-alkoxy-exchange. An electrochemical synthesis of N-acyl-2-alkoxyproline derivatives was reported by HORIKAWA et al. *(477)*.

ad V.C:

For a recent synthesis of α-oxoacid amides from nitriles using methyl methylthiomethyl sulfoxide see ref. *(478)*. Further interconversions between α-acylamino-α-hydroxy acid derivatives and the corresponding α-acylamino-α-aryl(alkyl)thio-, α-arylamino- and α-alkoxy compounds were reported *(479, 480, 511)*.

ad VI:

An α,β-dehydrovaline derivative was used in an intramolecular 2 + 2 photochemical cycloaddition reaction yielding a new β-lactam *(481)*.

In several papers MATSUMURA et al. studied reactions of β,β-dichloro-acylenamino-nitriles and their corresponding esters *(536—540)*.

ad VI.A.2.5:

Several authors reported further asymmetric hydrogenations of dehydropeptide derivatives: A Japanese group was able to show that not only an L-proline unit gives high optical induction in heterogeneous hydrogenation reactions of cyclic dehydrodipeptides, but that Ala, Val, Phe or Lys (ε Ac) do so as well, the highest effect being observed with L-valine (482). Pieroni and coworkers reported hydrogenations of acyclic dehydrophenylalanine containing peptides to be asymmetric (483). [For studies on conformation and chiroptical properties of dehydropeptides published by the same group see ref. (484 and (485).]

Further several papers were published dealing with homogeneous catalytic reductions of unsaturated amino acid derivatives using chiral phosphine Rhodium and Ruthenium complexes (486, 487, 488, 493, 512—529). See also the review of Valentine and Scott (530).

Stille and Achiwa employed further polymer-attached optically active phosphine ligands in Rh-catalyzed heterogeneous hydrogenation reactions of dehydro amino acid compounds (489, 490, 531).

ad VI.B:

Aminals and hemiaminals of glyoxylic acid esters were employed in Diels-Alder-reactions with cyclohexa-1,3-diene (491, 492), the intermediate imino acid derivative acting as dienophile. Several authors dealt with asymmetric hydrogenation of optically active Schiff bases of α-keto acids (532—535). For prior publications in this field these papers should be consulted.

References

1. Aberhart, D. J., and L. J. Lin: Studies on the Biosynthesis of β-Lactam Antibiotics. Part I. Stereospecific Syntheses of $(2RS,3S)$-[4,4,4-^{2}H$_3$]-, $(2RS,3S)$-[4-^{3}H]-, and $(2RS,3S)$-[4-^{13}C]-Valine. Incorporation of $(2RS,3S)$-[4-^{13}C]-Valine into Penicillin V. J. Chem. Soc., Perkin I 1974, 2320.

2. Aberhart, D. J.: Biosynthesis of β-Lactam Antibiotics. Tetrahedron 33, 1545 (1977).

3. Achiwa, K.: Asymmetric Hydrogenation with New Chiral Functionalized Bisphosphine-Rhodium Complexes. J. Amer. Chem. Soc. 98, 8265 (1976).

4. — New Chiral Phosphine-Rhodium Catalysts for Asymmetric Synthesis of (R)- and (S)-N-Benzyloxycarbonylalanine. Chemistry Letters 1977, 777.

5. Albers-Schönberg, G., B. H. Arison, and J. L. Smith: New β-Lactam Antibiotics; Structure Determination of Cephamycins A and B. Tetrahedron Letters 1972, 2911.

6. Ames, D. E., R. E. Bowman, J. F. Cavalla, and D. D. Evans: Griseoviridin. Part I. J. Chem. Soc. 1955, 4260.

7. Ames, D. E., and R. E. Bowman: Griseoviridin. Part II. J. Chem. Soc. 1955, 4264.

8. — — Griseoviridin. Part III. Degradation to 10-Amino-decanoic Acid, and Other Reactions. J. Chem. Soc. 1956, 2925.

9. ANATOL, J., and A. MEDÈTE: A New General Procedure for the Preparation of α-Oxocarboxylic Acids. Synthesis 1971, 538.

10. ANDERSON, L., and J. J. KELLEY: The Dephosphorylation of Casein by Alkalies. J. Amer. Chem. Soc. 81, 2275 (1959).

11. ANDERSON, B., D. C. HODGKIN, and M. A. VISWAMITRA: The Structure of Thiostrepton. Nature 225, 233 (1970).

12. APPEL, R., und A. HAUSS: Über einige Reaktionen des Triphenylphosphinimins und des Triphenylphosphinbromimins. Z. Anorg. Chem. 311, 290 (1961).

13. APPLEGATE, H. E., J. E. DOLFINI, M. S. PUAR, W. A. SLUSARCHYK, and B. TOEPLITZ: Synthesis of 7α-Methoxycephalosporins. J. Org. Chem. 39, 2794 (1974).

14. ARNSTEIN, H. R. V., and J. C. CRAWHALL: The Biosynthesis of Penicillin. 6. A Study of the Mechanism of the Formation of the Thiazolidine-β-Lactam Rings, Using Tritium-labelled Cystine. Biochem. J. 67, 180 (1957).

15. ARNSTEIN, H. R. V., and M. E. CLUBB: The Biosynthesis of Penicillin. 8. Investigation of Cyclic Cysteinylvaline Peptides as Precursors. Biochem. J. 68, 528 (1958).

16. ASQUITH, R. S., and P. CARTHEW: Synthesis and PMR Properties of Some Dehydroalanine Derivatives. Tetrahedron 28, 4769 (1972).

17. ASQUITH, R. S., K. W. YEUNG, and M. S. OTTERBURN: Synthesis, Identification and Properties of Some β-Aminoalanine Derivatives. Tetrahedron 33, 1633 (1977).

18. ATKINS, P. R., and I. T. KAY: Synthesis of (±)Versimide. J. Chem. Soc., Chem. Comm. 1971, 430.

19. AUGUSTIN, M.: Die Umsetzung des 2,5-Diketopiperazins mit Aldehyden und Nitrosoverbindungen. J. prakt. Chem. 32, 158 (1966).

20. AUSTEL, V., und W. STEGLICH: Reaktionen von Aminosäuren mit Trifluoressigsäureanhydrid III. 4^1-Substituierte 2-Trifluormethyl-4-(3,3,3-trifluor-2-trifluoracetoxypropyliden)-2-oxazolin-5-one. Chem. Ber. 108, 2361 (1975).

21. BALDWIN, J. E., S. B. HABER, and J. KITCHIN: Dehydropeptides Related to β-Lactam Antibiotics: A Schema for the Biosynthesis of Penicillins and Cephalosporins. J. C. S. Chem. Comm. 1973, 790.

22. BALDWIN, J. E., J. LÖLIGER, W. RASTETTER, N. NEUSS, L. L. HUCKSTEP, and N. DE LA HIGUERA: Use of Chiral Isopropyl Groups in Biosynthesis. Synthesis of (2RS,3S)-[4-^{13}C]-Valine. J. Amer. Chem. Soc. 95, 3796 (1973).

23. BALDWIN, J. E., F. J. URBAN, R. D. G. COOPER, and F. L. JOSE: Direct 6-Methoxylation of Penicillin Derivatives. A Convenient Pathway to Substituted β-Lactam Antibiotics. J. Amer. Chem. Soc. 95, 2401 (1973).

24. BALDWIN, J. E., A. AU, M. CHRISTIE, S. B. HABER, and D. HESSON: Stereospecific Conversion of Peptides into β-Lactams. J. Amer. Chem. Soc. 97, 5957 (1975).

25. BALTAZZI, E.: The Chemistry of 5-Oxazolones. Quart. Rev. (London) 9, 150 (1955).

26. BAPAT, J. B., D. ST. C. BLACK, and R. F. BROWN: Cyclic Hydroxamic Acids. Adv. Heterocyclic Chem. 10, 199 (1969).

27. BAUTE, R., G. DEFFIEUX, M.-A. BAUTE, M.-J. FILLEAU, and A. NEVEU: Un nouveau métabolite fongique du groupe des epidithio-3,6-dioxo-2,5-piperazines: l'epicorazine A, isolée d'une souche d'Epicoccum nigrum link (adelomycetes). Tetrahedron Letters 1976, 3943.

28. BAYER, E., und W. PARR: Eliminierung von Schwefelwasserstoff aus Ferredoxin und Cysteinmethylester. Angew. Chem. 78, 824 (1966); Angew. Chem. Int. Ed. Engl. 5, 840 (1966).

29. BEHRENS, O. K., D. G. DOHERTY, and M. BERGMANN: Resolution of d,l-Phenylalanine by Asymmetric Enzymic Synthesis. J. Biol. Chem. 136, 61 (1940).

30. BEHRINGER, H.: Synthese des Cystins. Chem. Ber. 81, 326 (1948).

31. BEHRINGER, H., and E. FACKLER: Eine einfache Synthese der racemischen Mercaptursäuren. Ann. 564, 73 (1949).

32. Belitz, H. D.: Eine Synthese von β-Methyllanthionin. Tetrahedron Letters **1967**, 749.

33. Ben-Ishai, D., J. Altmann, and Z. Bernstein: The Reactions of Ureas with Glyoxylic Acid and Methyl Glyoxylate. Tetrahedron **33**, 1191 (1977).

34. Benn, M. H., and R. E. Mitchell: Attempts on the Synthesis of the Penicillin Ring System by Transannular Reactions. Canad. J. Chem. **50**, 2195 (1972).

35. Benoiton, L., R. W. Hanson, and H. N. Rydon: Polypeptides. Part VIII. Synthesis of Oxazoline Peptides. J. Chem. Soc. **1964**, 824.

36. Berg, D. H., R. P. Massing, M. M. Hoehn, L. D. Boeck, and R. L. Hamill: A 30641, a New Epidithiodiketopiperazine with Antifungal Activity. J. Antibiot. **29**, 394 (1976).

37. Bergmann, M., und A. Miekeley: Derivate des d,l-Serins. Über neuartige Anhydride des Glycylserins. Z. physiol. Chem. **140**, 128 (1924).

38. Bergmann, M., A. Miekeley und E. Kann: Verwandlung des Serins in Brenztraubensäure und in Alanin. Z. physiol. Chem. **146**, 247 (1925).

39. Bergmann, M., E. Kann und A. Miekeley: Dehydrierung von Asparagin und Verwandlung in Brenztraubensäure. Ann. **449**, 135 (1926).

40. Bergmann, M., and F. Stern: Dehydrierung von Aminosäuren (Alanin, Phenylalanin, Tyrosin). Ann. **448**, 20 (1926).

41. Bergmann, M., F. Stern und C. Witte: Über neue Verfahren der Synthese von Dipeptiden und Dipeptid-Anhydriden. Ann. **449**, 277 (1926).

42. Bergmann, M., und H. Köster: Synthese argininhaltiger Dipeptide: Isomere Phenylalanyl-arginine und ihre Umwandlung in Phenylalanyl-ornithin. Z. physiol. Chem. **167**, 91 (1927).

43. Bergmann, M., und A. Miekeley: Zur Kenntnis des Abbaues der Aminosäuren: Serin als Dehydrierungsmittel. Ann. **458**, 40 (1927).

44. Bergmann, M., L. Zervas und V. Du Vigneaud: d-Tyrosyl-d-arginin und sein Anhydrid. Ber. **62**, 1905 (1929).

45. Bergmann, M., und K. Grafe: Zur Kenntnis der Peptidbindung. Z. Physiol. Chem. **187**, 183 (1930).

46. — — Synthese eines Peptids und anderer Abkömmlinge der α-Aminoacrylsäure aus Brenztraubensäure. Z. Physiol. Chem. **187**, 187 (1930).

47. — — Verbindungen der Brenztraubensäure mit Aminosäuren. Z. Physiol. Chem. **187**, 196 (1930).

48. Bergmann, M., V. Schmitt und A. Miekeley: Über Peptide dehydrierter Aminosäuren, ihr Verhalten gegen pankreatischen Fermenten und ihre Verwendung zur Peptidsynthese. Z. physiol. Chem. **187**, 264 (1930).

49. Bergmann, M., und H. Schleich: Über die enzymatische Spaltung dehydrierter Peptide. Auffindung einer Dehydropeptidase. Z. physiol. Chem. **205**, 65 (1931).

50. Bergmann, M., L. Zervas und F. Lebrecht: Dehydrierung von Aminosäuren, Übergang zur Pyrrolreihe. Chem. Ber. **64**, 2315 (1931).

51. Bergmann, M.: Aufgaben der Synthese für die Erforschung der Eiweißstoffe und ihrer Fermente. Naturwiss. **20**, 941 (1932).

52. Bergmann, M., und H. Schleich: Weiteres über Dehydropeptidasen: Über die enzymatische Angreifbarkeit von Verbindungen aus Brenztraubensäure und Aminosäure. Z. physiol. Chem. **207**, 235 (1932).

53. Bergmann, M., O. K. Behrens, and D. G. Doherty: Asymmetric Course of the Enzymic Synthesis of Peptide Bonds. J. Biol. Chem. **124**, 7 (1938).

54. Bergmann, M., and J. E. Tietzmann: Transformation of an Acyl Diketopiperazine. J. Biol. Chem. **155**, 535 (1944).

55. Birkinshaw, J. H., M. G. Kalyanpur, and C. E. Stickings: Studies in the Biochemistry of Micro-organisms. 113. Pencolide, a Nitrogen-containing Metabolite of Penicillium Multicolor Grigorieva-Manilova and Poradielova. Biochem. J. **86**, 237 (1963).

56. BIRNBAUM, G. I., and S. R. HALL: Structure of the Antibiotic Griseoviridin. J. Amer. Chem. Soc. **98**, 1926 (1976).

57. BLAKE, K. W., and P. G. SAMMES: Geometrical Isomerism and Tautomerism of 3-Arylidene-6-methyl-piperazine-2,5-diones. J. Chem. Soc. (C) **1970**, 980.

58. BLONDEAU, P., R. GAUTHIER, C. BERSE, and D. GRAVEL: Synthesis of Some Stable 7-Halo-1,4-thiazepines. Potential Substituted Penam Precursors. Can. J. Chem. **49**, 3866 (1971).

59. BODANSZKY, M., J. IZDEBSKI, and I. MURAMATSU: The Structure of the Peptide Antibiotic Stendomycin. J. Amer. Chem. Soc. **91**, 2351 (1969).

60. BODANSKY, M., G. G. MARCONI, and A. BODANSKY: The Structure of Stendomycicidine. J. Antibiot. **22**, 40 (1969).

61. BODANSKY, M., J. A. SCOZZIE, and I. MURAMATSU: Dehydroalanine Residues in Thiostrepton. J. Antibiot. **23**, 9 (1970).

62. BODANSKY, M., and A. BODANSKY: Addition of Thioglycolic Acid to Stendomycin. J. Antibiot. **27**, 312 (1974).

63. BOGGS, N. T., R. E. GAWLEY, K. A. KOEHLER, and R. G. HISKEY: Synthesis of DL-γ-Carboxyglutamic Acid Derivatives. J. Org. Chem. **40**, 2850 (1975).

64. BOHAK, Z.: N^ε-(DL-2-Amino-2-carboxyethyl)-L-lysine, a New Amino Acid Formed on Alkaline Treatment of Proteins. J. Biol. Chem. **239**, 2878 (1964).

65. BOSE, A. K., and M. S. MANHAS: Cephalosporins, Penicillins and Other β-Lactams. J. Heterocyclic Chem. **13**, 43 (1976).

66. BOWIE, J. H., D. A. COX, A. W. JOHNSON, and G. THOMAS: Viomycin. Tetrahedron Letters **1964**, 3305.

67. BOWIE, J. H., A. W. JOHNSON, and G. THOMAS: The Chromophore of Viomycin. Tetrahedron Letters **1964**, 863.

67A. BRAZHNIKOWA, M. G., M. K. KUDINOVA, N. P. POTAPOVA, T. M. FILIPPOVA, E. BOROWSKI, J. ZIELINSKI, and J. GOLIC: Structure of the Antibiotic Madumicin. Bioorg. Khim. **1975**, 1383. Chem. Abstr. **84**, 140654a (1976).

68. BREITHOLLE, E. G., and C. H. STAMMER: The Synthesis and Reactions of Dehydro Phenylalanine Anilide. Tetrahedron Letters **1975**, 2381.

69. — — Synthesis of Some Dehydrophenylalanine Peptides. J. Org. Chem. **41**, 1344 (1976).

70. BREMNER, D. H., M. M. CAMPBELL, and G. JOHNSON: Conversion of 6α-Alkoxyformamidopenicillanates into 6α-Aminopenicillanates, and the Formation of 6-Spiropenicillanates. J. Chem. Soc., Chem. Commun. **1976**, 293.

71. — — — Transformations of Penicillins: New Methods of Formation and Reactions of 6,6-Disubstituted Penams and 7,7-Disubstituted Cephems. J. Chem. Soc., Perkin I **1976**, 1918.

72. — — — Transformation of Penicillins: Reactions of Penam S-Oxides with N-Chloro-N-sodio-carbamates. J. Chem. Soc., Perkin I **1977**, 1943.

73. BRINK, A. J., and A. JORDAAN: Synthesis of Branched-Chain Sugars by Reaction of Glycosuloses with α-Metalated Isocyanoacetic Esters. Carbohydrate Research **34**, 1 (1974).

74. BROSIO, E., F. CONTI, R. DEL GIUDICE, A. DI NOLA, and D. GATTEGNO: The Conformation of Viomycin in Solution. Gazz. Chim. Ital. **107**, 139, 1977.

75. BROWN, A. G., and T. G. SMALE: Assignment of the Stereochemistry of α-Benzamido- and α-Phthalimido-crotonates using Nuclear Magnetic Resonance Spectroscopy. J. Chem. Soc., Chem. Comm. **1969**, 1489.

76. BROWN, A. G.: Versimide, a Metabolite of Aspergillus Versicolor. J. Chem. Soc. (C) **1970**, 2572.

77. BROWN, A. G., and T. C. SMALE: Synthesis of (±)-Versimide (Methyl-α-(methylsuccimido) acrylate) and Related Compounds. J. Chem. Soc. Perkin I **1972**, 65.

78. Bycroft, B. W., D. Cameron, L. R. Croft, A. W. Johnson, and T. Webb: Viomycin. Further Degradative Studies. Tetrahedron Letters 1968, 2925.
79. Bycroft, B. W., D. Cameron, L. R. Croft, A. Hassanali-Walji, A. W. Johnson, and T. Webb: The Chromophore and Partial Structure of Viomycin. Tetrahedron Letters 1968, 5901.
80. Bycroft, B. W.: Structural Relationships in Microbial Peptides. Nature 224, 595 (1969).
81. Bycroft, B. W., D. Cameron, A. Hassanali-Walji, and A. W. Johnson: Synthesis of a Model Relating to the Chromophores of Capreomycin and Viomycin. Tetrahedron Letters 1969, 2539.
82. Bycroft, B. W., D. Cameron, L. R. Croft, A. Hassanali-Walji, A. W. Johnson, and T. Webb: The Total Structure of Viomycin, a Tuberculostatic Peptide Antibiotic. Experientia 1971, 27, 501.
83. Bycroft, B. W.: The Crystal Structure of Viomycin, a Tuberculostatic Antibiotic. J. Chem. Soc., Chem. Comm. 1972, 660.
84. Bycroft, B. W., L. R. Croft, A. W. Johnson, and T. Webb: Viomycin Part I. The Structure of the Guanidine-Containing Unit. J. Chem. Soc. Perkin I 1972, 820.
85. Bycroft, B. W., D. Cameron, L. R. Croft, A. Hassanali-Walji, A. W. Johnson, and T. Webb: Viomycin Part II. The Structure of the Chromophore. J. Chem. Soc. Perkin I 1972, 827.
86. Bycroft, B. W., and G. R. Lee: Efficient Assymmetric Synthesis of α-Amino Acids from α-Keto Acids and Ammonia with Conservation of the Chiral Reagent. J. C. S. Chem. Comm. 1975, 988.
87. — — Efficient Asymmetric Synthesis of α-Amino Acids from α-Keto Acids and Ammonia with Conservation of the Chiral Reagent. J. Chem. Soc. Chem. Comm. 1975, 988; Addendum: J. Chem Soc. Chem Comm. 1976, 616.
88. Bycroft, B. W., and I. J. King: Revised Constitution, Absolute Configuration and Conformation of Griseoviridin, a Modified Cyclic Peptide Antibiotic. J. Chem. Soc. Perkin I, 1976, 1996.
88A. Bycroft, B. W., and R. Pinchin: Structure of Althiomycin, A Highly Modified Peptide Antibiotic. J. Chem. Soc. Chem. Comm. 1975, 121.
89. Cama, L. D., W. J. Leanza, T. R. Beattie, and B. G. Christensen: Substituted Penicillin and Cephalosporin Derivatives. I. Stereospecific Introduction of the C-6(7) Methoxy Group. J. Amer. Chem. Soc. 94, 1408 (1972).
90. Cama, L. D., and B. G. Christensen: Substituted Penicillins and Cephalosporins. VII. A Stereospecific Introduction of the C-6(7) Methoxy Group. Tetrahedron Letters 1973, 3505.
91. Cardillo, R., C. Fuganti, D. Ghiringhelli, and P. Grasselli: Stereochemical Course of the α,β-Desaturation of L-Tryptophan in the Biosynthesis of Crypto-echinuline A in Aspergillus Amstelodami. J. Chem. Soc. Chem. Comm. 1975, 778.
92. Carter, H. E., P. Handler, and D. B. Melville: Azlactones I. Preparation of Benzoyl-α-aminocrotonic Acid Azlactone and the Conversion of Allo-Threonine to Threonine. J. Biol. Chem. 129, 359 (1939).
93. Carter, H. E., and C. M. Stevens: Azlactone Formation in Glacial and in Aqueous Acetic Acid and Preparation of Benzoyl-α-amino-crotonic Acid Azlactone II. J. Biol. Chem. 133, 117 (1940).
94. Carter, H. E., and W. C. Risser: Preparation of Benzoyl-α-aminocinnamic acid Azlactones I and II. The Use of β-Phenylethylamine in the Purification of α-Amino-β-methoxy(hydroxy)acids. J. Biol. Chem. 139, 255 (1941).
95. Carter, H. E.: Azlactones: In Org. Reactions 3, 198ff., New York: J. Wiley & Sons, Inc., 1946.
96. Carter, J. H., R. H. du Bus, J. R. Dyer, J. C. Floyd, K. C. Rice, and P. D. Shaw:

Biosynthesis of Viomycin I. Origin of α,β-Diamino-propionic Acid and Serine. Biochemistry **13**, 1221 (1974).

97. — — — — — — Biosynthesis of Viomycin II. Origin of β-Lysine and Viomycidine. Biochemistry **13**, 1227 (1974).

98. CHATTERJEE, R., A. H. COOK, I. HEILBRON, and A. L. LEVY: Studies in the Azole Series. Part VII. A New Route to α-Amino-β-mercapto Acids. J. Chem. Soc. **1948**, 1337.

99. CHEMIAKIN, M. M., and V. K. ANTONOV: A New Method of Synthesis of α-Substituted α-Acylamino Acids. Doklady Akad. Nauk. S.S.S.R. **129**, 349 (1959). Chem. Abstr. **54**, 7633d.

100. CHEMIAKIN, M. M., E. S. TCHAMAN, L. I. DENISOVA, G. A. RAVDEL, and W. J. RODIONOW: Synthèses et propriétés des α-aminoacides α-substitués. Bull. Soc. Chim. France **1959**, 530.

101. CHEUNG, Y., and C. WALSH: Stereospecific Synthesis of Isotopically Labeled Serine at Carbon 3 and Stereochemical Analysis of D-Serine Dehydrase Reaction. J. Amer. Chem. Soc. **98**, 3397 (1976).

102. CHIGIRA, Y., M. MASAKI, and M. OHTA: Syntheses and Reactions of N(Phenylpyruvoyl) Amino Acids. Bull. Chem. Soc. Jap. **42**, 224 (1969).

103. CHOU, T. S.: Sulfenic Acid Trimethylsilylesters. A Convenient Protection for a Reactive Functionality. Tetrahedron Letters **1974**, 725.

104. COFFEN, D. L., D. A. KATONAK, N. R. NELSON, and F. D. SANCILIO: A Short Synthesis of Aromatic Analogues of the Aranotines. J. Org. Chem. **42**, 948 (1977).

105. COOPER, R. D. G.: Structural Studies on Penicillin Derivatives. VIII. A Possible Model Biosynthetic Route to Penams and Cephems. J. Amer. Chem. Soc. **94**, 1018 (1972).

106. CORNFORTH, J. W.: Oxazoles and Oxazolones. In: H. T. CLARKE, J. R. JOHNSON, and SIR R. ROBINSON, The Chemistry of Penicillin, p. 730ff. Princeton: Princeton University Press. 1949.

107. — Oxazole and its Derivatives. In: R. C. ELDERFIELD, Heterocyclic Compounds, Vol. 5, p. 336ff., New York: J. Wiley & Sons Inc., 1957.

108. DANG, T. P., and H. B. KAGAN: The Asymmetric Synthesis of Hydratropic Acid and Amino Acids by Homogeneous Catalytic Hydrogenation. J. Chem. Soc. Chem. Comm. **1971**, 481.

109. DANG, T. P., J. C. POULIN et H. B. KAGAN: Reduction asymetrique catalysée par des complexes de metaux de transition. III. Diphosphines chirales derivées de l'Isopropylidene dihydroxy-2,3-bis(diphenylphosphino)-1,4-butane (DIOP). J. Organomet. Chem. **91**, 105 (1975).

110. DAVIES, J. S., and M. N. IBRAHIM: Asymmetric Hydrogenation of Model Dehydrovalyl Peptides. Tetrahedron Letters **1977**, 1453.

111. DAVIS, L., and D. E. METZLER: Pyridoxal-Linked Elimination and Replacement Reactions. In: The Enzymes, Vol. VII, PAUL D. BOYER, Third Ed. New York and London: Academic Press. 1972.

112. DELPIERRE, G. R., F. W. EASTWOOD, G. E. GREAM, D. G. I. KINGSTON, P. S. SARIN, Lord TODD, and D. H. WILLIAMS: Antibiotics of the Ostreogrycin Complex. Part II. Structure of Ostreogrycin A. J. Chem. Soc. (C) **1966**, 1653.

113. DEPAIRE, H., J.-P. THOMAS, and A. BRUN: The Structure Relationship between the Antibiotics Nosiheptide and Thiostrepton. Tetrahedron Letters **1977**, 1403.

114. DOHERTY, D. G., J. E. TIETZMAN, and M. BERGMANN: Peptides of Dehydrogenated Amino Acids. J. Biol. Chem. **147**, 617 (1943).

114A. DURANT, F., G. EVRARD, J. P. DECLERCQ, and G. GERMAIN: Virginiamycin. Factor, M-Dioxane (of Virginiamycin). Cryst. Struct. Commun. **1974**, 503. Chem. Abstr. **81**, 128072g (1974).

115. Dyer, J. R., G. K. Kellogg, R. F. Nassar, and Wm. E. Streetman: Viomycin II. The Structure of Viomycin. Tetrahedron Letters 1965, 585.

116. Eiger, J. Z., and J. P. Greenstein: Addition Products of Dehydropeptides. Arch. Biochem. 19, 467 (1948).

117. Erlenmeyer, E., und E. Frühstück: Über Phenyl-α-amidomilchsäure (Phenylserin). Ann. 284, 36 (1895).

118. Faleev, N. G., Yu. N. Belokon, V. M. Belikov, and I. M. Mel'nikova: Oxidative Deamination of the Alanine Ligand by Air Oxygen in Stereochemically Inert Bis-(N-salicylidene alaninato)cobaltate(III) Complexes. J. C. S. Chem. Commun. 1975, 85.

119. Fallona, M. C., T. C. Morris, P. de Mayo, T. Money, and A. Stoessel: Griseoviridin. J. Amer. Chem. Soc. 84, 4162 (1962).

120. Fallona, M. C., P. de Mayo, T. C. McMorris, T. Money, and A. Stoessel: Mold Metabolites. II. The Structure of Griseoviridin. Canad. J. Chem. 42, 371 (1964).

121. Farlow, M. W.: Cysteine and Cystine. U.S. 2,406,362 (1946); Chem. Abstr. 40, 7233 (1946).

122. — A New Synthesis of Cystine. J. Biol. Chem. 176, 71 (1948).

123. Fawcett, P. A., J. J. Usher, and E. P. Abraham: Proceedings of the Second International Symposium on the Genetics of Industrial Microorganisms (K. O. McDonald, ed.), p. 129 ff. New York: Academic Press. 1975.

124. Fayos, J., D. Lokensgard, J. Clardy, R. J. Cole, and J. W. Kirksey: Structure of Verruculogen, a Tremor Producing Peroxide from Penicillium verruculosum. J. Amer. Chem. Soc. 96, 6785 (1974).

125. Fiaud, J. C., et H. B. Kagan: Une Nouvelle Synthèse d'α Amino Acides. Synthèse Asymétrique de l'Alanine. Tetrahedron Letters 1970, 1813.

126. Filler, R.: Recent Advances in Oxazolone Chemistry. In: Advances in Heterocyclic Chemistry 4, p. 75 ff. (A. R. Katritzky. A. J. Boulton, and J. M. Lagowski, eds.). New York and London: Academic Press. 1965.

127. Filler, R., and Y. S. Rao: New Developments in the Chemistry of Oxazolones. In: Advances in Heterocyclic Chemistry 21, 175 (A. R. Katritzky and A. J. Boulton eds.). New York, San Francisco, London: Academic Press. 1977.

128. Firestone, R. A., and B. G. Christensen: Functionalization of Penicillins at Carbon 6 via N-Acylimines. 6-Hydroxypenicillin. Substituted Penicillins and Cephalosporins. VIII. J. Org. Chem. 38, 1436 (1973).

129. Fischer, G., G. Oehme und A. Schellenberger: Zur Theorie der α-Ketosäuren: Beziehungen zwischen Struktur und UV-Spektren von α-Ketosäuren und verwandten α-Dicarbonylverbindungen. Tetrahedron 27, 5683 (1971).

130. Fisher, C., and H. S. Mosher: Asymmetric Homogenous Hydrogenation with Phosphine-Rhodium-Complexes Chiral Both at Phosphorus and Carbon. Tetrahedron Letters 1977, 2487.

131. Forster, M. O., and W. B. Saville: Isolation of Picrorocellin from Rocella fuciformis. J. Chem. Soc. 121, 816 (1922).

132. Fry, E. M.: Oxazolines, J. Org. Chem. 14, 887 (1949).

133. Fryzuk, M. D., and B. Bosnich: Asymmetric Synthesis. Production of Optically Active Amino Acids by Catalytic Hydrogenation. J. Amer. Chem. Soc. 99, 6262 (1977).

134. Fu, S.-C. J., and J. P. Greenstein: Saturation of Acetyldehydroalanine with Benzylamine. J. Amer. Chem. Soc. 77, 4412 (1955).

135. Fukuyama, T., S. Nakatsuka, and Y. Kishi: A New Synthesis of Epidithiapiperazinediones. Tetrahedron Letters 1976, 3393.

136. Gallina, C., F. Petrini, and A. Romeo: Synthesis of Derivatives of 2-Aminoproline and 5-Aminoproline. J. Org. Chem. 35, 2425 (1970).

137. GALLINA, C., C. MARTA, C. COLOMBO, and A. ROMEO: Capreomycidine and 3-Guanidinoproline from Viomycidine. Tetrahedron **27**, 4681 (1971).

138. GALLINA, C., and A. LIBERATORI: A New Synthesis of 1-Acetyl-3-arylidene (alkylidene) piperazine-2,5-diones. Tetrahedron Letters **1973**, 1135.

139. GALLINA, C., M. MANESCHI, and A. ROMEO: Synthesis of 2-Alkoxy-2-acylaminopropionic Acids by Alkoxymercuration-Demercuration of 2-Acylaminoacrylic Acids. J. Chem. Soc. Perkin I **1973**, 1134.

140. GALLINA, C., and A. LIBERATORI: Condensation of 1,4-Diacetylpiperazine-2,5-dione with Aldehydes. Tetrahedron **30**, 667 (1974).

141. GAWRON, O., and G. ODSTRCHEL: Kinetic Studies on the Alkaline Decomposition of Cystine Derivatives and Peptides. J. Amer. Chem. Soc. **89**, 3263 (1967).

142. GELBARD, G., H. B. KAGAN et R. STERN: Catalyse asymetrique avec des Complexes chiraux de Rhodium-DIOP — V. Effets des substituants lors de la Reduction d'acides N-Acylamino cinnamiques. Tetrahedron **32**, 233 (1976).

143. GINSBURG, S., and I. B. WILSON: Factors Affecting the Competitive Formation of Oxazolines and Dehydroalanines from Serine Derivatives. J. Amer. Chem. Soc. **86**, 4716 (1964).

144. GIVOT, J. L., T. A. SMITH, and R. H. ABELES: Studies on the Mechanism of Action and the Structure of the Electrophilic Center of Histidine Ammonia Lyase. J. Biol. Chem. **244**, 6341 (1969).

145. GLASER, R., and J. BLUMENFELD: Inhibition of the Asymmetric Hydrogenation of Z-Methyl-α-acetamidocinnamate Catalyzed by DIOP-Rhodium Catalyst in the Presence of Z-Adamantyl- or Bornyl-α-acetamidocinnamate. Tetrahedron Letters **1977**, 2525.

146. GLASER, R., and S. GERESH: Structural Requirements in Chiral Diphosphine-Rhodium-Complexes. VII. Use of Z-Methyl-α-acylaminocinnamates as Structural Probes for DIOP-Rhodium(I) Complexes. Tetrahedron Letters **1977**, 2527.

146 A. GORDON, E. M., H. W. CHANG, and C. M. CIMARUSTI: Sulfenyl Transfer Rearrangement of Thiooximes: A Novel Conversion of Cephalosporins to 7-γ-Methoxycephalosporins. J. Amer. Chem. Soc. **99**, 5504 (1977).

147. GRAVEL, D., R. GAUTHIER, and C. BERSE: A General Synthesis of α-Acylaminoacrylic Esters. J. C. S. Chem. Comm. **1972**, 1322.

148. GRIGG, R., and J. KEMP: New Selected Dehydroamino-acid Esters and Triazolidines. J. Chem. Soc. Chem. Comm. **1977**, 125.

149. GREENSTEIN, J. P.: Dehydropeptidases. In: Advances in Enzymology VIII. New York: Interscience Publishers. Inc. 1948.

150. GREENSTEIN, J. P., and M. WINITZ: Chemistry of the Amino Acids, Vol. 2, p. 843. New York-London-Sydney: J. Wiley and Sons. Inc. 1961.

151. GROSS, E., and J. L. MORELL: The Presence of Dehydroalanine in the Antibiotic Nisin and Its Relationship to Activity. J. Amer. Chem. Soc. **89**, 2791 (1967).

152. GROSS, E., J. L. MORELL, and L. C. CRAIG: Dehydroalanyllysine: Identical COOH-terminal Structures in the Peptide Antibiotics Nisin and Subtilin. Proc. of the National Academy of Sciences **62**, 952 (1969).

153. GROSS, E., and J. L. MORELL: Nisin. The Assignment of Sulfide Bridges of β-Methyllanthionine to a Novel Bicyclic Structure of Identical Ring Size. J. Amer. Chem. Soc. **92**, 2919 (1970).

154. — — The Structure of Nisin. J. Amer. Chem. Soc. **93**, 4634 (1971).

155. GROSS, E., H. H. KILTZ und L. C. CRAIG: Subtilin II. — Die Aminosäurezusammensetzung des Subtilins. Z. Physiol. Chem. **354**, 799 (1973).

156. GROSS, E., H. H. KILTZ und E. NEBELIN: Subtilin VI. Struktur des Subtilins. Z. Physiol. Chem. **354**, 810 (1973).

157. Gross, E., and H. H. Kiltz: The Number and Nature of α,β-Unsaturated Amino Acids in Subtilin. Biochem. Biophys. Res. Comm. **50,** 559, 1973.

158. Gross, E., K. Noda und B. Nisula: Festphasensynthese von Peptiden mit carboxy-terminalen Amidgruppen. — Thyrotropin-freisetzendes Hormon (TRF). Angew. Chem. **85,** 672 (1973); Angew. Chem. Internat. Ed. Engl. **12,** 664 (1973).

159. Gross, E., K. Noda, and S. Matsuura: The Utility of α,β-Unsaturated Amino Acids in Peptide Synthesis. II. The Synthesis of Peptides *via* α,β-Unsaturated Amino Acids. Peptides 1974, Proc. of the 13th Europ. Peptide Symposium, p. 403. New York: J. Wiley & Sons. 1975.

160. Gross, E.: Subtilin and Nisin: The Chemistry and Biology of Peptides with α,β-Unsaturated Amino Acids. In: Peptides: Chemistry, Structure and Biology Proceedings of the 4th American Peptide Symp. **1975,** p. 31. Ed. by R. Walter and J. Meienhofer.

161. Gross, E., and S. Matsuura: α,β-Unsaturated and Thioether Amino Acids in Peptide Synthesis. In: Peptides, Chemistry, Structure and Biology Proceedings of the 4th American Peptide Symp. **1975,** p. 351.

162. Gross, H., J. Gloede, J. Keitel und D. Kunath: α-Aminosäuren und Derivate. II. Isonitrilreaktionen mit Glyoxylsäurederivaten. J. Prakt. Chem. 4. Reihe **37,** 192 (1968).

163. Hamill, R. L., and W. M. Stark: Antibiotic A-2315 from Actinoplanes philippinensis. U.S. Appl. **276,** 546 (1972); Chem. Abstr. **81,** 2390 (1974).

164. Hanson, K. R., and E. A. Havir: L-Phenylalanine Ammonia-Lyase. IV. Evidence that the Prostetic Group Contains a Dehydroalanyl Residue and Mechanism of Action. Arch. Biochem. and Biophys. **141,** 1 (1970).

165. Haskell, T. H., S. A. Fusari, R. P. Frohardt, and Q. R. Bartz: The Chemistry of Viomycin. J. Amer. Chem. Soc. **74,** 599 (1952).

166. Häusler, J., und U. Schmidt: Über Pyruvoylaminosäuren. Chem. Ber. **107,** 145 (1974).

167. — — Hydroxylsubstituierte Cyclodipeptide durch Ringschluß von Pyruvoylaminosäure-amiden. Chem. Ber. **107,** 2804 (1974).

168. Häusler, J., R. Jahn und U. Schmidt: Radikalisch und photochemisch initiierte Oxidation von Aminosäurederivaten. Chem. Ber. **111,** 361 (1978).

169. Häusler, J., und U. Schmidt: Ringschlüsse von Pyruvoylpeptiden und Dehydro-peptiden. Mh. Chem. **109,** 147 (1978).

170. Hawkes, G. H., and E. W. Randall: High Field ^{15}N Nuclear Magnetic Resonances Spectroscopy of Peptides. Assignments in Viomycin Sulphate. J. Chem. Soc. Chem. Comm. **1977,** 546.

171. Hayashi, T., T. Mise, S. Mitachi, K. Yamamoto, and M. Kumada: Asymmetric Hydrogenation Catalyzed by a Chiral Ferrocenylphosphine-Rhodium-Complex. Tetrahedron Letters **1976,** 1133.

172. Hayashi, T., M. Tanaka, and I. Ogata: Asymmetric Hydrogenation by Rhodium Complex with d-*trans*-1,2-bis-(Diphenylphosphinoxy)-cyclopentane as a Chiral Ligand. Tetrahedron Letters **1977,** 295.

173. Hedy, P. H., E. B. Hodge, V. V. Young, R. L. Harried, G. A. Brewer, W. F. Phillips, W. F. Runge, H. E. Stavely, A. Pohland, H. Boaz, and H. R. Sullivan: Structure and Reaction of Cycloserine. J. Amer. Chem. Soc. **77,** 2345 (1955).

174. Helbling, A. M., und M. Viscontini: 96. Naturstoffe aus Mikroorganismen. Synthese von racemischen Proferrorosamin und Ferrorosamin. Helv. Chim. Acta **59,** 938 (1976).

175. Hellmann, H., und E. Folz: Über den Mechanismus der Reaktionen von quartären Ammoniumsalzen mit Alkalicyanid. Chem. Ber. **88,** 1944 (1955).

176. — — Kondensationen mit Dimethylaminomethyl-acetamino-malonester-jodmethylat.

Ein Beitrag zum Mechanismus der Kondensationsreaktionen quartärer Ammoniumsalze. Chem. Ber. **89**, 2000 (1956).

177. HELLMANN, H., K. TEICHMANN und F. LINGENS: α-Acylaminoacrylester aus Acylaminomalonestern. Chem. Ber. **91**, 2427 (1958).

178. HENERY-LOGAN, K. R., and C. G. CHEN: Synthesis of Oxygen Analogues of the Penicillins. I. Photocyclisation of 2-Oxoamides to 3-Carbomethoxy-6-hydroxypenams. Tetrahedron Letters **1973**, 1103.

179. HERBST, R. M.: The Condensation of α-Keto Acids and Amides. II. Pyruvic Acid and Acetamide. J. Amer. Chem. Soc. **61**, 483 (1939).

180. HISKEY, R. G., R. A. UPHAM, G. M. BEVERLY, and W. C. JONES, JR.: Sulfur-Containing Polypeptides X. A Study of β-Elimination of Mercaptides from Cysteine Peptides. J. Org. Chem. **35**, 513 (1970).

181. HOPPE, D.: 3-Acyl-2-thioxo-1,3-oxazolidin-4-carbonsäureester und ihre Umwandlung in α-(N-Acylamino)-acrylsäureester. Angew. Chem. **85**, 659 (1973); Int. Edit. **12**, 656 (1973).

182. — 2-Alkylthio-2-oxazolin-4-carbonsäureester und ihre Umwandlung in α-(N-Alkylthiocarbonylamino)-acrylsäureester. Angew. Chem. **85**, 660 (1973); Int. Edit. **12**, 658 (1973).

183. — Metallierte Stickstoffderivate der Kohlensäure in der organischen Synthese. X. 2-Alkylthio-2-oxazolin-4-carbonsäureester und ihre baseninduzierte Ringöffnung zu 2-⟨N-[(Alkylthio)carbonyl]amino⟩ acrylsäureestern. Ann. **1976**, 1843.

184. HOPPE, D., und M. KLOFT: Metallierte Stickstoff-Derivate der Kohlensäure in der organischen Synthese. XI. α-Isothiocyanatoacrylsäureester. II. Zur Synthese von α-Thiocyanatoacrylsäureestern aus Isothiocyanatoessigsäureestern und Carbonylverbindungen. Ann. **1976**, 1850.

185. HOPPE, D., und R. FOLLMANN: Metallierte Stickstoff-Derivate der Kohlensäure in der organischen Synthese IX. α-(N-alkoxycarbonyl)-aminoacrylsäureester (N-Acyl-α,β-dehydroaminosäureester) durch baseninduzierte Ringöffnung von 3-Alkoxycarbonyl-2-thioxo-oxazolidin-4-carbonsäureestern. Chem. Ber. **109**, 3062 (1976).

186. HORIKAWA, H., T. IWASAKI, K. MATSUMOTO, and M. MIYOSHI: A New Synthesis of 2-Alkoxy- and 2- Acetoxy-2-amino Acids by Anod. Oxidation. Tetrahedron Letters **1971**, 191.

187. HUANG, F. C., J. A. CHAN, C. J. SIH, P. FAWCETT, and E. P. ABRAHAM: The Nonparticipation of α,β-Dehydrovalinyl Intermediates in the Formation of δ-(L-α-Aminoadipyl)-L-cysteinyl-D-valine. J. Amer. Chem. Soc. **97**, 3858 (1975).

188. INOUE, S., J. MURATA, N. TAKAMATSU, H. NAGANO, and Y. KISHI: Synthetic Studies on Echinulins and Related Natural Products. Part 5. Isolation, Structure and Synthesis of Echinulin-Neoechinulin Type Alkaloids Isolated from Aspergillus Amstelodami. Yakugaku Zasshi **97**, 576 (1977).

189. INOUE, S., N. TAKAMATSU, K. HASHIZUME, and Y. KISHI: Synthetic Studies on Echinulin and Related Products. Part 6. Structure and Synthesis of Aurechinulin. Yakugaku Zasshi **97**, 582 (1977).

190. ITO, Y., M. OKANO, and R. ODA: Studies on Isocyanides III. The Addition of N,N-Dialkyl Chlorides to Isocyanides. Tetrahedron **22**, 447 (1966).

191. IZUMI, R., T. NODA, T. ANDO, T. TAKE, and A. NAGATA: Studies on Tuberactinomycin. III. Isolation and Characterisation of Two Minor Components, Tuberactinomycin B and Tuberactinomycin O. J. Antibiot. **25**, 201 (1972).

192. JEN, T., J. FRAZEE, and J. R. E. HOOVER: A Stereospecific Synthesis of C-6(7) Methoxypenicillin and -cephalosporin Derivatives. J. Org. Chem. **38**, 2857 (1973).

193. KAGAN, H. B., and T. P. DANG: Asymmetric Catalytic Reduction with Transition Metal Complexes. I. A Catalytic System of Rhodium (I) with (−)-2,3-O-Isopropylidene-2,3-dihydroxy-1,4-bis(diphenylphosphino)butane, a New Chiral Diphosphine. J. Amer. Chem. Soc. **94**, 6429 (1972).

194. Kagan, H. B., N. Langlois, and T. P. Dang: Réduction asymétrique catalysée par des complexes de métaux de transition. IV. Synthèse d'Amines chirales au moyen d'un complexe de Rhodium et d'Isopropylidene Dihydroxy-2,3-bis-(diphenyl-phosphino)-1,4-butane (DIOP). J. Organomet. Chem. **90**, 353 (1975).

195. Kaneda, A., and R. Sudo: The Preparation of α-Amino-α-benzylmercaptopropionic Acid Derivatives. Bull. Chem. Soc. Jap. **43**, 2159 (1970).

195A. Katrukha, G. S., S. N. Maevskaja, and A. B. Silaev: Use of Chemical Methods for the Analysis of Amino Acid Sequences in the Polypeptide Antibiotic A-128-OP. Structure of the Antibiotic A-128-OP. Ref. Dokl. Soobshch.-Mendeleevsh. S'ezd Obshch. Prikl. Khim. **1975**, 60. Chem. Abstr. **88**, 191 458e (1978).

196. Kil'disheva, O. V., L. P. Rasteikene, and I. L. Knunyants: Transformation of Mercapto Amino Acids. IV. 2,3-Dihalo-2-acylaminopropionic Acids. Bull. Acad. Sci. USSR. Div. Chem. Sci. **1955**, 231; Chem. Abstr. **50**, 4914 (1956).

197. Kil'disheva, O. V., M. G. Lin'kova, and I. L. Knunyants: Transformations of the Mercapto Amino Acids V. 3-Halo-2-hydroxy-2-(acylamino)propionic Acids and Their Derivatives. Bull. Acad. Sci., U.S.S.R., Div. Chem. Sci. **1955**, 241. Chem. Abstr. **50**, 4914i (1956).

198. — — — 3-Halo-2-(acylamino)acrylic Acids and Their Derivatives. Izvest. Akad. Nauk. S.S.S.R., Otdel. Khim. Nauk **8**, 282 (1955); Bull. Acad. Sci. U.S.S.R., Div. Chem. Sci. **1955**, 251; Chem. Abstr. **50**, 4915 (1956).

199. Kil'disheva, O. V., M. G. Lin'kova, V. M. Savosina, and I. L. Knunyants: α,β-Disubstituted α-Acylaminocarboxylic Acids II. New Method of Formation of Oxazole-4-carboxylic Acids. Izvest. Akad. Nauk. S.S.S.R., Otdel. Khim. Nauk. **1958**, 1348; Chem. Abstr. **53**, 7140i (1959).

200. Kiltz, H. H., and E. Gross: Subtilin III. — Enzymatische Fragmentierung mit Trypsin und Thermolysin. Z. Physiol. Chem. **354**, 802 (1973).

201. — — Subtilin IV. — Sequenz und Sulfidbrückenzuordnung in heterodetem bicyclischen Peptid der Aminosäurereste 20—29. Z. Physiol. Chem. **354**, 805 (1973).

202. Kingston, D. G. J., Lord Todd, and D. H. Williams: Antibiotics of the Ostreogrycin Complex. Part III. The Structure of Ostreogrycin A. Evidence Based on Nuclear Magnetic Double Resonance Experiments and High-resolution Mass Spectrometry. J. Chem. Soc. (C) **1966**, 1669.

202A. Kingston, D. G. I., P. S. Sarin, Lord Todd, and D. H. Williams: Antibiotics of the Ostreogrycin Complex. IV. The Structure of Ostreogrycin G. J. Chem. Soc. (C) **1966**, 1856.

203. Kirby, G. W., and S. Narayanaswami: Biosynthetic Incorporation of Stereoselectively Labelled [β-³H]-Tyrosine into Mycelianamide. J. Chem. Soc. Chem. Comm. **1973**, 322.

204. — — Stereochemical Studies on the Biosynthesis of the α,β-Didehydro-amino-acid Units of Mycelianamide, Cyclopenin and Cyclopenol. J. Chem. Soc. Perkin I **1976**, 1564.

205. Kisfaludy, L., A. Patthy, and M. Löw: A Beta-Elimination Reaction between Cysteine Derivatives Containing Free SH Group and Dicyclohexylcarbodiimide. Acta Chim. Acad. Sci. Hung. **59**, 159 (1969).

206. Kishi, Y., T. Fukuyama, and S. Nakatsuka: A New Method for the Synthesis of Epidithiodiketopiperazines. J. Amer. Chem. Soc. **95**, 6490 (1973).

207. — — — A Total Synthesis of Dehydrogliotoxin. J. Amer. Chem. Soc. **95**, 6492 (1973).

208. Kishi, Y., S. Nakatsuka, T. Fukuyama, and M. Havel: A Total Synthesis of Sporidesmin A. J. Amer. Chem. Soc. **95**, 6493 (1973).

209. Kitagawa, T., T. Miura, K. Fujiwara, and H. Taniyama: The Total Structure of Viomycin by Sequential Analysis. Chem. Pharm. Bull. **20**, 2215 (1972).

210. KITAGAWA, T., T. MIURA, and H. TANIYAMA: Characterization of Viomycin and Its Acyl Derivatives. Chem. Pharm. Bull. **20**, 2176 (1972).

211. KITAGAWA, T., T. MIURA, S. TANAKA, and H. TANIYAMA: Relationships between Antimicrobial Activities and Chemical Structures of Reduced Products of Viomycin. J. Antibiot. **26**, 528 (1973).

212. KITAGAWA, T., T. MIURA, Y. SAWADA, K. FUJIWARA, R. ITO, and H. TANIYAMA: Studies on Viomycin. VII. Oxidative Modifications of Viomycin. Chem. Pharm. Bull. **22**, 1827 (1974).

213. KITAGAWA, T., T. MIURA, K. MORI, H. TANIYAMA, K. KAWANO, and Y. KYOGOKU: Studies on Viomycin X. Carbon-13-nuclear Magnetic Resonance Studies on Viomycin and Its Related Compounds. Chem. Pharm. Bull. **25**, 280 (1977).

214. KLUENDER, H., C. H. BRADLEY, C. H. SIH, P. FAWCETT, and E. P. ABRAHAM: Synthesis and Incorporation of (2S,3S)-[4-^{13}C] Valine into β-Lactam Antibiotics. J. Amer. Chem. Soc. **95**, 6149 (1973).

215. KLUENDER, H., F. C. HUANG, A. FRITZBERG, H. SCHNOES, C. J. SIH, P. FAWCETT, and E. P. ABRAHAM: Studies in the Incorporation of (2S,3R)-[4,4,4-^{2}H$_3$]Valine and (2S,3S)-[4,4,4-^{2}H$_3$]Valine into β-Lactam Antibiotics. J. Amer. Chem. Soc. **96**, 4054 (1974).

216. KNOWLES, W. S., and M. J. SABACKY: Catalytic Asymmetric Hydrogenation of β-Substituted α-(Acylamido)acrylic Acids and/or Their Salts. Ger. Offen. 2,123,063. Chem. Abstr. **76**, 60074f (1972).

217. KNOWLES, W. S., M. J. SABACKY, and B. D. VINEYARD: 3-(3,4-Dihydroxyphenyl)-L-alanine. Ger. Offen. 2,210,938; Chem. Abstr. **77**, 165 073d (1972).

218. — — — Asymmetric Hydrogenation Yields α-Amino Acids. Chem. Technol. **1972**, 2 (10), 590.

219. — — — Catalytic Asymmetric Hydrogenation. J. Chem. Soc. Chem. Comm. **1972**, 10.

220. KNOWLES, W. S., and M. J. SABACKY: Catalytic Asymmetric Hydrogenation of α-(Acylamino)acrylic and -cinnamic Acids. Brit. 1,349,895 (cl. C 07c), 10 Apr 1974, Appl. 17,256/71, 26 May 1971; Chem. Abstr. **81**, 25942 (1974).

221. KNOWLES, W. S., M. J. SABACKY, and B. D. VINEYARD: α-Amino Acids by Asymmetric Hydrogenation. Adv. Chem. Ser. **1974**, 132.

222. KNOWLES, W. S., M. J. SABACKY, B. D. VINEYARD, and D. J. WEINKAUFF: Asymmetric Hydrogenation with a Complex of Rhodium and a Chiral Bisphosphine. J. Amer. Chem. Soc. **97**, 2567 (1975).

223. KOBAYASHI, T., K. IINO, and T. HIRAOKA: A Novel Synthetic Route to 7α-Methoxy-cephalosporins. J. Amer. Chem. Soc. **99**, 5505 (1977).

224. KONCEWICZ, M., P. MATHIAPARANAM, T. F. UCHYTIL, L. SPARAPANO, J. TAM, D. H. RICH, and R. D. DURBIN: Sequence and Optical Configuration of the Amino Acids in Tentoxin. Biochem. Biophys. Res. Comm. **53**, 653 (1973).

225. KOPPEL, G. A., and R. E. KOEHLER: Functionalisation of C$_{6(7)}$ of Penicillins and Cephalosporins. A One-Step Stereoselective Synthesis of 7-α-Methoxycephalosporins C. J. Amer. Chem. Soc. **95**, 2403 (1973).

226. KURITA, H., Y. CHIGIRA, M. MASAKI, and M. OHTA: Synthesis of 4-Alkylidene- and 4-Aralkylidene-2-chloromethyl-5-oxazolones and N-(Chloroacetyl)dehydroamino Acids. Bull. Chem. Soc. Japan **41**, 2758 (1968).

227. KRAFT, W. M., and R. M. HERBST: The Condensation of Carbonyl Compounds with Amides. Aliphatic Aldehydes and Pyruvic Acid with Aliphatic Carbamates. J. Org. Chem. **10**, 483 (1945).

228. KOTULA, Z., E. ZYBURA, and Z. KOWSZYK-GINDIFER: Abstracts, 7th International Symposium on the Chemistry on Natural Products, Riga Latvia, **1970**, p. 629.

229. — — — De-β-lysylviomycin. I. Isolation and Physicochemical Properties. Act. Pol. Pharm. **30**, 431 (1973).

230. Lee, S., H. Aoyagi, Y. Shimohigashi, and N. Izumiya: Syntheses of Cyclotetra-depsipeptides, AM-Toxin I and its Analogs. Tetrahedron Letters 1976, 843.

231. Leonard, N. J., and G. E. Wilson, Jr.: Stereospecific Synthesis and Oxidative Transformation of a Synthetic 1,4-Thiazepine from D-Penicillamine. Tetrahedron Letters 1964, 1465.

232. Leonard, N. J., and G. E. Wilson: The Synthesis and Oxidative Rearrangement of Some 1,4-Thiazepines Related to the Penicillins. J. Amer. Chem. Soc. 86, 5307 (1964).

233. Leonard, N. J., and R. Y. Ning: The Synthesis and Stereochemistry of Substituted 1,4-Thiazepines Related to Penicillins. J. Org. Chem. 31, 3928 (1966).

234. Levi, A., G. Modena, and G. Scorrano: Asymmetric Reduction of Carbon-Nitrogen. Carbon-Oxygen, and Carbon-Carbon Double Bonds by Homogenous Catalytic Hydrogenation. J. Chem. Soc. Chem. Comm. 1975, 6.

235. Liesch, J. M., and K. L. Rinehart, Jr.: Berninamycin. 3. — Total Structure of Berninamycin A. J. Amer. Chem. Soc. 99, 1645 (1977).

236. Love, A. L., and R. K. Olsen: Orientation in Electrophilic Addition Reactions to 2-Acetamidoacrylic Acid Derivatives. J. Org. Chem. 37, 3431 (1972).

237. Lucente, G., G. M. Lucente, and A. Romeo: Synthesis of Various Compounds Derived from α,α-Diaminopropionic Acid. Ann. Chim. (Rome) 56, 572 (1966).

238. Lucente, G., C. Gallina, and A. Romeo: Some Azlactones. Ann. Chim. (Rome) 56, 1192 (1966).

239. Lucente, G., P. Pantanella, and A. Romeo: Synthesis of 2-Alkoxy-2-phenyl-acetamidocarboxylic Acids. J. Chem. Soc. (C) 1967, 1264.

240. Lucente, G., A. Romeo, and G. Zanotti: A New Route to 2-Acylamido-2-hydroxy-carboxylic Esters. Chem. and Ind. 1968, 1602.

241. Lucente, G., and D. Rossi: Synthesis of 2-Alkoxy- and 2-Hydroxy-2-acylamino esters. Chem. and Ind. 1973, 324.

242. Machin, P. J., and P. G. Sammes: Pyrazine Chemistry. Part VI. Addition of Sulphur Nucleophiles across Dehydrocyclodipeptides. J. C. S. Perkin Transactions I 1974, 698.

243. — — Pyrazine Chemistry. Part VII. Oxidations of Piperazine-2,5-diones and Derivatives. J. Chem. Soc. Perkin I 1976, 624.

244. — — Pyrazine Chemistry. Part VIII. Oxidation involving 3-Arylmethylene-piperazine-2,5-diones. J. C. S. Perkin I 1976, 628.

245. Maclaren, J. A., W. E. Savige, and J. M. Swan: Amino Acids and Peptides IV, Intermediates for the Synthesis of Certain Cystine-containing Peptide Sequences in Insulin. Austr. J. Chem. 11, 345 (1958).

246. Maclaren, J. A.: Amino Acids and Peptides V., The Alkaline Saponification of N-Benzyloxycarbonyl Peptide Esters. Austral. J. Chem. 11, 360 (1958).

247. Marchand, J., M. Païs, X. Monseur, and F.-X. Jarreau: Alcaloïdes Peptidiques — VII. Les Lasiodines A et B, Alkaloïdes du Lasiodiscus Marmoratus C. H. Wright (Rhamnacées). Tetrahedron 25, 937 (1969).

248. Marchelli, R., A. Dossena, and G. Casnati: Biosynthesis of Neochinulin by Aspergillus Amstelodami from cyclo-L-[U-^{14}C]Alanyl-L-[5,7-^{3}H$_2$]tryptophyl. J. Chem. Soc. Chem. Comm. 1975, 779.

249. Marchelli, R., A. Dossena, A. Pochini, and E. Dradi: The Structures of Five New Didehydropetides related to Neochinulin, Isolated from Aspergillus Amstelodami. J. Chem. Soc., Perkin Transactions I 1977, 713.

250. Märki, W., und R. Schwyzer: Herstellung von DL-N-Benzyloxycarbonyl-γ-carboxy-glutaminsäure-γ,γ'-di-t-butyl-α-methyl-ester, einem für die Peptidsynthese geeigneten Derivat der neuen Aminotricarbonsäure aus Prothrombin. Helv. Chimica Acta 58, 1471 (1975).

251. MARSCHALL, J. A., T. F. SCHLAF. and J. G. CSERNANSKY: A Convenient Synthesis of Diketopiperazines *via* Aminolysis of N-Pyruvoyl-α-amino Esters. Synth. Comm. **5**, 237 (1975).

252. MARTELL, A. E., and R. M. HERBST: Condensation of Amides with Carbonyl Compounds: Benzyl Carbamate with Aldehydes and α-Keto Acids. J. Org. Chem. **6**, 878 (1941).

253. MASAKI, M., C. SHIN, H. KURITA, and M. OHTA: Independent Isolation of a Primary Enamine and the Tautomeric Imine. J. C. S. Chem. Comm. **1968**, 1447.

254. MASSEN, J. A., T. A. J. W. WAJER, and T. J. DE BOER: C-Nitroso Compounds, Part VIII. A Synthesis of Aldimines from Phosphonate-Carbanions and Nitroso-Alkanes. Rec. Trav. Chim. **88**, 5 (1969).

255. MAUGER, A. B.: Peptide Antibiotic Biosynthesis: A New Approach. Experientia **24**, 1068 (1968).

256. MAYO, P. DE, and A. STOESSEL: Griseoviridin: The C_6-Fragment. Canad. J. Chem. **38**, 950 (1960).

257. McCAPRA, F., and M. ROTH: Cyclisation of a Dehydropeptide Derivative. A Model for Cypridina Luciferin Biosynthesis. J. C. S. Chem. Commun. **1972**, 894.

258. McGAHREN, W. J., G. O. MORTON, M. P. KUNSTMANN, and G. A. ELLESTAD: Carbon-13 Nuclear Magnetic Resonances Studies on a New Antitubercular Peptide Antibiotic LL-BM547ß. J. Org. Chem. **42**, 1282 (1977).

259. McHALE, D., P. MAMALIS, and J. GREEN: Partial Racemisation Accompanying the Acid Hydrolysis of Dibenzoyl-D-cystathionine and -lanthionine. J. Chem. Soc. **1960**, 2847.

260. McINNES, A. G., A. TAYLOR, and J. A. WALTER: The Structure of Chetomin. J. Amer. Chem. Soc. **98**, 6741 (1976).

261. MECHAM, D. K., and H. S. OLCOTT: Phosvitin, the Principal Phosphoprotein of Egg Yolk. J. Amer. Chem. Soc. **71**, 3670 (1949).

262. MEYER, W. L., G. E. TEMPLETON, and C. I. GRABLE: The Structure of Tentoxin. Tetrahedron Letters **1971**, 2357.

263. MEYER, W. L., L. F. KUYPER, R. B. LEWIS, G. E. TEMPLETON, and. S. H. WOODHEAD: Amino Acid Sequence and Configuration of Tentoxin. Biochem. Biophys. Res. Comm. **56**, 234 (1974).

264. MEYER, W. L., L. F. KUYPER, D. W. PHELPS, and A. W. CORDES: Structure of the Cyclic Tetrapeptide Tentoxin. Crystal and Molecular Structure of the Dihydro Derivative. J. Chem. Soc. Chem. Comm. **1974**, 339.

265. MODERHACK, D., and G. ZINNER: 2-(N-Alkyl-hydroxylamino)säureamide als potentielle 2-Oxo-säureamide. Chemiker Zeitung **98**, 110 (1974).

266. MORELL, J. L., P. FLECKENSTEIN, and E. GROSS: Stereospecific Synthesis of (2S,3R)-2-Amino-3-mercaptobutyric Acid — an Intermediate for Incorporation into β-Methyllanthionine-Containing Peptides. J. Org. Chem. **42**, 355 (1977).

267. MORIN, R. B., and E. M. GORDON: Chemistry of Dehydropeptides. Formation of Dehydropeptides by Oxidation of Peptide Oxazolones. Tetrahedron Letters **1973**, 2163.

268. MORRIS, H. R., M. R. THOMPSON, and A. DELL: Synthesis and Proof of Structure of the New Amino Acid in Prothrombin. Biochem. and Biophys. Research Comm. **62**, 856 (1975).

269. MUKERJEE, A. K., and A. K. SINGH: Reactions of Natural and Synthetic β-Lactams. Synthesis **1973**, 547.

270. MÜLLER, L.: Über Reaktionen von 4^1-bromierten 2-Trifluormethyl-Δ^3-oxazolonen-(5) mit Aminen und Sulfoxiden. Dissertation Technische Universität München, 1971.

271. NAGARAJAN, R., L. D. BOECK, M. GORMAN, R. L. HAMILL, C. E. HIGGENS, M. M. HOEHN, W. M. STARK, and J. G. WHITNEY: β-Lactam Antibiotics from Streptomyces. J. Amer. Chem. Soc. **93**, 2308 (1971).

272. Nakagawa, Y., T. Tsuno, K. Nakajima, M. Iwai, H. Kawai, and K. Okawa: Studies on Hydroxy Amino Acids IV. Syntheses of Several Peptides Containing Aziridinecarboxylic Acid Derived from the Corresponding Hydroxy Amino Acid Derivatives. Bull. Chem. Soc. Jap. **45**, 1162 (1972).

273. Nakajima, K., H. Kawai, M. Takai, and K. Okawa: Studies on Hydroxy Amino Acids. VI. Formation of the Oxazoline Derivatives from N-Acyl-β-hydroxy Amino Acid Peptides. Bull. Chem. Soc. Jap. **50**, 917 (1977).

274. Nakatsuka, S., H. Tanino, and Y. Kishi: Biogenetic-Type Synthesis of Penicillin-Cephalosporin Antibiotics. I. A Stereocontrolled Synthesis of the Penam- and Cephem-Ring Systems from an Acyclic Tripeptide Equivalent. J. Amer. Chem. Soc. **97**, 5008 (1975).

275. — — — Biogenetic-Type Synthesis of Penicillin-Cephalosporin Antibiotics. II. An Oxidative Cyclisation Route to β-Lactam Thiazoline Derivatives. J. Amer. Chem. Soc. **97**, 5010 (1975).

276. Nakayama, M., G. Maeda, T. Kaneko, and H. Katsura: Asymmetric Reduction of Some Dehydrophenylalanyl Peptides. Bull. Chem. Soc. Jap. **44**, 1150 (1971).

277. Nebelin, E., und E. Gross: Subtilin V. — Sequenz und Sulfidbrückenzuordnung im heterodeten tricyclischen Peptid der Aminosäurereste 3—19. Z. Physiol. Chem. **354**, 807 (1973).

278. Nicolet, B. H.: The Mechanism of Sulfur Lability in Cysteine and Its Derivatives. II. The Addition of Mercaptan to Benzoylaminocinnamic Acid Derivatives. J. Biol. Chem. **95**, 389 (1932).

279. — The Special Reactivity of Peptides. Science **81**, 181 (1935).

280. Noda, T., T. Take, and A. Nagata: Chemical Studies on Tuberactinomycin. III. The Chemical Structure of Viomycin (Tuberactinomycin B). J. Antibiot. **25**, 427 (1972).

281. Noda, Y., K. Takai, T. Tokuyama, S. Narumiya, H. Ushiro, and O. Hayaishi: Enzymatic Oxidation of Acetyltryptophanamide- and Tryptophan-Containing Peptides. Formation of Dehydrotryptophan. J. Biol. Chem. **252**, 4413 (1977).

282. Öhler, E., H. Poisel, F. Tataruch, und U. Schmidt: Synthese des Epidithio-L-prolyl-L-prolinanhydrids. Chem. Ber. **105**, 635 (1972).

283. Öhler, E., F. Tataruch, und U. Schmidt: Nucleophile Einführung von Schwefelfunktionen über Sulfone und Hydroxyderivate cyclischer Dipeptide (Dioxopiperazine). Chem. Ber. **106**, 165 (1973).

284. — — — Über die Einführung von Sauerstofffunktionen in Prolyl-prolinanhydrid mit Bleitetraacetat: Ein neuer Weg zum Epidisulfid des Prolyl-prolinanhydrids. Chem. Ber. **106**, 396 (1973).

285. Öhler, E., und U. Schmidt: Hydroxylsubstituierte Cyclodipeptide durch Ringschluß von Pyruvoylaminosäureamiden. II. Zweifacher Ringschluß. Chem. Ber. **108**, 2907 (1975).

286. — — Über Dehydroaminosäuren IV. Ringschlüsse an Dehydropeptiden. Chem. Ber. **110**, 921 (1977).

287. Öhler, E., E. Prantz, und U. Schmidt: Über Dehydroaminosäuren. XIV. Biomimetische Versuche zur Cysteinbildung. — Addition von SH-Verbindungen an Dehydroaminosäuren. Chem. Ber. **111**, 1058 (1978).

287A. Öhler, E., und U. Schmidt: Schiffsche Basen von Dehydroaminosäuren aus 2-Aryl-4-thiazolidincarbonsäuren. — Erste Synthese eines N-Aryliden-dehydroalaninesters. Chem. Ber. **112**, 107 (1979).

288. Okawa, K., T. Kinutani, and K. Sakai: Studies on Hydroxy Amino Acids. I. A New Synthesis of Aziridine Derivative from β-Hydroxy-α-Amino Acid. Bull. Chem. Soc. Jap. **41**, 1353 (1968).

289. Okawa, K., K. Nakajima, T. Tanaka, and Y. Kawana: Studies on Hydroxy Amino

Acids V. Synthesis and N-Acylation of 3-Methyl-L-azylylglycine Benzyl Ester. Chemistry Letters **1975**, 591.

290. OKUNO, T., Y. ISHITA, K. SAWAI, and T. MATSUMOTO: Characterisation of Alternariolide, a Host-specific Toxin Produced By Alternaria Mali. Chemistry Letters **1974**, 635.

291. OKUNO, T., Y. ISHITA, A. SUGAWARA, Y. MORI, K. SAWAI, and T. MATSUMOTO: Structure of the Biological Active Cyclopeptides Produced by Alternaria Mali Roberts. Tetrahedron Letters **1975**, 335.

292. OLSEN, R. K., and A. J. KOLAR: N-Acylimines as Intermediates in Reactions of α-Substituted α-Amino Acids and Dehydroamino Acids. Tetrahedron Letters **1975,** 3579.

293. OTTENHEIJM, H. C. J., T. F. SPANDE, and B. WITKOP: Approaches to Analogs of Anhydrogliotoxin. J. Amer. Chem. Soc. **95,** 1989 (1973).

294a. OTTENHEIJM, H. C. J., N. P. E. VERMEULEN, und L. J. F. M. BREUER: Modellversuche zur Synthese von Anhydrogliotoxin-Analoga: Eine bequeme Synthese von Thiazolo-indolon-Derivaten. Liebigs Ann. Chem. **1974,** 206.

294b. OTTENHEIJM, H. C. J., A. D. POTMAN, and T. VAN VROONHOVEN: Approaches to Analogues of Anhydrogliotoxin IV. Synthesis and Reactions of 2-Mercapto-2-amino-propionic Acid Derivatives. Rec. Trav. Chim. Pays-Pas **94,** 135 (1975).

294c. OTTENHEIJM, H. C. J., J. A. M. HULSHOF, and R. J. F. NIVARD: Approaches to Analogs of Anhydrogliotoxin. 3. Synthesis of a Desthiomethylene Analog. J. Org. Chem. **40,** 2147 (1975).

295a. OTTENHEIJM, H. C. J., G. P. C. KERKHOFF, J. W. H. A. BIJEN, and T. F. SPANDE: A three-step Synthesis of a Gliotoxin Analogue with Anti-reverse Transcriptase Activity. J. C. S. Chem. Comm. **1975,** 768.

295b. OTTENHEIJM, H. C. J., J. D. M. HERSCHEID, G. P. C. KERKHOFF, and T. F. SPANDE: Approaches to Analogs of Dehydrogliotoxin VI. An Efficient Synthesis of a Gliotoxin Analog with Anti-reverse Transcriptase Activity. J. Org. Chem. **41,** 3433 (1976).

296. PASCARD, C., A. DUCRUIX, J. LUNEL, and T. PRANGÉ: Highly Modified Cysteine-Containing Antibiotics. Chemical Structure and Configuration of Nosiheptide. J. Amer. Chem. Soc. **99,** 6418 (1977).

297. PATCHORNIK, A., M. SOKOLOVSKY, and T. SADEH: Proc. of the 5th Internat. Congr. Biochemistry, Moscow **1961,** 11.

298. PATCHORNIK, A., and M. SOKOLOVSKY: Oxidative Cleavage of Dehydroalanine (α-Amino-Acrylic Acid) Peptides. Bull. Research Council of Israel, Proc. 30th Meeting of the Israel Chem. Soc. **11A,** 80 (1962).

299. — — Non-Enzymatic Cleavage of Peptide Chains at the Cysteine and Serine Residues. In: Peptides: Proceedings of the Fifth European Symposium. Oxford, p. 253. September 1962. Pergamon Press, 1963.

300. — — Nonenzymatic Cleavages of Peptide Chains at the Cysteine and Serine Residues through their Conversion into Dehydroalanine. I. Hydrolytic and Oxidative Cleavage of Dehydroalanine Residues. J. Amer. Chem. Soc. **86,** 1206 (1964).

301. PATEL, S. M., J. O. CURRIE, JR., and R. K. OLSEN: The Synthesis of N-Acyl-α-mercaptoalanine Derivatives. J. Org. Chem. **38,** 126 (1973).

302. PHOTAKI, I.: Transformation of Serine to Cysteine. β-Elimination Reactions in Serine Derivatives. J. Amer. Chem. Soc. **85,** 1123 (1963).

303. PHOTAKI, I., and V. BARDAKOS: Transformation of L-Serine to L-Cysteine. Experientia **1965,** 371.

304. — — Transformation of L-Serine Peptides to L-Cysteine Peptides. J. Amer. Chem. Soc. **87,** 3489 (1965).

305. — — Transformation of β-Chloro-L-alanine Peptides into L-Cysteine Peptides. J. C. S. Chem. Comm. **1974,** 818.

306. Pieroni, O., G. Montagnoli, A. Fissi, S. Merlino, and F. Ciardelli: Structure and Optical Activity of Unsaturated Peptides. J. Amer. Chem. Soc. **97**, 6820 (1975).

307. Poisel, H., und U. Schmidt: Syntheseversuche in der Reihe der 3,6-Epidithio-2,5-dioxopiperazin-Antibiotika Gliotoxin, Sporidesmin, Aranotin und Chaetocin. II. Chem. Ber. **104**, 1714 (1971).

308. — — Über die elektrophile Einführung von Alkylgruppen und Schwefelfunktionen in den 2,5-Dioxopiperazinkern. Chem. Ber. **105**, 625 (1972).

309. — — Asymmetrische Induktion bei Reaktionen von Aminosäuren und Peptiden, I. Asymmetrische Synthese aromatischer α-Aminosäuren und N-Methyl-α-Aminosäuren. — Synthese von L-Dopa. — Über die katalytische Hydrierung ungesättigter Cyclodipeptide. Chem. Ber. **106**, 3408 (1973).

310. — — Über Dehydroaminosäuren II, Dehydroaminosäuren aus Aminosäuren. Chem. Ber. **108**, 2547 (1975).

311. — — Über Dehydroaminosäuren III. Additionen an α-Iminocarbonsäuren. Chem. Ber. **108**, 2917 (1975).

312. — — Synthese von α,β-Dehydroaminosäureestern und N-tert. Butoxycarbonyl-α,β-dehydroaminosäuren. Angew. Chem. **88**, 295 (1976). Angew. Chem. Int. Ed. Engl. **15**, 294 (1976).

313. Poisel, H.: Über Dehydroaminosäuren VII, Synthese von Dehydroaminosäureestern. Chem. Ber. **110**, 942 (1977).

314. — Über Dehydroaminosäuren VIII, N-Acyl-α,β-dehydroaminosäuren durch Umlagerung von N-Acyl-α-iminosäuren. Chem. Ber. **110**, 948 (1977).

314A. — α-Ketoester aus α-Aminosäureestern. Chem. Ber. **111**, 3136 (1978).

314B. Poisel, H., and U. Schmidt: Unpublished Results.

315. Pojer, P. M., and I. D. Rae: Synthesis of 2-Benzamido-2-mercaptopropanoic Acid. Tetrahedron Letters **1971**, 3077.

316. — — Synthesis of 2-Benzamido-2-mercaptopropanoic Acid from 4-Methyl-2-phenyl-2-oxazolin-5-one. Austral. J. Chem. **25**, 1737 (1972).

317. Prangé, T., A. Ducruix, C. Pascard, and J. Lunel: Structure of Nosiheptide, a Polythiazole-Containing Antibiotic. Nature **265**, 189 (1977).

318. Price, V. E., and J. P. Greenstein: A New Synthesis of Chloroacetyldehydroalanine. Arch. Biochem. **14**, 249 (1947).

319. — — Acetylated Dehydroamino Acids. Arch. Biochem. **18**, 383 (1948).

320. — — N-Acetylated and N-Methylated Glycyldehydroalanine. J. Biol. Chem. **173**, 337 (1948).

321. Rambacher, P.: Weitere einfache Cystin-Synthesen aus α-Acetaminoacrylsäure oder Serin und Thioharnstoff. Chem. Ber. **101**, 3433 (1968).

322. Rao, Y. S., and R. Filler: Geometric Isomers of 2-Aryl(aralkyl)-4-arylidene (alkylidene)-5(4H)-oxazolones. Synthesis **1975**, 749.

323. Ratcliffe, R. W., and B. G. Christensen: Total Synthesis of β-Lactam Antibiotics III. (±-Cefoxitin.) Tetrahedron Letters **1973**, 4653.

324. Reusser, F.: Mode of Action of Berninamycin. An Inhibitor of Protein Biosynthesis. Biochemistry **8**, 3303 (1969).

325. Rich, D. H., and P. Mathiaparanam: Synthesis of the Cyclic Tetrapeptide Tentoxin. Effect of an N-Methyldehydrophenylalanyl Residue on Conformation of Linear Tetrapeptides. Tetrahedron Letters **1974**, 4037.

326. Rich, D. H., J. P. Tam, P. Mathiaparanam, J. A. Grant, and C. Mabuni: General Synthesis of Didehydro-amino-acids and Peptides. J. C. S. Chem. Comm. **1974**, 897.

327. Rich, D. H., and J. P. Tam: Synthesis of Didehydropeptides from Peptides Containing 3-Alkylthio-Amino Acid Residues. Tetrahedron Letters **1975**, 211.

328. Rich, D. H., J. P. Tam, P. Mathiaparanam, and J. Grant: Selective N-Methylation of Dehydroamino Acids and Peptides. Synthesis **1975**, 402.

329. RICH, D. H., and J. P. TAM: A Method for Introducing Secondary Amide Bonds into Strained Cyclic Peptides. Tetrahedron Letters **1977**, 749.

330. RICHARDS, K. D., A. J. KOLAR, A. SRINIVASAN, R. W. STEPHENSON, and R. K. OLSEN: The Reaction of Dialkylcopper Lithium Reagents with 3-Halo-2-acylaminoacrylic Acids. J. Org. Chem. **41**, 3674 (1976).

331. RILEY, G., J. H. TURNBULL, and W. WILSON: O-Phosphorylserine Derivatives. Chem. and Ind. **1953**, 1181.

332. — — — Synthesis of Some Phosphorylated Amino-hydroxy-acids and Derived Peptides related to the Phosphoproteins. J. Chem. Soc. **1957**, 1373.

333. RIORDAN, J. M., and C. H. STAMMER: The Direct Conversion of N-Acyl-α-amino Acids into N-Acyl-α,β-unsaturated α-Amino Acids. Tetrahedron Letters **1971**, 4969.

334. — — Synthesis of Unsaturated Azlactones from N-Acylamino Acids. J. Org. Chem. **39**, 654 (1974).

335. — — o-Chloranil Oxidation of Azlactones. Tetrahedron Letters **1976**, 1247.

336. RIORDAN, J. M., M. SATO, and C. H. STAMMER: p-Chloranil-Azlactone Adducts and Their Conversion to Unsaturated Amino Acid Derivatives. J. Org. Chem. **42**, 236 (1977).

337. ROTHSTEIN, E.: Experiments in the Synthesis of Derivatives of α-Aminoacrylic Acid from Serine and N-Substituted Serines. J. Chem. Soc. **1949**, 1968.

338. SAEGUSA, T., N. TAKA-ISHI, and Y. ITO: The Thermal Rearrangement and Degradation of 2,3-Bis(alkylimino)oxetane. Bull. Chem. Soc. Jap. **44**, 1121 (1971).

339. SAITO, T., Y. SUGIMURA, Y. IWANO, K. IINO, and T. HIRAOKA: A New Synthetic Route to 7α-Methoxy-cephalosporins. J. Chem. Soc., Chem. Commun. **1976**, 516.

340. SAKAKIBARA, S.: Studies on Dehydroalanine Derivatives. IV. Synthesis of N-Phthaloyl-Dehydroalanine by Thermal Decomposition of Cysteine and Serine in the Presence of Phthalic Anhydride. Bull. Chem. Soc. Jap. **34**, 171 (1961).

341. SAMMES, P. G.: Naturally Occurring 2,5-Dioxopiperazines and Related Compounds. In: Fortschritte der Chemie organischer Naturstoffe (W. HERZ, H. GRISEBACH, G. W. KIRBY, eds.), Vol. 32, p. 51. Wien-New York: Springer. 1975.

342. — Recent Chemistry of the β-Lactam Antibiotics. Chem. Rev. **1976**, 113.

343. SASAKI, T.: Über die Kondensation von Glycinanhydrid mit Aldehyden. Eine neue Synthese von d,l-Phenylalanin und d,l-Tyrosin. Chem. Ber. **54**, 163 (1921).

344. SASAKI, T., und T. HASHIMOTO: Über die Kondensation einiger Dipeptidanhydride mit Benzaldehyd. Chem. Ber. **54**, 168 (1921).

345. SATO, T., and T. HINO: Decarboxylative C-S Bond Formation. Synthesis of 1,4-Dimethyl-3,6-epidithio-2,5-piperazinedione and Related Compounds. Chem. Pharm. Bull. **24**, 285 (1976).

346. SAVARD, K., E. M. RICHARDSON, and G. A. GRANT: Synthesis of a New α-Amino Acid, S-Methyl-β,β-dimethylcysteine. Can. J. Res. **24B**, 28 (1946).

346A. SCHMIDT, H., and W. STEGLICH: Private Communication.

347. SCHMIDT, U., A. PERCO, und E. ÖHLER: Über Dehydroaminosäuren, I. Ringschluß des Z-Dehydroalanyl-L-prolin-N-methylamids zum Z-Amino-cyclodipeptid durch Amidaddition an die C=C-Doppelbindung. Chem. Ber. **107**, 2816 (1974).

348. SCHMIDT, U., und E. ÖHLER: Optische Induktion bei der biomimetischen Cystein-bildung. Angew. Chem. **88**, 54 (1976), Angew. Chem., Int. Ed. Engl. **15**, 42 (1976).

349. SCHMIDT, U., und J. HÄUSLER: Radikalische Oxidation von Aminosäurederivaten. Angew. Chem. **88**, 538 (1976); Angew. Chem. Int. Ed. Engl. **15**, 497 (1976).

350. SCHMIDT, U., und E. ÖHLER: Einfache Synthese von α,β-Dehydroaminosäureestern. Angew. Chem. **89**, 344 (1977). Angew. Chem. Int. Ed. Engl. **16**, 327 (1977).

351. SCHMIDT, U., J. HÄUSLER, E. OHLER, und H. POISEL: Bemerkungen zur ,,Three-step Synthesis of a Gliotoxin Analogue..." von H. C. J. OTTENHEIJM et al. Chem. Ber. **110**, 3722 (1977).

352. Schmidt, U., und E. Prantz: Doppelbindungsaktivierung in Dehydroaminosäuren: Ein Modell pyridoxalhaltiger Enzyme in Eliminierungs-Additionsreaktionen. Angew. Chem. **89,** 345 (1977). Chem. Int. Ed. Engl. **16,** 328 (1977).

353. Schöberl, A.: Über die Anlagerung von Sulfhydrylcarbonsäuren an ungesättigte Säuren und über eine neue Synthese von Lanthionin. Chem. Ber. **80,** 379 (1947).

354. Schöberl, A., und A. Wagner: Eine neue Synthese von Cystein und Cystin. Naturwissenschaften **34,** 189 (1947).

355. Schöberl, A.: Neue Synthesen schwefelhaltiger Aminosäuren. Angew. Chem. **60A,** 308 (1948).

356. Schöberl, A., und A. Wagner: Über eine neue Synthese von Cystathionin. Naturwiss. **37,** 113 (1950).

357. Schöberl, A., und G. Täuber: Über die Synthese der optisch aktiven Diastereomeren Cystathionin und Allocystathionin und über Methoden zu deren Trennung. Ann. **599,** 23 (1956).

358. Schöberl, A., und A. Wagner: Untersuchungen zur Frage der Lanthionin-Bildung aus Wolle und Cystin. Z. Physiol. Chem. **304,** 97 (1956).

359. Schöberl, A., M. Rimpler, und K. H. Magosch: Notiz zur Synthese von D-Cystin und seinen Derivaten. Chem. Ber. **102,** 1767 (1969).

360. Schöllkopf, U., F. Gerhart, R. Schröder, und D. Hoppe: β-Substituierte α-Formylaminoacrylsäureäthylester aus α-metallierten Isocyanessigestern und Carbonylverbindungen. (Formylaminomethylierung von Carbonylverbindungen.) Liebigs Ann. Chem. **766,** 116 (1972).

361. Sen, L. C., E. Gonzalez-Flores, R. E. Feeney, and J. R. Whitaker: Reactions of Phosphoproteins in Alkaline Solutions. J. Agric. Food Chem. **25,** 632 (1977); Chem. Abstr. **87,** 6361m (1977).

362. Sheehan, J. C., and R. E. Chandler: A Sterically Controlled Synthesis of Amino Acids. J. Amer. Chem. Soc. **83,** 4795 (1961).

363. Sheehan, J. C., D. Mania, S. Nakamura, J. A. Stock, and K. Maeda: The Structure of Telomycin. J. Amer. Chem. Soc. **90,** 462 (1968).

364. Shemin, D., and R. M. Herbst: The Condensation of α-Keto Acids and Acetamide. J. Amer. Chem. Soc. **60,** 1954 (1938).

365. Shiba, T., S. Nomoto, T. Teshima, and T. Wakamiya: Revised Structure and Total Synthesis of Capreomycin. Tetrahedron Letters **1976,** 3907.

366. Shin, C., M. Masaki, and M. Ohta: The Synthesis of 3-Isopropylidene-2,5-dioxopiperazines. Bull. Chem. Soc. Japan **39,** 858 (1966).

367. Shin, C., Y. Chigira, M. Masaki, and M. Ohta: Total Synthesis of Albonoursin. Tetrahedron Letters **1967,** 4601.

368. Shin, C., M. Masaki, and M. Ohta: The Synthesis of 3-Isopropylidene-2,5-piperazinediones. J. Org. Chem. **32,** 1860 (1967).

369. Shin, C., Y. Chigira, M. Masaki, and M. Ohta: Synthesis of Albonoursin. Bull. Chem. Soc. Jap. **42,** 191 (1969).

370. Shin, C., M. Masaki, and M. Ohta: The Synthesis and Reaction of α,β-Unsaturated α-Nitrocarboxylic Esters. Bull. Chem. Soc. Japan **43,** 3219 (1970).

371. Shin, C., H. Ando, and J. Yoshimura: The Reaction of α-Oxo Acids with N-Phenyltriphenylphosphinimine. Bull. Chem. Soc. Jap. **44,** 474 (1971).

372. Shin, C., M. Fujii, and J. Joshimura: The General Synthesis of 3-Alkylidene-2,5-piperazinediones. Tetrahedron Letters **1971,** 2499.

373. Shin, C., M. Masaki, and M. Ohta: The Independent Isolation of a Primary Enamine and the Tautomeric Imine. Bull. Chem. Soc. Jap. **44,** 1657 (1971).

374. Shin, C., K. Nanjo, and J. Yoshimura: Cyclisation Reaction of N-(Haloacetyl)- or N-(Phtaloylglycyl)hydroxyaminoacid Esters with Ammonia. Chemistry Letters **1973,** 1039.

375. Shin, C., K. Sato, A. Ohtsuka, K. Mikami, and J. Yoshimura: α,β-Unsaturated Carboxylic Acid Derivatives. IV. General Synthesis of Unsaturated Unsymmetric 3,6-Disubstituted-2,5-piperazinediones. Bull. Chem. Soc. Japan **46**, 3876 (1973).

376. Shin, C., K. Nanjo, and J. Yoshimura: A Facil Synthesis of α,β-Unsaturated β-Bromo-N-acyl-α-amino Acids. Tetrahedron Letters **1974**, 521.

377. Shin, C., K. Nanjo, E. Ando, and J. Yoshimura: α,β-Unsaturated Carboxylic Acid Derivatives VI, New Synthesis of N-Acyl-α-dehydroamino Acid Esters. Bull. Chem. Soc. Japan **47**, 3109 (1974).

378. Shin, C., K. Nanjo, T. Nishino, Y. Sato, and J. Yoshimura: α,β-Unsaturated Carboxylic Acid Derivatives. VIII. The Synthesis and Reaction of Esters of N-Acyl-N-bromo-α-dehydroamino Acid. Bull. Chem. Soc. Jap. **48**, 2492 (1975).

379. Shin, C., Y. Sato, and J. Yoshimura: α,β-Unsaturated Carboxylic Acid Derivatives. XI. Convenient Synthesis of tert.-Butyl 2-Alkoxy- and Hydroxy-2-acetylamino-3-mono- or 3,3-dihaloalkanoates. Bull. Chem. Soc. Japan **49**, 1909 (1976).

380. Shin, C., Y. Yonezawa, and J. Yoshimura: A General Synthesis of α,β-Unsaturated α-Amino Acid Ethyl Esters. Chemistry Letters **1976**, 1095.

381. Shin, C., M. Hayakawa, K. Mikami, and J. Yoshimura: Synthesis and Configurational Assignments of Albonoursin and Its Three Geometric Isomers. Tetrahedron Letters **1977**, 863.

382. Shive, W., and G. W. Shive: The Condensation of Pyruvic Acid and Formamide. J. Amer. Chem. Soc. **68**, 117 (1946).

383. Sicher, J., M. Svoboda, and J. Farkas: The Synthesis and Configuration of the Two Racemic β-Phenylcysteines. Coll. Czech. Chem. Comm. **20**, 1439 (1955).

384. Sjöberg, B., H. Thelin, L. Nathorst-Westfelt, E. E. van Tamelen, and E. R. Wagner: On the Role of "Cyclic Cysteinylvaline" in Penicillin Biosynthesis. Tetrahedron Letters **1965**, 281.

385. Slusarchyk, W. A., H. E. Applegate, P. Funke, W. Koster, M. S. Puar, M. Young, and J. E. Dolfini: Synthesis of 6-Methoxythiopenicillins and 7-Heteroatom-Substituted Cephalosporins. J. Org. Chem. **38**, 943 (1973).

386. Smale, T. C., and S. Bailey: Insecticidal Amidoacrylate Derivatives. Brit. 1,354,571 (1974); Chem. Abstr. **81**, 63353k (1974).

387. Smrt, J., J. Beranek, and J. Sicher: Esters of N-Substituted Ethyleniminecarboxylic Acids. U.S. 2,958,691 (Nov. 1, 1960), Chem. Abstr. **55**, 10468i (1961).

388. Snow, J. T., J. W. Finley, and M. Friedman: Relative Reactivities of Sulfhydryl Groups with N-Acetyl Dehydroalanine and N-Acetyl Dehydroalanine Methyl Ester. Int. J. Peptide Protein Res. **8**, 57 (1976).

389. Snyder, H. R., and J. A. MacDonald: A Synthesis of Tryptophan and Tryptophan Analogs. J. Amer. Chem. Soc. **77**, 1257 (1955).

390. Sokolovsky, M., M. Wilchek, and A. Patchornik: The Formation of Dehydroalanine Derivatives from S-DNP Cysteine Peptides. Proc. 30th Meeting of the Israel Chem. Soc. **11A**, 79 (1962).

391. Sokolovsky, M., T. Sadeh, and A. Patchornik: Nonenzymatic Cleavages of Peptide Chains at the Cysteine and Serine Residues through their Conversion to Dehydroalanine (DHAL). II. The Specific Chemical Cleavage of Cysteinyl Peptides. J. Amer. Chem. Soc. **86**, 1212 (1964).

392. Spitzer, W. A., and T. Goodson: The Synthesis of S-Methyl and O-Methyl β-Lactam Antibiotics. Tetrahedron Letters **1973**, 273.

393. Srinivasan, A., K. D. Richards, and R. K. Olsen: Comments on Assignment of Stereochemistry to 2-Acylaminocrotonates. Tetrahedron Letters **1976**, 891.

394. Srinivasan, A., R. W. Stephenson, and R. K. Olsen: Synthesis of Dehydroalanine Peptides from β-Chloroalanine Peptide Derivatives. J. Org. Chem. **42**, 2253 (1977).

395. Srinivasan, A., R. W. Stephenson, and R. K. Olsen: Conversion of Threonine Derivatives to Dehydroamino Acids by Elimination of β-Chloro and O-Tosyl Derivatives. J. Org. Chem. **42**, 2256 (1977).

396. Steglich, W., und R. Hurnaus: Über den Verlauf der Bergmannschen Azlaktonsynthese, 2-Alkyliden-pseudo-oxazolone-(5). Tetrahedron Letters **1966**, 383.

397. Steglich, W., H. Tanner, und R. Hurnaus: 2-Dichlormethylen-pseudooxazolone-(5). Chem. Ber. **100**, 1824 (1967).

398. Steglich, W.: Fortschritte in der Chemie der Oxazolinone-(5). In: Fortschritte chem. Forsch. Bd. 12/1, p. 77. Berlin-Heidelberg-New York: Springer. 1970.

399. Strunz, G. M., and M. Kakushima: Total Synthesis of (±)-Hyalodendrin. Experientia **30**, 719 (1974).

400. Sugimura, Y., K. Iino, Y. Iwano, T. Saito, and T. Hiraoka: A Novel Synthesis of 7-Methoxycephalosporins and 6-Methoxypenicillins. Tetrahedron Letters **1976**, 1307.

401. Süs, O.: Über die Anlagerung von Schwefelverbindungen an die β,β-Dimethylacrylsäure. — Synthesen der β,β-Dimethyl-α-aminopropionsäure, dl-Penicillamin. Ann. **559**, 92 (1948).

402. — Synthetische Versuche in der Penicillinreihe V. Versuche zum Aufbau von Zwischenprodukten der Biosynthese bei dem Penicillin G. Ann. **569**, 153 (1950).

403. — α-Amino-β,β-dimethyl-β-(arylmethylthio)propionic Acids. Ger 831,997 (1952). Chem. Abstr. **47**, 2201 (1953).

404. — Aliphatic Amino Thio Carboxylic Acids. Ger 831, 998 (1952). Chem. Abstr. **47**, 6978 (1953).

405. Sutherland, J. K.: The Proton-Magnetic-Resonance Spectrum of Pencolide. Biochem. J. **86**, 243 (1963).

406. Swan, J. M.: Thiols, Disulphides and Thiosulphates: Some New Reactions and Possibilities in Peptide and Protein Chemistry. Nature **1957**, 643.

407. — Mechanism of Alkaline Degradation of Cystine Residues in Proteins. Nature **179**, 965 (1957).

408. Takaishi, N., H. Imai, C. A. Bertelo, and J. K. Stille: Transition Metal Catalyzed Asymmetric Organic Syntheses *via* Polymer Bound Chiral Ligands. Synthesis of R Amino Acids and Hydratropic Acid by Hydrogenation. J. Amer. Chem. Soc. **98**, 5400 (1976).

408A. Takita, Y., Y. Muraoka, T. Yoshioka, A. Fujii, K. Maeda, and H. Umezawa: Chemistry of Bleomycin. IX. Structures of Bleomycin and Phleomycin. J. Antibiot. **25**, 755 (1972).

409. Tanaka, M., and J. Ogata: Asymmetric Hydrogenation by a Chiral Diphosphinite Rhodium Complex. J. Chem. Soc. Chem. Comm. **1975**, 735.

410. Tatsuoka, S., M. Murakami, and T. Tamura: Reactions of β-Hydroxyvaline and Related Compounds. J. Pharm. Soc. Jap. **70**, 230 (1950).

411. Teshima, T., S. Nomoto, T. Wakamiya, and T. Shiba: Chemical Studies on Tuberactinomycins. X. Total Synthesis of Tuberactinomycin O. Tetrahedron Letters **1976**, 2343.

412. Testa, B., and P. Jenner: Oxidation of Nitrogen-Containing Functional Groups. In: "Drug Metabolism. Chemical and Biochemical Aspects", p. 61ff. Marcel Dekker Inc., 1976.

413. Theodoropoulos, D., I. L. Schwartz, and R. Walter: Synthesis of Selenium-Containing Peptides. Biochemistry **6**, 3927 (1967).

414. Tolosa, E. A., R. N. Maslowa, E. V. Goryachenkowa, I. H. Willhardt, and A. E. Braunstein: Isotopic Hydrogen Exchange in Reactions Catalysed by Cysteine Lyase and Serine Sulphhydrase. Eur. J. Biochem. **53**, 429 (1975).

415. Tori, K., K. Tokura, K. Okabe, M. Ebata, H. Otsuka, and G. Lukacs: Carbon-13 NMR Studies of Peptide Antibiotics, Thiostrepton and Siomycin A: The Structure and Relationship. Tetrahedron Letters **1976**, 185.

416. TROWN, P. W.: Antiviral Activity of N,N'-Dimethyl-3,6-epidithiopiperazine-2,5-dione. A Synthetic Compound Related to the Gliotoxins, LL S88α and β-Chetomin, and the Sporidesmins. Biochem. Biophys. Res. Commun. **33**, 402 (1968).

417. TSHAMANN, E. S., and M. M. SHEMJAKIN: α-Substituted α-Amino Acids Series. I. Synthesis and Properties of the Simplest α-Hydroxy-α-acylamino Carboxylic Acids. Zhur. Obscheĭ Khim. **25**, 1360 (1955); Chem. Abstr. **50**, 4913 (1956).

418. UENO, T., T. NAKASHIMA, Y. HAYASHI, and H. FUKAMI: Isolation and Structure of AM-Toxin III, a Host-specific Phytotoxic Metabolite Produced by *Alternaria Mali.* Agric. Biol. Chem. **39**, 1115, 2081 (1975).

419. UGI, I., und U. FETZER: Isonitrile III. Die Addition von Carbonsäurechloriden an Isonitrile. Chem. Ber. **94**, 1116 (1961).

419A. UMEZAWA, H.: Natural and Artificial Bleomycins: Chemistry and Antitumor Activities. Pure and Applied Chem. **28**, 665 (1971).

420. WAKAMIYA, T., T. SHIBA, T. KANEKO, H. SAKAKIBARA, and T. TAKE: Chemical Studies on Tuberactinomycin. I. Structure of Tuberactidine, Guanidino Amino Acid Component. Tetrahedron Letters **1970**, 3497.

421. WAKAMIYA, T., T. SHIBA, T. KANEKO, Y. YOSHIOKA, T. AOKI, K. NAKATSU, T. NODA, T. TAKE, A. NAGATA, and J. ABE: The Chemical Structures of Tuberactinomycines. 15th Symposium on the Chemistry of Natural Products, p. 16. Nagoya 1971.

422. WAKAMIYA, T., T. SHIBA, and T. KANEKO: Chemical Studies on Tuberactinomycin. IV. Chemical Structure of γ-Hydroxy-β-lysine. Bull. Chem. Soc. Japan **45**, 3668 (1972).

423. WAKAMIYA, T., T. SHIBA, T. KANEKO, H. SAKAKIBARA, T. NODA, and T. TAKE: Chemical Studies on Tuberactinomycin. V. Structures of Guanidino Amino Acids. Bull. Chem. Soc. Japan **46**, 949 (1973).

424. WAKAMIYA, T., and T. SHIBA: Chemical Studies on Tuberactinomycin. VI. — The Absolute Configuration of γ-Hydroxy-β-lysine in Tuberactinomycins A and N. J. Antibiot. **27**, 900 (1974).

425. WAKAMIYA, T., T. TESHIMA, I. KUBOTA, T. SHIBA, and T. KANEKO: Chemical Studies on Tuberactinomycin. VII. Synthesis of γ-Hydroxy-β-lysine. Bull. Chem. Soc. Japan **47**, 2292 (1974).

426. WAKAMIYA, T., and T. SHIBA: Chemical Studies on Tuberactinomycin. VIII. Isolation of Tuberactinomycin N and Conversion of Tuberactinomycin N to O. J. Antibiot. (Tokyo) **28**, 292 (1975).

427. — — Chemical Studies on Tuberactinomycin. IX. Nuclear Magnetic Resonance Studies on Tuberactinomycins and Tuberactinamin N. Bull. Chem. Soc. Japan **48**, 2502 (1975).

427A. WALKER, J., A. OLESKER, L. VALENTE, R. RABANAL and G. LUCACS: Total Structure of the Polythiazole-containing Antibiotic Micrococcin P. A ^{13}C Nuclear Magnetic Resonance Study. J. Chem. Soc., Chem. Commun. **1978**, 256.

428. WALTER, R., and J. ROY: Selenomethionine, a Potential Catalytic Antioxidant in Biological Systems. J. Org. Chem. **36**, 2561 (1971).

429. WARNHOFF, E. W.: Peptide Alkaloids. In: Fortschritte der Chemie organischer Naturstoffe (W. HERZ, H. GRISEBACH, A. I. SCOTT, eds.), Vol. 28, p. 162. Wien-New York: Springer. 1970.

430. WEINER, W., W. N. WHITE, D. G. HOARE, and D. E. KOSHLAND, Jr.: The Formation of Anhydrochymotrypsin by Removing the Elements of Water from the Serine at the Active Site. J. Amer. Chem. Soc. **88**, 3851 (1966).

431. WEINSTEIN, B., K. G. WATRIN, H. J. LOIE, and J. C. MARTIN: Amino Acids and Peptides. 44. Synthesis of DL-γ-Carboxyglutamic Acid, a New Amino Acid. J. Org. Chem. **41**, 3634 (1976).

432. WEYGAND, F., W. STEGLICH, und H. TANNER: Eine neue Methode zur Umwandlung von α-Aminosäuren in α-Ketosäuren. Liebigs Ann. **658**, 128 (1962).

433. Weygand, F., W. Steglich, D. Mayer, und W. von Phillipsborn: 2-Trifluormethyl-pseudooxazolone-(5). Chem. Ber. **97**, 2023 (1964).

434. Wickner, R. B.: Dehydroalanine in Histidine Ammonia Lyase. J. Biol. Chem. **244**, 6550 (1969).

435. Wieland, T., G. Ohnacker, und W. Ziegler: Aminosäuresynthesen mit α-Acylamino-acrylestern. Chem. Ber. **90**, 194 (1957).

436. Wieland, T., K. H. Shin, und B. Heinke: Synthese einiger Pyruvoylaminosäuren nach der Phsophoroxychlorid-Methode. Chem. Ber. **91**, 483 (1958).

437. Wohl, A., und L. H. Lips: Über Amide der Brenztraubensäure. Ber. **40**, 2313 (1907).

438. Woodard, J. C., D. D. Short, C. E. Strattan, and J. H. Duncan: Synthesis and Properties of N^{ε}-(DL-2-Amino-2-carboxyethyl)-L-lysine, Lysinoalanine. Food. Cosmet. Toxicol. **15**, 109 (1977); Chem. Abstr. **87**, 68619k (1977).

439. Wolfe, S., R. N. Bassett, S. M. Caldwell, and F. I. Wasson: Reversal of the Anhydropenicillin Rearrangement. J. Amer. Chem. Soc. **91**, 7205 (1969).

440. Yamazaki, M., K. Sasago, and K. Miyaki: The Structure of Fumitremorgin B (FTB), a Tremorgenic Toxin from *Aspergillus fumigatus* Fres. J. Chem. Soc. Chem. Commun. **1974**, 408.

441. Yanagisawa, H., M. Fukushima, A. Ando, and H. Nakao: A Novel General Method for Synthesizing 7α-Methoxycephalosporins. Tetrahedron Letters **1975**, 2705.

442. — — — — A Novel Simple Synthesis of 7α-Substituted Cephalosporins. Tetrahedron Letters **1976**, 259.

443. Yoshimoto, M., S. Ishihara, E. Nakayama, E. Shoji, H. Kuwano, and N. Soma: Studies on β-Lactam Antibiotics II. A New Synthesis of 1,2-Secopenicillin and its Conversion to the Cephem Nucleus. Tetrahedron Letters **1972**, 4387.

444. Yoshimura, J., Y. Sugiyama, K. Matsunari, and H. Nakamura: Addition of Methanesulfenyl Chloride and Sulfur Chloride to 1,4-Dimethyl-3,6-dimethylene-2,5-piperazinedione and Substitution of the Adducts. Bull. Chem. Soc. Jap. **47**, 1215 (1974).

445. Yoshimura, J., H. Nakamura, and K. Matsunari: Synthesis of 3,6-Dialkyl-1,4-dimethyl-3,6-epithio- and -3,6-epidithio-2,5-piperazinediones. Bull. Chem. Soc. Jap. **48**, 605 (1975).

446. Yoshioka, H., T. Aoki, H. Goko, K. Nagatsu, T. Noda, H. Sakakibara, T. Take, A. Nagata, J. Abe, T. Wakamiya, T. Shiba, and T. Kaneko: Chemical Studies on Tuberactinomycin. II. The Structure of Tuberactinomycin O. Tetrahedron Letters **1971**, 2043.

447. Zervas, L., und I. Photaki: Über Cystein- und Cystinpeptide. Chimia **14**, 375 (1960).

448. Zervas, L., und N. Ferderigos: Umwandlung von Cysteinylserin in Lanthionin. Experientia **29**, 262 (1973).

449. — — On Lanthionine and Cyclolanthionyl. Israel J. Chem. **12**, 139 (1974).

450. Zioudrou, C., M. Wilchek, and A. Patchornik: Conversion of the L-Serine Residue to an L-Cysteine Residue in Peptides. Biochemistry **1965**, 1811.

451. Zoller, U., and D. Ben-Ishai: Amidoalkylation of Mercaptans with Glyoxylic Acid Derivatives. Tetrahedron **31**, 863 (1975).

452. Shimohigashi, Y., S. Lee, T. Kato, and N. Izumiya: Synthesis of Cyclotetradepsi-peptides, AM-Toxin II and Its Analog. Chemistry Letters **1977**, 1411.

453. Shimohigashi, Y., S. Lee, T. Kato, N. Izumiya, T. Ueno, and H. Fukami: Synthesis and Necrotic Activity of Dihydro-AM-toxin I. Agric. Biol. Chem. **41**, 1533 (1977).

454. Shimohigashi, Y., S. Lee, H. Aoyagi, T. Kato, and N. Izumiya: Cyclic Peptides. I. Synthesis of AM-Toxin Analog Containing O-Methyl-L-tyrosine. Int. J. Pept. Protein Res. **1977**, 197.

454A. — — — — — Cyclic Peptides. Part 3. Synthesis of AM-Toxin I. Int. J. Pept. Protein Res. **1977**, 323.

455. SHIMOHIGASHI, Y., S. LEE, T. KATO, and N. IZUMIYA: Cyclic Peptides. IV. Synthesis of Diastereomeric Dihydro-AM-toxin I and Its Analogs. Bull. Chem. Soc. Japan **51**, 584 (1978).

456. UENO, T., T. NAKASHIMA, M. UEMOTO, H. FUKAMI, S. LEE, and N. IZUMIYA: Mass Spectrometry of Alternaria Mali Toxins and Related Cyclodepsipeptides. Biomed. Mass Spectrom. **1977**, 134.

457. NOMOTO, S., T. TESHIMA, T. WAKAMIYA, and T. SHIBA: The Revised Structure of Capreomycin. J. Antibiot. **1977**, 955.

458. — — — — Total Synthesis of Capreomycin. Tetrahedron **34**, 921 (1978).

459. PALLAI, P., T. WAKAMIYA, and E. GROSS: Studies on the Synthesis and Biology of Nisin: Ring A. Pept., Proc. Am. Pept. Symp., 5th. **1977**, 205.

460. RICH, D. H., and P. K. BHATNAGAR: Isolation and Conformational Analysis of Two Conformers of β-Methylalanine[1]-tentoxin. J. Amer. Chem. Soc. **100**, 2218 (1978).

461. RICH, D. H., and P. K. BHATNAGAR: Conformational Studies of Tentoxin by Nuclear Magnetic Resonance Spectroscopy. Evidence for a New Conformation For a Cyclic Tetrapeptide. J. Amer. Chem. Soc. **100**, 2212 (1978).

462. RICH, D. H., P. BHATNAGAR, P. MATHIAPARANAM, J. A. GRANT, and J. P. TAM: Synthesis of Tentoxin and Related Dehydro Cyclic Tetrapeptides. J. Org. Chem. **43**, 296 (1978).

463. WAKAMIYA, T., T. TESHIMA, H. SAKAKIBARA, K. FUKUKAWA, and T. SHIBA: Chemical Studies on Tuberactinomycin. XI. Semisyntheses of Tuberactinomycin Analogs with Various Amino Acids in the Branched Part. Bull. Chem. Soc. Japan **50**, 1984 (1977).

464. TESHIMA, T., S. NOMOTO, T. WAKAMIYA, and T. SHIBA: Chemical Studies on Tuberactinomycin. XII. Syntheses and Antimicrobial Activities of [Ala³,Ala⁴]-, [Ala³]-. and [Ala⁴]-Tuberactinomycin O. Bull. Chem. Soc. Japan **50**, 3372 (1977).

465. NOMOTO, S., and T. SHIBA: Chemical Studies on Tuberactinomycin. XIII. Modification of β-Ureidodehydroalanine Residue in Tuberactinomycin N. J. Antibiot. **1977**, 1008.

466. TESHIMA, T., S. NOMOTO, T. WAKAMIYA, and T. SHIBA: Chemical Studies on Tuberactinomycin. XV. Total Synthesis of Tuberactinomycin O. J. Antibiot. **1977**, 1073.

467. KONISHI, M., K. SAITO, K. NUMATA, T. TSUNO, K. ASAMA, H. TSUKIURA, T. NAITO, and H. KAWAGUCHI: Tallysomycin, a New Antitumor Complex Related to Bleomycin. II. Structure Determination of Tallysomycin. J. Antibiot. **1977**, 789.

468. FEREZOU, J. P., C. RICHE, A. QUESNEAU-THIERRY, C. PASCARD-BILLY, M. BARBIER, J. F. BOUSQUET, and G. BOUDART: Structures of Two Toxins Isolated from Cultures of the Fungus Phoma Lingam Tode: Sirodesmin PL and Deacetylsirodesmin PL. Nouv. J. Chim. **1977**, 327.

469. MCGOWAN, D. A., U. JORDIS, D. K. MINSTER, and S. M. HECHT: A Biomimetic Synthesis of the Bisthiazole Moiety of Bleomycin. J. Amer. Chem. Soc. **99**, 8078 (1977).

470. BYCROFT, B. W., and M. S. GOWLAND: The Structures of the Highly Modified Peptide Antibiotics Micrococcin P_1 and P_2. J. Chem. Soc. Chem. Comm. **1978**, 256.

471. BYCROFT, B. W.: Configurational and Conformational Studies on the Group A Peptide Antibiotics of the Mikamycin (Streptogramin, Virginiamycin) Family. J. Chem. Soc., Perkin I **1977**, 2464.

472. RICH, D., and J. P. TAM: Synthesis of Dehydro Amino Acids and Peptides by Dehydrosulfenation. Rate Enhancement Using Sulfenic Acid Trapping Agents. J. Org. Chem. **42**, 3815 (1977).

473. SHIN, C., M. HAYAKAWA, T. SUZUKI, A. OHTSUKA, and J. YOSHIMURA: α,β-Unsaturated Carboxylic Acid Derivatives. XIII. The Synthesis and Configuration of Alkyl 2-Acylamino-2-alkenoates and Their Cyclized 2,5-Piperazinedione Derivatives. Bull. Chem. Soc. Japan **51**, 550 (1978).

474. Saito, T., and T. Hiraoka: Reactions of Iminophosphoranes of α-Amino Acid Derivatives with Dimethyl Acetylendicarboxylate. An Application to a Synthesis of 7α-Methoxycephalosporins. Chem. Pharm. Bull. **25**, 1645 (1977).

475. Taylor, A. W., and G. Burton: Formation and 6α-Substitution of 6β-(2-Carboxy)-ketenimino Penicillins. Tetrahedron Letters **1977**, 3831.

476. Tajima, K.: Autoxidation of 4H-5-Oxazolone. Chemistry Letters **1977**, 279.

477. Horikawa, H., T. Iwasaki, K. Matsumoto, and M. Miyoshi: Electrochemical Synthesis of N-Acetyl-2,3-substituted Pyrroles. J. Org. Chem. **43**, 335 (1978).

478. Ogura, K., N. Katoh, I. Yoshimura, and G. Tsuchihashi: New Synthesis of α-Keto Acid Derivatives From Nitriles Using Methyl Methylthiomethyl Sulfoxide. Tetrahedron Letters **1978**, 375.

479. Matthies, D., und E. D. Setiakusuma: Reaktionen mit Arylglyoxal-thiocarbonsäure-amid. Arch. Pharm. (Weinheim) **310**, 996 (1977).

480. Matthies, D.: Herstellung und Umsetzung von N-Acyl-α-chloroglycinen. Synthesis **1978**, 53.

481. Sen, P. K., C. J. Veal, and D. W. Young: Photochemical Synthesis of a Novel β-Lactam. J. Chem. Soc. Chem. Comm. **1977**, 678.

482. Izumiya, N., S. Lee, T. Kanmera, and H. Aoyagi: Asymmetric Hydrogenation of α,β-Dehydroamino Acid Residue in Cyclic Dipeptides. J. Amer. Chem. Soc. **99**, 8346 (1977).

483. Pieroni, O., D. Bacciola, A. Fissi, R. A. Felicioli, and E. Balestreri: Asymmetric Hydrogenation of Unsaturated Peptides. Int. J. Peptide Protein Res. **10**, 107 (1977).

484. Pieroni, O., A. Fissi, S. Merlino, and F. Ciardelli: Chiroptical Properties and Conformation of Dehydrophenylalanine Peptides. Isr. J. Chem. **15**, 22 (1976/77).

485. Pieroni, O., A. Fissi, and G. Montagnioli: Unsaturated Amino Acid Residues as Probes for the Conformation of Polypeptides in Solution. Biopolymers **16**, 1677 (1977).

486. Glaser, R., M. Twaik, S. Geresh, and J. Blumenfeld: Structural Requirements in Chiral Diphosphine-Rhodium Complexes. VIII. Asymmetric Hydrogenation of N-Acetyldehydroamino Acids with Rhodium(I) Complexes Containing Chiral Carboxylic Analogues of DIOP. Tetrahedron Letters **1977**, 4635.

487. Glaser, R., J. Blumenfeld, and M. Twaik: Structural Requirements in Chiral Diphosphine-Rhodium Complexes. X. Asymmetric Hydrogenation of Z-N-Acetyl-dehydroamino Acids and Esters with (1R,2R)-Trans-1,2-bis (diphenylphosphinome-thyl)cyclobutane/Rhodium(I) Complexes. Tetrahedron Letters **1977**, 4639.

488. Pracejus, G., and H. Pracejus: Chiral Amino and Diamino Phosphines as Ligands for Asymmetric Hydrogenation Catalysts. Tetrahedron Letters **1977**, 3497.

489. Masuda, T., and J. K. Stille: Transition Metal Catalyzed Asymmetric Organic Syntheses via Polymer-Attached Optically Active Phosphine Ligands. Synthesis of R-Amino Acids by Hydrogenation with a Polymer Catalyst Containing Optically Active Alcohol Sites. J. Amer. Chem. Soc. **100**, 268 (1978).

490. Takaishi, N., H. Imai, C. A. Bertelo, and J. K. Stille: Transition Metal Catalyzed Asymmetric Organic Syntheses via Polymer-Attached Optically Active Phosphine Ligands. Synthesis of R-Amino Acids and Hydratropic Acid by Hydrogenation. J. Amer. Chem. Soc. **100**, 264 (1978).

491. Baxter, A. J. G., and A. B. Holmes: Synthetic Studies in the Piperidine Alkaloid Field. Part 1. The 2-Azabicyclo[2.2.2]octan-5-one Approach to Prosopine. J. Chem. Soc., Perkin I **1977**, 2343.

492. Krow, G. R., C. Johnson, and M. Boyle: Heterodienophiles. Part 9. On the

Preference For Exo-Orientation in Aldimine Cycloadditions. Tetrahedron Letters **1978**, 1971.

493. FRYZUK, M. D., and B. BOSNICH: Asymmetric Synthesis. An Asymmetric Homogeneous Hydrogenation Catalyst Which Breeds Its Own Chirality. J. Amer. Chem. Soc. **100**, 5491 (1978).

494. HENSENS, O. D., and G. ALBERS-SCHÖNBERG: Total Structure of the Peptide Antibiotic Components of Thiopeptin by ^{1}H and ^{13}C-NMR Spectroscopy. Tetrahedron Letters **1978**, 3649.

495. OLESKER, A., L. VALENTE, L. BARATA, G. LUKACS, W. E. HULL, K. TORI, K. TOKURA, K. OKABE, M. EBATA, and H. OTSUKA: Natural Abundance ^{15}N Nuclear Magnetic Resonance Spectroscopic Evidence For the Structural Relationship Between The Peptide Antibiotics Thiostrepton and Siomycin A. J. Chem. Soc. Chem. Comm. **1978**, 577.

496. MICHEL, K. H., M. O. CHANEY, N. D. JONES, M. M. HOEHN, and R. NAGARAJAN: Epipolythiopiperazinedione Antibiotics From Penicillium Turbatum. J. Antibiot. **27**, 57 (1974).

496A. MINSTER, D. K., U. JORDIS, D. L. EVANS, and S. M. HECHT: Thiazoles From Cysteinyl Peptides. J. Org. Chem. **43**, 1624 (1978).

497. BREWER, D., A. G. MCINNES, D. G. SMITH, A. TAYLOR, J. A. WALTER, H. R. LOOSLI, and Z. L. KIS: Sporidesmins. Part 16. The Structure of Chaetomin, A Toxic Metabolite of Chaetomium Cochlides, By Nitrogen-15 and Carbon-13 Nuclear Magnetic Resonance Spectroscopy. J. Chem. Soc., Perkin I **1978**, 1248.

498. ABE, H., T. TAKAISHI, T. OKUDA, K. AOE, and T. DATE: Methanolysis Products of Sulfomycin I. Tetrahedron Letters **1978**, 2791.

499. KONNO, S., and C. H. STAMMER: A New Dehydropeptide Synthesis. Direct Oxidation of a Dipeptide Azlactone. Synthesis **1978**, 598.

500. BARRETT, G. C., L. A. CHOWDHURY, and A. A. USMANI: Formation of "Dehydropeptides" from Peptides. A Model System Establishing a Mechanism For the Biogenesis of Peptide Amides and α-Keto Acids. Tetrahedron Letters **1978**, 2063.

501. WOJCIECHOWSKA, H., R. PAWLOWICZ, R. ANDRUSZKIEWICZ, J. GRZYBOWSKA: Conversion of Protected Serine and Threonine to Corresponding Dehydroamino Acids Under Mild Conditions. Tetrahedron Letters **1978**, 4063.

502. WATANABE, K., und H. KLOSTERMEYER: Bildung von Dehydroalanin, Lanthionin und Lysinoalanin beim Erhitzen von β-Lactoglobulin A. Z. Lebensm. Unters.-Forsch. **164**, 77 (1977).

503. SUZUKI, M., K. NUNAMI, and N. YONEDY: Synthesis of Cycloalk-1-enylglycines. J. Chem. Soc. Chem. Comm. **1978**, 270.

504. TATSUMOTO, K., and A. E. MARTELL: Catalysis of the β-Elimination of O-Phosphoserine and β-Chloroalanine by Pyridoxal and Zink(II) Ion. J. Amer. Chem. Soc. **99**, 6082 (1977).

505. — — Reaction Kinetics of the Metal Ion Catalyzed β-Phenylserine-Pyridoxal Model System. J. Amer. Chem. Soc. **100**, 5549 (1978).

506. KEITH, D. D., R. YANG, J. A. TORTORA, and M. WEIGELE: Synthesis of DL-2-Amino-4-(2-aminoethoxy)-*trans*-but-3-enoic Acid. J. Org. Chem. **43**, 3713 (1978).

507. GREENLEE, W. J., D. TAUB, and A. A. PATCHETT: A General Synthesis of α-Vinyl-α-amino Acids. Tetrahedron Letters **1978**, 3999.

508. KOLASA, T., and A. CHIMIAK: Unambiguous Synthesis of N-Hydroxypeptides. Tetrahedron **33**, 3285 (1977).

509. CAMA, L. D., and B. G. CHRISTENSEN: Total Synthesis of β-Lactam Antibiotics IX. (±)-1-Oxabisnorpenicillin G. Tetrahedron Letters **1978**, 4233.

510. CONWAY, T. T., G. LIM, J. L. DOUGLAS, M. MENARD, T. W. DOYLE, P. RIVEST, D. HORNING, L. R. MORRIS, and D. CIMON: Nuclear Analogs of β-Lactam Antibiotics. VIII. Synthesis of 3-Acetoxymethyl-Δ^3-O-2-isocephems. Canad. J. Chem. **56**, 1335 (1978).

511. Schouteeten, A., Y. Christidis et G. Mattioda: Les N-acylhémiaminals de l'acide glyoxylique et leur utilisation en synthèse. Bull. Soc. Chim. France **1978**, 248.

512. Achiwa, K.: Homogeneous Catalytic Asymmetric Hydrogenation of (Z)-2-Aceta-mido-3-methyl-fumaric Acid Ester, a Tetrasubstituted Olefin. Tetrahedron Letters **1978**, 2583.

513. Brown, J. M., and P. A. Chaloner: The Mechanism of Asymmetric Hydrogenation Catalyzed by Rhodium(I) DIPAMP-Complexes. Tetrahedron Letters **1978**, 1877.

514. Brown, J. M., and P. A. Chaloner: Mechanism of Asymmetric Hydrogenation Catalysed by Rhodium(I) trans-4,5-Bis-(diphenylphosphinomethyl)-2,2-dimethyldi-oxolan (DIOP) Complexes. J. Chem. Soc. Chem. Comm. **1978**, 321.

515. Cullen, W. R., and Y. Sugi: Asymmetric Hydrogenation Catalyzed by Diphos-phinite Rhodium Complexes Derived From A Sugar. Tetrahedron Letters **1978**, 1635.

516. Fiorini, M., G. M. Giongo, F. Marcati, and W. Marconi: Asymmetric Hydro-genation by Chiral Aminophosphine-Rhodium Complexes. J. Molecular Catal. **1**, 451 (1976).

517. Glaser, R.: Prediction of Chirality of Major Product by Models of DIOP-Rho-dium(I) Complexes For Asymmetric Hydrogenation and Hydrosilylation. Tetrahedron Letters **1975**, 2127.

518. Glaser, R., S. Geresh, and J. Blumenfeld: Structural Requirements in Chiral Diphosphine-Rhodium Complexes. III. Small Scale Method For Fresh Preparation of Cationic DIOP-Rhodium Complexes and Comparison with Neutral DIOP-Rhodium Complexes. J. Organomet. Chem. **112**, 355 (1976).

519. Cullen, W. R., and E-Shan Yeh: Asymmetric Hydrogenation Using Ferrocenyl-phosphine Rhodium(I) Cationic Complexes. J. Organomet. **139**, C13 (1977).

520. Glaser, R., and M. Twaik: Structural Requirements in Chiral Diphosphine-Rhodium Complexes. II. NMR-Determination of E-Z-Geometry in Prochiral Substrates Used in Asymmetric Hydrogenation Reactions. α-Acetamidocinnamic Acids, Esters and Parent Azlactones. Tetrahedron Letters **1976**, 1219.

521. Glaser, R., S. Geresh, J. Blumenfeld, and M. Twaik: Structural Requirements in Chiral Diphosphine-Rhodium Complexes. XI. Asymmetric Homogeneous Hydro-genation of Z-α-Acylaminocinnamic Acids and Esters with (1S,2S)-trans-1,2-Bis-(diphenylphosphinomethyl) Cyclohexane/Rhodium(I) Complexes. Tetrahedron **34**, 2405 (1978).

522. Grubbs, R. H., and R. A. DeVries: Asymmetric Hydrogenation By an Atropisomeric Diphosphinite Rhodium Complex. Tetrahedron Letters **1977**, 1879.

523. Hanaki, K., K. Kashiwabara, and J. Fujita: Asymmetric Hydrogenation of α-Acylaminoacrylic Acids by the Rhodium(I) Complex of (1R,2R)-Bis(N-diphenyl-phosphinomethylamino)-cyclohexane. Chemistry Letters **1978**, 489.

524. James, B. R., D. K. W. Wang, and R. F. Voigt: Catalytic Asymmetric Hydrogen-ation Using Ruthenium(II) Chiral Phosphine Complexes. J. Chem. Soc. Chem. Comm. **1975**, 574.

525. James, B. R., R. S. MacMillan, and K. J. Reimer: Catalytic Asymmetric Synthesis Using Ruthenium Complexes Containing Sulfoxide Ligands. J. Mol. Catal. **1**, 439 (1975/76).

526. Morrison, J. D., W. F. Masler, and M. K. Neuberg: Asymmetric Homogeneous Hydrogenation. Adv. Catal. **25**, 81 (1976).

527. Tamao, K., H. Yamamoto, H. Matsumoto, N. Miyake, T. Hayashi, and M. Kumada: Optically Active 2,2'-Bis-(diphenylphosphinomethyl)-1,1'-binaphthyl: A New Chiral Bidentate Phosphine Ligand For Transition-Metal Complex Catalyzed Asymmetric Reactions. Tetrahedron Letters **1977**, 1389.

528. VILÍM, J., and J. HETFLEJŠ: Kinetics of Enantioselective Hydrogenation of α-Acetyl-aminocinnamic Acid Catalysed by A Rhodium Complex. Coll. Czech. Chem. Comm. **43**, 122 (1978).

529. VINEYARD, B. D., W. S. KNOWLES, M. J. SABACKY, G. L. BACHMAN, and D. J. WEIN-KAUFF: Asymmetric Hydrogenation. Rhodium Chiral Bisphosphine Catalyst. J. Amer. Chem. Soc. **99**, 5946 (1977).

530. VALENTINE, D. Jr., and J. W. SCOTT: Asymmetric Synthesis. Synthesis **1978**, 329.

531. ACHIWA, K.: Catalytic Asymmetric Hydrogenations With Polymer Supported Chiral Pyrrolidinephosphine-Rhodium Complexes. Chemistry Letters **1978**, 905.

532. YAMADA, S., and S. HASHIMOTO: Asymmetric Transamination From Amino Acids(I). Asymmetric Synthesis of Amino Acid By Chemical Transamination From Optically Active Amino Acids to α-Keto Acid. Tetrahedron Letters **1976**, 997.

533. HARADA, K., and I. NAKAMURA: The Formation of Schiff Base From Dimethyl α,β-Dibromosuccinate and an Asymmetric Synthesis of Aspartic Acid. Chemistry Letters **1978**, 9.

534. HARADA, K., and Y. KATAOKA: The Temperature Dependence of Hydrogenolytic Asymmetric Transamination Between Esters of Optically Active Phenylglycine and Pyruvic Acid. Chemistry Letters **1978**, 791.

535. HARADA, A., and Y. KATAOKA: Asymmetric Synthesis of Alanine by Hydrogenolytic Asymmetric Transamination. Tetrahedron Letters **1978**, 2103.

536. MATSUMURA, K., T. SARAIE, and N. HASHIMOTO: Studies of Nitriles. Part 7. Synthesis and Properties of 2-Amino-3,3-dichloroacrylonitrile (ADAN). Chem. Pharm. Bull. **24**, 912 (1976).

537. MATSUMURA, K., T. SARAIE, and N. HASHIMOTO: Studies of Nitriles. Part 8. Reactions of N-Acyl Derivatives of 2-Amino-3,3-dichloroacrylonitrile (ADAN) with Amines. (1). A New Synthesis of 2-Substituted-5-(substituted amino)-oxazole-4-carbonitriles and 4-N-acylcarboxamides. Chem. Pharm. Bull. **24**, 924 (1976).

538. MATSUMURA, K., H. SHIMADZU, O. MIYASHITA, and N. HASHIMITO: Studies of Nitriles. Part 9. Reactions of 2-Acylamino-3,3-dichloroacrylic Amide and -N-Acyl-amide with Aliphatic Amines. (2). Syntheses of Some α,α-Diamino-Acid Derivatives. Chem. Pharm. Bull. **24**, 941 (1976).

539. MATSUMURA, K., O. MIYASHITA, H. SHIMADZU, and N. HASHIMOTO: Studies of Nitriles. Part 10. Synthesis and Reactions of 2-Acylamino-3,3-bis(substituted mercapto)-acrylonitriles and Their Derivatives. A New Synthesis of 2-Substituted-5-(substituted mercapto)oxazole-4-carbonitriles and Their Derivatives. Chem. Pharm. Bull. **24**, 948 (1976).

540. MATSUMURA, K., M. KURITANI, H. SHIMADZU, and N. HASHIMOTO: Studies of Nitriles. Part 11. Preparation and Chemistry of Schiff Base of ADAN. 2-Amino-3,3-dichloroacrylnitrile. A Highly Effective Conversion into 2-Substituted-4(5)-chloro-imidazole-5(4)-carboldehydes. Chem. Pharm. Bull. **24**, 960 (1976).

(Received February 23, 1978)

Author Index

Page numbers printed in *italics* refer to References

Subject Index

By

A. SIEGEL, Wien

Fortschritte der Chemie organischer Naturstoffe

Progress in the Chemistry of Organic Natural Products

All Volumes and Cumulative Index 1—20 available / Alle Bände und Generalregister 1—20 lieferbar.

Price reduction for subscribers / Preisermäßigung für Subskribenten: 10%.

Special price reduction (20% of the list price) for the Vols. 1—20 plus Cumulative Index. / Vorzugspreis (20% Nachlaß) bei Bezug der Bände 1—20 inklusive Generalregister.

Volume 33: 48 figures. VIII, 581 pages. 1976. ISBN 3-211-81 357-8.

> *Contents:* L. MINALE, G. CIMINO, S. DE STEFANO, and G. SODANO, Natural Products from Porifera – R. M. COATES, Biogenetic-Type Rearrangements of Terpenes – K. L. RINEHART, JR., and L. S. SHIELD, Chemistry of the Ansamycin Antibiotics – A. FONTANA and C. TONIOLO, The Chemistry of Tryptophan in Peptides and Proteins – P. HEMMERICH, The Present Status of Flavin and Flavocoenzyme Chemistry – Author Index – Subject Index.

Volume 34: 63 figures. X, 620 pages. 1977. ISBN 3-211-81415-9.

> *Contents:* C. R. ENZELL, I. WAHLBERG, and A. J. AASEN, Isoprenoids and Alkaloids of Tobacco – A. R. PINDER, The Chemistry of the Eremophilane and Related Sesquiterpenes – D. GROSS, Phytoalexine und verwandte Pflanzenstoffe – K. H. OVERTON and D. J. PICKEN, Studies in Secondary Metabolism with Plant Tissue Cultures – D. P. CHAKRABORTY, Carbazole Alkaloids – J. JACOB, Bürzeldrüsenlipide – W. VOELTER, Hypothalamus-Regulationshormone – Author Index – Subject Index.

Volume 35: VIII, 589 pages. 1978. ISBN 3-211-81460-4.

> *Contents:* O. R. GOTTLIEB, Neolignans – K. HERRMANN, Hydroxyzimtsäuren und Hydroxybenzoesäuren enthaltende Naturstoffe in Pflanzen – G. PATTENDEN, Natural 4-Ylidenebutenolides and 4-Ylidenetetronic Acids – R. D. H. MURRAY, Naturally Occurring Plant Coumarins – G. OHLOFF, Recent Developments in the Field of Naturally-Occurring Aroma Components – Author Index – Subject Index.

Volume 36: 11 figures. VII, 425 pages. 1979. ISBN 3-211-81472-8.

> *Contents:* F. W. WEHRLI and T. NISHIDA, The Use of Carbon-13 Nuclear Magnetic Resonance Spectroscopy in Natural Products Chemistry – G. OHLOFF and I. FLAMENT, The Role of Heteroatomic Substances in the Aroma Compounds of Foodstuffs – A. J. WEINHEIMER, C. W. J. CHANG, and J. A. MATSON, Naturally Occurring Cembranes – Author Index – Subject Index.

Springer-Verlag Wien · New York